London,
2010

Dear Tino!

Thank you so much for coming to London for our civil wedding!

Happy reading....

Love,
Fiona

Environmental Finance and Socially Responsible Business in Russia

Legal and Practical Trends

Environmental Finance and Socially Responsible Business in Russia

Legal and Practical Trends

edited by

Wybe Th. Douma and Fiona M. Mucklow

T·M·C·Asser Press

Cover photograph: Anjoum Noorani, 2009

ISBN 978-90-6704-330-4

All rights reserved.
© 2010, T·M·C·ASSER PRESS, The Hague, The Netherlands and the Authors

<www.asserpress.nl>

No part of the material protected by this copyright notice may be reproduced or utilized in any form or by any means, electronic or mechanical, including photocopying, recording, or by any information storage and retrieval system, without written permission from the copyright owner.

PRINTED IN THE NETHERLANDS

FOREWORD

We all have to look carefully at investment ideas, whether in good or difficult times. Corporations everywhere are under pressure to behave responsibly. Despite what some believe, the vast majority are naturally disposed to want to do that: it is in their longer term interests, and after all, who would of themselves wish to be irresponsible? But context determines behaviour.

And the context is changing. The legislative constraints on corporate behaviour, and the social expectations of business, have evolved fast in recent years. The papers in this volume look at the way that process has affected Russia.

The result is illuminating. Russia, like other formerly Communist countries, has had to undergo far greater changes than any experienced by the outside developed world. The very nature of the modern Russian corporation has been a new and rapidly changing element of Russian society. The relationship of private business with the state has been subject to sometimes dramatic change in the past few years. Property rights are not as firm as elsewhere. The execution of tax law, the regularity of the judicial system, the transparency of administration, and corruption are all matters for concern.

But none of this should disguise the fact that there is measurable progress too. Many Russian corporations increasingly see the case for accountable behaviour, and the connection between that and their interest in long term investment, whether domestic or foreign.

Both Russian and outside investors have a direct interest in the way that the assumptions and beliefs that surround Russian business will develop. That is of course a matter of knowing what Russian law may say, and how it is applied. The papers which follow explain that. But they go further, because they deal too with the wider context. It is essential for all investors, and perhaps particularly foreign investors, to understand that context, and to track its evolution. We can all have our different hopes and fears, but we need to know what is really going on before we can judge how best to promote the one and guard against the other.

That should be enough to justify close consideration of the material set out in this volume. But I would commend it for a wider reason too: Russia is an essential part of Europe, and her fate and prosperity matter to all of us on this continent, and beyond it. Russia's future will depend on responsible and effective investment being undertaken now. That future will be more hopeful if both Russian and foreign investment is for the long term, and undertaken with the wider interests of society fully in mind.

London, November 2009

Andrew Wood

Former British Ambassador to the Russian Federation

ACKNOWLEDGEMENTS

The editors would like to thank the many people who have been instrumental in making this book come to fruition, including all the dedicated and patient authors, as well as Laura Brank at Dechert LLP, Irina Tribushnaya, Nadia Walton-Sharp, Pam Marsh, Isabelle Pearson, Shirley McGee, Russian translators Natalia Kolochkova, Elena Balagurova, and Natalia Darchieva, the Russo-British Chamber of Commerce, the United Nations Development Programme in Russia, WWF Russia, Trevor Barton at The PBN Company, Anjoum Noorani for the book cover photograph of Lake Baikal, the colleagues, contributors and friendships that stemmed from the first Harmonisation of environmental standards TACIS project (in which the consortium partners ERM, Milieu Ltd. and DHV cooperated), Karin Peters-Moes, Patrick Mucklow for his help with websites construction, and many others, who all in their own way contributed to this book becoming reality.

Most importantly, however, Fiona Mucklow would like to thank her parents and Nicholas Cheremeteff, and Wybe Douma his mum and Dima Avilov, for all their boundless love, friendship, support and encouragement to follow our paths.

The Hague, December 2009

SUMMARY OF CONTENTS

LIST OF ABBREVIATIONS

A

AA	Assigned Amount
AAU	Assigned Amount Units
AIE	Accredited Independent Agency
AIJ	Activities implemented jointly

B

BAT	Best Available Techniques
BREF	BAT Reference Documents

C

CDM	Clean Development Mechanism
CEB	Council of Europe Development Bank
CHPP	Combined heat and power plant
CIS	Commonwealth of Independent States
COP	Conference of the Parties
CSR	Corporate Social Responsibility
CTE	Committee on Trade and Environment

D

DEA	Danish Energy Agency
DFID	Department for International Development (Great Britain)

E

EBRD	European Bank for Reconstruction and Development
EIA	Environmental Impact Assessment
EIB	European Investment Bank
EIC	Environmental Investment Center
EM	Environmental Monitoring
EMAS	EU Eco-Management and Audit Scheme
EMEs	Emerging Market Economies
EMP	Environmental Management Plan
EMS	Environmental Management System
EQS	Environment Quality Standards
ERPA	Emission Reduction Purchase Agreement
ERU	Emissions Reduction Unit
ERUPT	Emission Reduction Unit Procurement Tender
ET	Emissions Trading
EU	European Union

F

FAS	Federal Anti-Monopoly Service (Russian Federation)
FDI	Foreign Direct Investment
FSC	Forest Stewardship Council

G

GDP	Gross Domestic Product
GEF	Global Environment Facility
GHG	Greenhouse gas
GIS	Green Investment Scheme
GOF	Global Opportunities Fund (now referred to as the SPF)
GRI	Global Reporting Initiative

J

JI	Joint Implementation
JISC	Joint Implementation Supervisory Committee

K

KNPP	Kola Nuclear Power Plant
KP	Kyoto Protocol to the United Nations Framework Convention on Climate Change

L

LULUCF	Land Use, Land Use Change and Forestry

M

MED	Ministry of Economic Development (Russian Federation)
MFI	Multilateral Financial Institution
MNR	Ministry of Natural Resources and Environment (Russian Federation)
MSC	Marine Stewardship Council
$MtCO_2e$	Million tonnes of CO_2 equivalent

N

NEFCO	Nordic Environment Finance Corporation
NGO	Non-Governmental Organisation
NIB	Nordic Investment Bank
NIS	Newly Independent States
NPAF	National Pollution Abatement Facility (Russian Federation)

O

OECD	Organisation for Economic Co-Operation and Development

P

PDD	Project Design Document
PRI	Principles for Responsible Investing

Q

QMS	Quality management system

R

RBCC	Russo-British Chamber of Commerce
REEEP	Renewable Energy and Energy Efficiency Partnership
REC CEE	Regional Environmental Centre for Central and Eastern Europe
RF	Russian Federation
RREC	Russian Regional Environment Center
RSPP or RUIE	Russian Union of Industrialists and Entrepreneurs

S

SER	State Environmental Review
SEI	Statement of Environmental Impact
SPF	Strategic Programme Fund

T

TACIS	EU Technical Assistance to the CIS
TED	Tenders Electronic Daily
ToR	Terms of reference
TPP	Tidal power plants

U

UN	United Nations
UNDP	United Nations Development Programme
UN ECE	United Nations Economic Commission for Europe
UNESCO	United Nations Educational, Scientific and Cultural Organization
UNFCCC	United Nations Framework Convention on Climate Change

W

WBCSD	World Business Council for Sustainable Development
WTO	World Trade Organisation

1

ENVIRONMENTAL FINANCE AND SOCIALLY RESPONSIBLE BUSINESS IN RUSSIA – AN INTRODUCTION

Fiona Mucklow[1] and Wybe Th. Douma[2]

As the world economies are experiencing a global financial crisis, matters such as environmentally responsible investing or environmental finance, socially responsible activity and sustainability are increasingly moving to the forefront of mainstream topics of discussion. Not merely of interest to politicians, but also increasingly addressed by businesses globally, environmentally and socially sound investment and business activity appears to be moving to the core of creating a new post-recession world order.

Much attention is paid to the new direction that the USA under President Obama is taking, and where China is heading. However, it is necessary, in addition, to obtain a better understanding of what is happening within Russia. Russia is a key player in many respects. It is the world's third largest emitter of greenhouse gases[3] after China and the USA and, thus, could play a vital role in the fight against global climate change. It is also the largest trading partner of the European Union. With Russia's recently stalled economy after almost a decade of exponential growth, the issues of legal certainty (rule of law), environmentally responsible (or 'green') finance and socially responsible business activity, too, are of increasing importance within the framework of Russia's future sustainable development and its role in the international arena.

ENVIRONMENTAL FINANCE IN RUSSIA – TIDES OF CHANGE?

Since embarking on the road of transition to a market economy in 1992, Russia has taken many steps within the sphere of liberalisation, stabilisation, private-sector development and redefining the role of the state. However, and as Philip Hanson (Chapter 1) illustrates, risks and challenges continue to exist, including in the area of environmental finance.

[1] Associate Researcher, Asser Institute in The Hague, The Netherlands; Email: *f.mucklow@asser.nl*. The authors may also be contacted at *info@EnvironmentalFinanceInRussia.co.uk* and further information may be available on <www.EnvironmentalFinanceInRussia.co.uk> or <www. EnvironmentalFinance InRussia.ru>.

[2] Senior Researcher European Law and International Trade Law, Asser Institute in The Hague, The Netherlands; Email: *W.T.Douma@asser.nl*.

[3] Russia has approximately 1,700 millon tonnes of emissions per year, see Energy Information Administration, *International Carbon Dioxide Emissions and Carbon Intensity*, at <http://www.eia.doe.gov/emeu/international/carbondioxide.html>.

W.Th. Douma and F.M. Mucklow, eds., Environmental Finance and Socially Responsible Business in Russia
© 2010, T·M·C·ASSER PRESS, The Hague, The Netherlands and the Authors

Environmental finance[4] has emerged globally, in particular since the beginning of the 21st century, as a market-based solution to environmental problems. It relies on a limited set of financial products and instruments, including local capital markets, carbon trading, as well as tariffs for environmental services and intergovernmental transfers. However, the reach of environmental finance extends not just to environmental matters, but also to related socio-economic issues. It encompasses voluntary initiatives (e.g., reporting and certification), environmental management systems ('EMS'), corporate social responsibity ('CSR') and touches the financial services sector, insurance business and many other sectors.

Since 2003, innovative policy instruments (such as industry rating, EMS and corporate reporting) have been adopted or further promoted, and traditional instruments (e.g., environmental quality standards, permitting, and environmental liability) have been under reform – and environmental finance in Russia illustrates this dynamic and some of its limitations.[5]

Thus, establishing a modern EMS that matches new economic and social realities is one of the challenges that Russia faces.[6] Many environmental finance projects have been funded with international involvement, for example within the ambit of the European Bank for Reconstruction ('EBRD'), the Global Environment Facility ('GEF'), the World Bank, the Nordic Investment Bank ('EIB') or the Nordic Environment Finance Corporation ('NEFCO'). However, as this book investigates and aims to exemplify, local efforts, albeit some in their nacency, are underway, too.[7]

As Daria Ratsiborinskaya (Chapter 4) and other authors in this volume discuss, Russia has dedicated major efforts to the development of its environmental legal frame-

---

[4] This term encompasses all market-based instruments designed 'to deliver environmental quality and to transfer environmental risk', see Labatt, S., and White, R.R., *Environmental Finance: A Guide to Risk Assessment and Financial Products* (2002). See also editions of the *Environmental Finance* magazine, at (<www.environmental-finance.com>); Labatt, S. and White, R.R., *Carbon Finance: The Financial Implications of Climate Change* (2007). Other useful publications on business and environment are by Willard, B., *The Sustainability Advantage: Seven Business Case Benefits of a Triple Bottom Line* (2002); Holliday, Jr., C.O., Schmidheiny, S., and Watts, P., *Walking the Talk: The Business Case for Sustainable Development* (2002); Schmidheiny, S., *Changing Course: A Global Perspective on Development and the Environment* (1992).

[5] OECD, *Mobilising Financial Resources for the Environment in Russia* (2008), see <http://www.oecd.org/dataoecd/23/5/41445931.pdf> ('OECD 2008 Report'), p. 6.

[6] See OECD 2008 Report, p. 6.

[7] See OECD 2008 Report, pp. 6-7: which provides an update of some of the developments in environmental expenditure and finance in Russia. It states that 'recent data collected in Russia and in EECCA countries indicate that environmental expenditure in Russia has increased to levels which compare with some CEE countries'. However, the report continues to comment that the level of expenditure remains low as a share of GDP, and expenditure is still largely concentrated on the water sector. The OECD 2008 Report further states that 'there are opportunities to increase both public and private expenditure volumes, to increase the share of investment in cleaner technologies, and to allocate resources to a wider variety of environmental domains'. The authors in the same report state that this requires making a better use of existing financing instruments and harnessing new sources of finance. The capacity to mobilise additional sources of finance for the environment and to make the best use of them can contribute to a number of issues which are high on the Russian agenda: sustained economic growth, more effective innovation policies, and tackling the health crisis (see OECD, *Economic Survey of the Russian Federation* (2006) and the OECD 2008 Report).

work. Russia has signed and ratified many multilateral environmental agreements and Russian environmental laws and regulations now address most of the priority environmental issues.[8] Unlike the USA, it ratified the Kyoto Protocol (in 2004, so enabling the KP to enter into force) under then President Vladimir Putin.

Nevertheless, and as has been widely publicised, environmental challenges still exist in Russia[9] – related in part to a (not unreasonable) focus on economic development, inconsistent implementation and enforcement of Russian environmental laws and perhaps also due to the main state body on environmental affairs, the State Committee on Environmental Protection ('Goskomekology') being closed in 2000, and its functions having been transferred to the Ministry of Natural Resources.[10]

Since the beginning of the 21st century, change in Russia, towards the creation of more 'green' and socially responsible public and private sectors, appears to be underway. It is notable that at the highest political levels, calls have been made to diversify investments in Russia, away from the traditional focus sectors connected to natural resources (such as oil, gas, forestry, minerals). Moreover, growing environmental consciousness was exemplified with a change in the planned route of the Eastern Siberia-Pacific Ocean ('ESPO')[11] oil pipeline being ordered, so as to minimise the risks to Lake Baikal,[12] the world's largest freshwater lake and a UNESCO World Heritage site. Towards the end of his presidency in 2008, then President Putin expressed that '[w]orking to protect nature must become the systematic, daily obligation of state authorities at all levels'.[13]

Russian President Medvedev has also expressed his support for environmental responsibility and stated that '...in the coming years the quality of the environment will

---

[8] See OECD 2008 Report, p. 6 and OECD, *Policies for a better environment. Progress in Eastern Europe, Caucasus and Central Asia* (2007) ('OECD 2007 Report').

[9] For example, OECD reports that the magnitude of these challenges can be illustrated by the fact that about 15% of Russia's territory suffers from exposure to high levels of ambient pollution. In many industrial centres (e.g., Dzerzhinsk, Irkutsk, Kemerovo, Krasnoyarsk, Novokuznetsk, Norilsk, and Cherepovts), the rates of morbidity and mortality are reported exceed 1.5-3 times the national average – see OECD 2008 Report, p. 9, and OECD, *Environmental Policy and Regulation in Russia. The Implementation Challenge* (2006) ('OECD 2006 Report'). See also Weiner, D.R., *A Little Corner of Freedom: Russian Nature Protection from Stalin to Gorbachev* (1999); Murray, F., *Ecological Disaster: Cleaning up the Hidden Legacy of the Soviet Regime* (1995); further material about Soviet environmental pollution see Feshbach, M. and Friendly, Jr., A., *Ecocide in the USSR: Health and Nature under Siege* (1992).

[10] The Yeltsin administration had demoted the Ministry of Ecology to Goskomekology in 1996. Recently, the Ministry of Natural Resources was renamed Ministry of Environmental Resources and Ecology ('MNR').

[11] See <www.ihsglobalinsight.com/SDA/SDADetail5684.htm> and <http://en.rian.ru/russia/20080303/100502195.html>; the construction of the 2,550 km ESPO pipeline by Transneft started in 2006; see also <www.hydrocarbons-technology.com/projects/espopipeline/>.

[12] Situated in south-east Siberia, the 3.15-million-ha Lake Baikal is the oldest (25 million years) and deepest (1,700 m) lake in the world. UNESCO reports that it contains 20% of the world's total unfrozen freshwater reserve. It is known as the 'Galapagos of Russia', its age and isolation have produced one of the world's richest and most unusual freshwater faunas, which is of exceptional value to evolutionary science, see <http://whc.unesco.org/en/list/754>.

[13] ABC News 8 February 2008, *Putin, Medvedev courting Russia's green vote*, at <www.abc.net.au/news/stories/2008/02/08/2157436.htm>, accessed 3 April 2009.

be a key factor in the competitiveness of each Russian region'.[14] At a meeting with the Minister of Natural Resources and Environment, Yury Trutnev, the new President also called on the government, civil society, and the business community to give priority to the improvement of the environment regardless of short-term issues. 'Our common goal – both that of the government, civil society, and the business community – is to continue working to make the environment healthier,' he said. 'Even when business sees its resources shrinking, it should not stop and should commission new environmentally friendly facilities, improve purification systems, reduce emissions, and so on.' 'All this work should continue non-stop, regardless of any external or internal reasons,' he stated.[15]

Old perceptions and new circumstances

The general perception amongst foreign investors is that Russia remains a challenging (though potentially very lucrative) market, which requires only minimal focus on environmental and social issues. 'Bad press', e.g., relating to the rule of law, transparency and corruption, has contributed towards the international business community being cautious about the viability of 'green' projects or focusing on taking seriously their social responsibilities in Russia – whilst domestic Russian companies are often sceptical of green initiatives or 'social responsibility' corporate policies due to perceived costs, lack of information and experience of successful environmental or socially responsible ventures.

This book hopes to adjust these perceptions and aims to exemplify that steps are being taken by Russia, as well as by businesses operating in Russia, in addressing environmental and social issues. The aim is to present a more accurate view of Russia's progress towards sustainable development and of recent Russian business practices. There are many opportunities and challenges that exist from a legal, as well as from a practical and policy perspective, however, the message is clear: environmentally and socially responsible investing and business activity is underway in Russia. Many real opportunities exist for businesses to participate in environmental and socially responsible projects in Russia.

The editors have chosen to combine the topics of 'environmental finance' and 'socially responsible business' in this book as they are of the opinion that these two topics are inextricably linked. Environmental finance or environmentally responsible investing cannot be effective without taking account of social issues. And businesses purporting to act in a socially responsible manner cannot be considered to be doing so effectively without undertaking environmental investments, in tandem. The one is not complete without the other.

Thus, ultimately, this book aims to contribute towards facilitating and promoting sustainable development in Russia. It is developed within the general spirit of the pro-

[14] 30 January 2008, <www.medvedev2008.ru/english_2008_01_30.htm>.

[15] See Interfax, 29 October 2008, at <http://www.cdi.org/russia/Johnson/2008-197-47.cfm>.

visions of the *Millennium Declaration*, the *Johannesburg Declaration on Sustainable Development*, which provide that 'the private sector, both large and small companies, have a duty to contribute to the evolution of equitable and sustainable communities and societies' (Article 27); as well as the provisions of the *World Summit on Sustainable Development Plan of Implementation* ('PoI'), which states that:

> 'Poverty eradication, changing unsustainable patters of production and consumption, and protecting and managing the natural resource base of economic and social development are overarching objectives of, and essential requirements for, sustainable development.' (Art. 2)

The PoI emphasizes, *inter alia*, the need to enhance corporate environmental and social responsibility and accountability (Article 17), and the need to actively promote corporate responsibility and accountability based on the Rio Principles, and support continuous improvement in corporate practices in all countries (Article 45ter).

Naturally, due to space constraints, and in no way for lack of their importance, numerous issues related to environmental finance and socially responsible business practices are not addressed (in full or at all) in this book – including, the development of the protection of human rights, such as indigenous peoples' rights, intellectual property rights issues, the rule of law, tax law and practice, transparency, corruption and other debilitating social matters, such as racism, alcoholism and sexism.

Over half of the authors presenting their views in this volume are from Russia, and most of the non-Russian authors have worked extensively in Russia. Together, the authors offer a practical account, and in some instances a reflective analysis, of the state of play of 'green' investments in Russia and how local and foreign businesses are operating in an increasingly more socially responsible manner. The result is a collection of chapters, each written in the style of the authors, focusing on a myriad of separate but interrelated issues. This provides for a varied introductory overview of where we stand today and how current and future generations of Russian and foreign businesses can continue to add value in an emerging market such as Russia.

Outline of the book

The contributions in this volume are grouped into three parts. The aim is to provide some structure in very much over-lapping topics and subject matters.

Part I acts as an overture, sketching the broad picture. On the one hand, Phil Hanson provides an introduction on the development of Russia's economy – from the turmoil in the post-communist days, through the exponential growth to the recent decline (Chapter 1). This account is followed by Godfrey Cromwell and Tatiana Guseva's chapters, which focus on the role that environmentally and socially responsible business and investing can play in Russia, taking into account its historical background, as well as its frail legal and social system (Chapters 2 and 3).

Part II focuses on the legal and policy aspects of investing and doing business in Russia. Daria Ratsiborinskaya starts off with a description of Russia's environmental law (Chapter 4). The role of NGO's in promoting green investments and their interaction with businesses in Russia is addressed by Olga Razbash (Chapter 5). Next, Dietrich Hahn and Michael Begak set out how the EU-TACIS harmonisation of environmental standards project in Russia raises new questions and issues. They discuss the aims of this project of approximating Russia's legal system with that of the EU where industrial permitting is concerned (Chapter 6). Thereafter, Soili Nystén-Haarala provides an insight into the status of governmental mechanisms in Russia, which are set to promote green business (Chapter 7). Evgeny Shvarts and Ivetta Gerasimchuk highlight how environmental policy and international competitiveness of the Russian economy can go hand in hand (Chapter 8). In Chapter 9, Tatiana Guseva discusses how environmental education and capacity building in Russia has developed. Ekaterina Khmeleva turns to an important potential instrument that could stimulate further greening of Russian society and business, namely green public procurement (Chapter 10). Ivetta Gerasimchuk discusses the interesting question as to whether Russia could become a "green investor" in Southern Africa (Chapter 11). In the final chapter of Part II, Wybe Douma reviews how the EBRD promotes the use of European Principles for the Environment and the application of Russian as well as European environmental law (including the system of using Best Available Techniques) when issuing permits to projects sponsored by this bank (Chapter 12).

Part III shifts the focus to practical trends and synergies. Here, the authors set out how businesses and NGOs have participated in 'green' projects (Part III.A) and how corporate social responsibility ('CSR') and philanthropy have developed in Russia (Part III.B).

Michael Yulkin discusses the involvement of Russian business in the Kyoto Protocol (Chapter 13), whilst Jonti Small, Trevor Barton and James Lanman focus on the changing context for green investment, including with respect to the Kyoto Protocol (Chapter 14). Yurij Maksimenko and Irina Gorkina discuss the practice and results of the Russian National Pollution Abatement Facility (Chapter 15). Whilst in Chapter 16, Sergey Dayman reviews how environmental management systems are increasingly becoming part of business operations in Russia, Nina Lesikhina and Michele Grønbech provide an interesting account of the opportunities of introducing renewable energy on the Kola Peninsula (Chapter 17). Finally, Jeffrey Roberts provides a personal account and reflections of his experiences of investments in the Murmansk Oblast and Pomor Region (Chapter 18).

In Chapter 19, the first in Part III.B, Alexey Kostin extensively discusses CSR in Russia and the role the Russian office of the United Nations Development Programme ('UNDP') played in this respect. This chapter is in slight contrast to a shorter contribution in Chapter 20, wherein Sergey Dayman sets out the role CSR plays and can play in a post-socialist state, such as Russia. Finally, Kit Lawry and Joanna Rogers debate the role of businesses in helping children in Russia within the framework of EasyChild, an international children's support organisation (Chapter 21).

The below provides a brief overview of the arguments raised in the chapters in this book.

Overview of chapters

In the introductory Chapter 1, Professor **Philip Hanson**, Associate Fellow at the Royal Institute of International Affairs, provides an economist's overview of how the Russian economy has developed since the 1990s until the present (end of 2009). He states that Russia's spectacular growth record (if vulnerable to the fluctuations in oil prices) has had a mixed impact on Russian society and the environment. Hanson highlights that there are risks ahead and that there remain environmental challenges to overcome.

Godfrey Cromwell, previously an Executive Director at the Russo-British Chamber of Commerce, discusses in Chapter 2 environmental and socially responsible business practices in Russia by asking the question 'why bother'? As is addressed in other chapters, too, focus is laid at first on the historical context. He argues that the Soviet legacy has placed Russia at a disadvantage with respect to, *inter alia*, environmental awareness and the awareness of the finiteness of resources. However, Cromwell explains that, although old attitudes are sometimes still prevalent, some change is underway. He states that as more investors and entrepreneurs are active in Russia, green investments and socially responsible business operations should not be a separate niche activity, but be a 'part of successful mainstream investment strategy'. He also raises important questions as regards the definition and what constitutes 'green' or 'socially responsible' investing and the costs associated with such activity. However, his conclusions are far from gloomy and he identifies many areas within which positive impacts can be achieved.

In the final chapter of Part I, Professor **Tatiana Guseva**, a distinguished scholar at D. Mendeleyev University of Chemical Technology (in Moscow, Russia), provides an insightful discussion of the needs and opportunities relating to socially responsible investing in Russia. She establishes the link between socially responsible investing and CSR – in that socially responsible investing also means investing in companies, which implement CSR within their operations. Drawing on the definition of CSR from the World Business Council for Sustainable Development ('WBCSD'), she concludes that CSR 'is about conducting business in a socially and environmentally responsible way, complemented by investment in communities for solid business case reasons'.

Within this framework, Guseva then reviews the impact and adoption of the concept of sustainable development in Russia. She points out that the concept was initially only accepted by environmentalists and for a long time remained misunderstood by the general public. Even today, she states, 'sustainability' is often mistakenly understood to mean 'stability', and the word 'development' is often omitted, as indeed seems to still be reflected in the "Social Charter" of the Russian Union of Industrialists and Entrepreneurs. She analyses the role of business in sustainable development within Agenda 21's emphasis "on the force which business can have in improving the efficiency of resource use, reducing risks and hazards, minimizing wastes and safeguarding environmental qualities". In this respect, she also discusses the nostalgia that still pervails on some levels for Soviet 'social-infrastructure', as well as the historical significance of the turmoil and economic and social restructuring of the early 1990s. She observes that the perception of many Russians is that businesses regard their environmental responsibility as merely an expensive 'end-of-pipe' activity and that 'safety' considerations are not adhered to properly. Amongst others points, she concludes that socially responsible investing in Russia 'needs to be developed transparently, based on lessons learnt internationally and by thoroughly considering national priorities and specific economic, social and environmental conditions of transition from an older totalitarian regime to a more democratically and sustainably managed country'.

Legal and Policy Perspectives

In Chapter 4, **Daria Ratsiborinskaya** provides an overview of current and some forthcoming Russian environmental law of importance to the business community, though not focusing on the more controversial topic of implementation and enforcement. Notably, she discusses in detail the institutional arrangements at the different levels (federal, regional and municipal), the judicial and arbitration system with regard to environmental law, the main pieces of environmental legislation (e.g., permitting, EIA and technical regulation), liability, economic and administrative enforcement instruments and environmental compliance control. At the end of the chapter, she deals with some of the new strategies developed to protect the environment in Russia and in that way touches upon some of the issues discussed in more detail in Chapter 6 by Dietrich Hahn and Michael Begak.

While not focusing on the issues relating to the 2006 Law on Non-Governmental and Non-Profit Organizations, in Chapter 5, **Olga Razbash**'s contribution describes the way in which public participation in Russia has developed. She describes how Non-Governmental Organisations ('NGOs') engage with Russian businesses and the authorities. In particular, the activities of the Russian Regional Environment Center ('RREC'), where Razbash works as the environmental law and public awareness expert, are used as a practical example of how environmental NGOs can interact and support businesses to engage in green investments and socially responsible business practices. It can be added that similar to the case with RREC, numerous other environmental NGOs (including WWF Russia) have been able to make a difference in Russia by advancing a meaningful dialogue between authorities, business and civil society and by realizing actual improvements in concrete plans, projects, legislation and the lives and health of individuals.

In Chapter 6, **Dietrich Hahn** and **Michael Begak**, the team leader and key expert, respectively, of the second phase of the EU-TACIS project on 'Harmonisation of Environmental Standards' ('HES II') describe the HES II project, which builds on the findings of the first 'Harmonisation of Environmental Standards' project ('HES I'). Within HES I, Russian and EU environmental law were compared in detail with the help of tables of concordance, and recommendations were formulated on areas where approximation of Russian law to EU practices could bring about improvements to the Russian system of permitting. Hahn and Begak explain that, in general, the partial harmonisation of Russian environmental law with relevant EU legislation supports the development of a joint EU-Russia economic space and international commercial activities generally. Furthermore, they stress the advantages which can result for Russian producers if they ensure that their production processes adhere to harmonised environmental standards in a transparent manner (which is increasingly expected by end consumers of goods). This is a similar argument to that raised by Shvarts and Gerasimchuk (in Chapter 8), who argue that on the public level, environmental policy can act as an instrument to improve Russia's international competitiveness. Hahn and Begak state that as a basis to promote such harmonization, EU experiences and examples relating to industrial environmental management serve as an excellent source of information and provide useful experiences, which can be applied to the Russian Federation. Moreover, the authors propose the introduction of a permitting system based on Best Available Techniques ('BAT') and Russian BAT Reference Documents ('BREFs'). They point out that such changes would only be possible if courageous institutional steps would also be taken.

In her chapter, **Soili Nystén-Haarala**, who is an established voice in Russian environmental matters and professor of civil (private) law at the University of Joensuu in Finland, reviews and discusses mechanisms to promote green business in Russia (Chapter 7). She highlights the need for further development of a 'green' business sector and green business practices in Russia. In this respect, Nystén-Haarala notes that the regulatory mechanisms to promote such a sector and practices may still be in their infancy, but that non-regulatory mechanisms are available to companies and investors. These include the voluntary certification of products and processes and cooperation activity with NGOs. Nystén-Haarala also provides a brief overview of the general legal framework, permitting, licensing and quality standards in Russia. She concludes that despite the integration of principles of international environmental law (e.g., the polluter pays principle), Russian environmental law has remained much the same since Soviet times. Strategies are discussed which companies can employ to promote green business or to undertake environmental investments in Russia. She notes, as do other authors in this publication, that the Medvedev administration (as did the Putin administration) encourages companies to participate in the building of Russian society. Thus, she says that 'essentially, socially responsible business investments and environmental investments are welcome and encouraged in Russia'.

In Chapter 8, **Evgeny Shvarts** and **Ivetta Gerasimchuk**, director of conservation policy and senior adviser to the trade and investment programme at WWF Russia, respectively, discuss how Russian environmental policy could act as an instrument to enhance Russia's international economic competitiveness. They identify core environment-related areas that they argue constrain Russia's international competitiveness and provide a series of solutions to address these constaints. They explain that Russia is an 'environmental super power' as it possesses approximately 9% of all 'biocapacity' on planet Earth. Furthermore, unlike EU member states, the USA, China and India, which are 'ecological debtors', Russia is an 'ecological creditor'. They argue that Russia should, therefore, aim to benefit from its natural and environmental capital perhaps following Brazil's example, which has used its biocapacity to its advantage. Shvarts and Gerasimchuk point out that Russia has had a tendency to deplete its natural resources and environmental capital. Thus, preservation and replenishing Russia's environmental capital is a key to becoming more competitive globally. They discuss further steps which need to be taken with respect to modernizing the environmental management system in Russia, as well as Russian environmental institutions. They conclude that 'green' economic development (or sustainable development) of Russia may become the most important factor in enabling the modernization of the Russian economy and assist in achieving the goals set out in the *2020 Concept of Long-Term Social and Economic Development of the Russian Federation*.

Professor **Tatiana Guseva** analytically reviews the evolution and current status of environmental education in Russia. Her first hand experience as a lecturing professor at the D. Mendeleyev University of Chemical Technology of Russia gives her an ideal platform from which to analyse this important issue. She reviews the traditional (USSR rooted) system of environmental eduction and points out how for decades the environmental educational curriculum taught by technical higher education establishments remained technocratic and emphasised the 'end-of-pipe' methods of environmental protection. Nevertheless, this situation resulted in a strong base of environmental scientists and engineers in Russia. Through the introduction of 'Methodological Councils' in the early 1990s, environmental education was improved. But given the elimination in 1999-2000 of the branch of 'Environmental Protection and Rational Use of Natural Resources' from the list of research areas (from which research students can obtain a doctorate), Guseva comments that, officially, this means that environmental sciences no longer exist as a discipline in Russia to this day. However, she concedes that some piecemeal research still survives in other formal research branches such as 'Ecology', 'Geoecology' or 'Economy of Nature Use'.

On a more positive note, Guseva states that with the increased attention given to energy (and environmental) efficiency and with Russia's adoption of market instruments of environmental regulation, this resulted in the development of new higher education programmes in areas such as in environmental auditing and environmental management. She highlights that although the concept of sustainable development is influencing environmental education, teaching methods are still more technical than holistic, or inter-disciplinary, in nature. Guseva argues that as a result, the concept of sustainable development is not understood well enough in Russia. She calls for more environmental capacity building in Russia targeted at producers (managers and engineers), regulators, consultants and NGOs. She also highlights the role that businesses take in supporting such environmental capactiy building.

In Chapter 10, **Ekaterina Khmeleva**, an environmental law programme officer at WWF Russia, explores how green (or environmentally responsible) public procurement laws and policies have developed in Russia. In comparision to EU laws on green public procurement (which she discusses in relative detail), Khemeleva points out that green public procurement has been slow to catch on in Russia, but that there are many opportunities for its development. She provides an overview of the relevant Russian laws and policies on public procurement and suggests avenues of reform which could be implemented. She highlights that the new *2020 Concept of Long-Term Social and Economic Development* (as approved in 2008) is a first step towards establishing a green public procurement system within Russia.

Ivetta Gerasimchuk, who is an environmental economist and senior advisor to the Trade and Investment Programme of WWF Russia, addresses environmental finance and socially responsible business in Russia from a different perspective than most of the other authors in this book (Chapter 11). She investigates how Russian companies and Russia engage in outward investment activities within Southern Africa. This chapter sets out how Russia-Southern African relations have been developing, in particular, since autumn 2006 following high-level Russian (business and political) delegations visiting South Africa, Namibia and Angola and since the conclusion of several bilateral treaties in 2007. As a result, the bulk of Russian investments in Southern Africa occurred during this period of 2006-2007. Gerasimchuk clarifies that Russia's revived interest towards Southern Africa stem from the interest in accessing natural resources (e.g., uranium, non-ferrous metals and diamonds) and from Russia's strategy in being a global guarantor of energy security. She highlights the socio-environmental potential of Russo-South African cooperation, but provides a list of practical measures which could be implemented by Russian companies in order to enhance their sustainability, both domestically and in Southern Africa. In this chapter, Gerasimchuk also highlights the opportunities that exist for Russian businesses to invest in renewable energy, climate change and joint implementation projects within Southern African countries.

Wybe Douma (Chapter 12) sheds light on the way in which companies operating in Russia on projects partially financed by the EBRD take the environmental and social aspects of their activities into account. He sets out how EBRD operates under its own 2008 Environmental and Social Policy and in accordance with the Declaration on European Principles for the Environment ('EPE') and thereby aims to ensure that its Russian clients subscribe to these principles, as well. The EPE Declaration was also signed and adopted by four other banks, namely the Council of Europe Development Bank ('CEB'), the EIB, NEFCO and the Nordic Investment Bank ('NIB'). Accordingly, projects financed by these multilateral financial institutions ('MFIs') in Russia should comply with the relevant EU environmental principles, practices and standards – and with regard to EU financings adhere to the European Neighbourhood Policy and the EU policy towards Russia – but remain subject to local conditions. Under the EPE, the signatory banks agree "to apply the European Principles for the Environment ('EPE'), with reference to such factors as the costs of application, the local conditions that prevail and the time frame for the phased application for implementing the EPE." In other words, local conditions (including economic and other conditions) can serve as a reason for a signatory bank, including EBRD, not to apply the EPE in full. Douma states that these MFIs are expected to build up the capacity in their countries of operation, in order to meet and enforce the requirements of the EPE. In practice, EBRD is, thus, able to ensure that projects are assessed in accordance with both Russian and EU standards and that the EU system of using Best Available Techniques ('BAT') for new projects is introduced in the Russian Federation. By chance, this is in line with the recommendations of the two EU-TACIS 'Harmonisation of Environmental Standards' projects.

Practical Trends and Synergies

The director general of Climate Change Global Services, **Michael Yulkin**, provides an interesting insight into how Russia has implemented the Kyoto Protocol (Chapter 13).[16] He sets out the background against which Russian regulations on joint implementation ('JI') have evolved in Russia and touches upon the new JI regulations of 28 October 2009 and the key role which Sberbank will be taking in the newly established bidding process. Yulkin discusses ways in which GHG emissions can be best controlled and emphasises the importance of Russian businesses taking the lead in the Kyoto Protocol (and post-2012) process. Numerous Russian companies are already involved in JI projects in Russia. These include Rosneft, SUEK, Euraz, Ilim Group, the Archangel'sk Paper and Pulp Mill ('APPM', which Yulkin discusses in more detail), and numerous others. He argues that these examples demonstrate how the implementation of GHG limitations and reduction measures do not prevent business development and production growth. Quite to the contrary, he argues: they act to assist in improving a company's operations, strengthen its competitiveness and sustainability.

Jonti Small, **Trevor Barton** and **James Lanman** (Chapter 14), all attached to The PBN Company,[17] discuss the opportunities for green projects and Western investments in Russia. They do so by first reviewing the significance of political statements by President Medvedev on overcoming 'legal nihilism' and looking at how environmental projects need to be considered in the context both of the domestic political agenda and Russia's ambitions in the international arena. After a brief review of some environmental projects to date, they turn to Russia's ratification of the Kyoto Protocol. They argue that the latter step, in theory, opened up a range of green opportunities for Western companies, but slow progress to date also demonstrates that implementation of green projects remains challenging in Russia. The chapter outlines the current situation with regard to environmental projects in Russia and looks ahead to future developments, particularly in the run-up to a possible new global climate change deal in Copenhagen.[18] These authors note that Russia's greater involvement in international

[16] See also Perelet, R., Pegov, S., and Yulkin, M., *Climate Change. Russia Country Paper* (2007/12), UNDP Russia, Human Development Report 2007-2008, see at <http://hdr.undp.org/en/reports/global/hdr2007-2008/papers/Perelet_Renat_Pegov_Yulkin.pdf>; see also Kokorin, A.O., Garnak, A, Gritsevich, I.G. and Safonov, G.V., *Developing Russia's economy and saving the climate*, Danish Energy Agency, Moscow, (2008), (at: http://www.wwf.ru/data/climate/cowi-210x297-eng_1.pdf); Kokorin, A. & Gritsevich, I. 'The Danger of Climate Change for Russia – Expected Losses and Recommendations', *Russian Analytical Digest*, No. 23, 19 June 2007; Korppoo, A., Karas, J., and Grubb, M., *Russia and the Kyoto Protocol – Opportunities and Challenges* (2005), London, Royal Institute of International Affairs, Chatham House.

[17] The PBN Company is a consultancy specialising in advising Western investors into Russia and the former Soviet Union on market entry and expansion.

[18] For information about Russia and the post-2012 deal, please see Korppoo, A., *The Russian Debate on Climate Doctrine Emerging Issues on the Road to Copenhagen* (2009), Briefing Paper 33, 5 June 2009, The Finnish Institute of International Affairs, see <www.upi-fiia.fi>; Korppoo, A., *Russia and the Post-2012 Climate Regime: Foreign Rather than Environmental Policy*, Briefing Paper 23, 24 November 2008,

organisations and its reassertion of national interests are both trends with potential implications for green projects in Russia. Small, Barton and Lanman conclude that it remains to be seen how the two – potentially contradictory – trends play out. They bring forward important nuances about doing business in Russia, warning that well-publicised cases like Shell and BP should not dominate the picture of business in Russia, as many Western investors that have operated very successfully and profitably in Russia seldom receive media coverage. This is certainly true of many environmental projects, of which they provide a fair set of examples.

In Chapter 15, **Yurij Maksimenko** and **Irina Gorkina** present an overview of the almost 14-year old National Pollution Abatement Facility ('NPAF'), which is a facility set up by the Russian government in cooperation with various other governments and international institutions within a Framework Environmental Programme ('EMP'). The authors, who were actively part of the NPAF for many years, describe how the aim of the EMP is to support and improve the implementation of Russian state environmental policy on all levels, whilst the aim of NPAF projects is to reduce environmental pollution and rehabilitation of natural resources in Russia. The authors present a practical account of the work of NPAF and some of the projects it was involved in, including the Arkhangelsk Pulp and Paper Mill and the Volgograd project. This account exemplifies how, in Russia, efforts to address environmental pollution have been underway with international support and cooperation, though it does not address how the activities have contributed to environmental standards in Russia. This latter point might be an interesting topic for a later edition of this book.

Sergey Dayman, who is senior consultant of the Sustainability Assurance and Advisory Services group with Ernst & Young in Moscow, gives a brief account of how EMS has developed in Russia (Chapter 16). He establishes that many Russian businesses have been implementing EMS since the late 1990s, which have also been influenced by management approaches in Western Europe, Japan and in the USA, as well as the ISO 14001 EMS standard. However, Dayman notes that the situation in Russia did not faciliate widespead implementation of EMS and that uptake of EMS has been slow by businesses. The lack of awareness of stakeholders, language difficulties (the ISO 14001 was first translated into Russian in 1998 and then a correct version published in late 2007) and difficulties in understanding the approach of EMS are reasons listed why uptake was slow. Multinational corporations operating in Russia lead the way but were soon followed by Russian companies. He notes that by 2004, more than 100 companies had their EMS certified and that many more are developing their EMS gradually. Dayman concludes that companies which implement EMS can reap many benefits and that many opportunities exist in EMS for foreign investors, as for Russian companies.

In Chapter 17, **Nina Lesikhina** and **Michele Hege Grønbech** of the Bellona Foundation provide an interesting review of the opportunities for renewable energy projects

---

The Finnish Institute of International Affairs, see <http://www.upi-fiia.fi/assets/events/UPI_Briefing_Paper_23_2008.pdf> and WWF and Allianz Partnership, *Russia: Energy Policies and Carbon Markets* (2007), at <http://assets.panda.org/downloads/13869__wwf_russian_brochure_6pp_version_low_resolution_.pdf>.

on the Kola Peninsula (in Northwest Russia). The authors highlight numerous reasons for developing renewable energy sources on the Kola Peninsula, which are mostly ecological, but also economic and socio-economic, political and practical. In this chapter, Lesikhina and Grønbech discuss the evaluations by the Bellona Foundation of the prospects of six renewable energy sources: solar, wind, hydro, tidal, wave and bioenergy resources. Currently, the region surrounding the Kola Peninsula relies mostly on nuclear energy and imported fossil fuel. With the development of renewable energy initiatives, including the Murmansk Oblast's creation of the multi-stakeholder Working Group on Renewable Energy in March 2007, the opportunities for businesses are vast. As the authors mention, before full-scale developments can occur, first the necessary technical and economic feasibility studies, pilot projects and infrastructure reinforcements need to take place. These again, providing ample opportunities for businesses to partipate in the development of renewable energy in Northwest Russia.

In the final chapter of Part III, **Jeffrey Roberts**, a geologist and director of Impivaara Securities, provides a personal account and some reflections of his experiences with investments in the Murmank Oblast and the Pomor region (Chapter 18). He highlights from a geological perspective how the Northwest region of Russia has much investment potential, in particular relating to mining (for example, apatite, platinum and phosphates), which in itself is an environmentally high-impact activity. Roberts argues that geographical and language barriers may be slowing the progress of 'frontier investments' in the Barents Sea region in Russia, despite bordering Finland and Norway with which there are many shared cultural (and geological) links. Other barriers posed are infrastructural and technical. He concludes that despite the region's relative remoteness (in geographic as well as economic terms), enormous opportunities exist for investments to thrive and mentions that such investments are, indeed, starting to occur.

CSR and philanthropy

The link between socially (and environmentally) responsible investing and CSR has been made by Tatiana Guseva. The editors, too, consider that CSR and socially responsible business is an integral part of environmental finance and 'green' projects, and *vice versa*. Within this framework, Alexey Kostin (Chapter 19) and Sergey Dayman (Chapter 20) discuss CSR from different perspectives. What pervades from the two chapters on CSR is that, in Russia, perceptions of what CSR constitutes differ widely and that although it had a slow start, many opportunities exist for CSR to become an integral part of Russian businesses.

Alexey Kostin, gives an informative account of how Russian companies are increasingly adopting CSR, corporate governance strategies and how ambitions of undertaking IPOs are providing further impetus to companies operating in Russia to implement transparent and 'good' corporate practices. He describes the role of non-financial reporting and the role of the Russian office of UNDP in promoting CSR within Russia.

Sergey Dayman reviews how CSR developed within the context of post-Soviet society. Dayman describes how various Russian companies responded to former President Putin's calls for Russian businesses to become 'socially responsible'. With businesses drawing on international experiences, first 'social reports' were published. However, Dayman highlights that, until recently, many such social reports 'bore more resemblance to PR-booklets, listing sponsorship or philanthropic activities', rather than actually providing detailed information about the company's economic, social and environmental policies, actions and impact(s) on society. Interestingly, Dayman argues that although companies make their operations more environmentally sensitive, they do not report on their improvements made in environmental performance. Similar to Tatiana Guseva (Chapter 3), Dayman mentions the difficulties caused by the diverging perceptions amongst Russian businesses of what 'sustainability' and sustainable development actually mean – and to distinguish these terms and concepts from the 'sustainability of the company'.

In the final chapter, **Kit Lawry** and **Joanna Rogers**, both (at the time of writing) involved in the work of EveryChild, discuss how businesses can be involved in child support organisations (Chapter 21). They highlight the focus of EveryChild, which aims at helping children through family-based care, rather than sending them to institutional orphanages where, often, the institutionalised children are neglected or, at worst, abused. They provide an interesting account of how businesses can channel their investments into the community in a more considered way than just through 'one-off' philanthropic donations or events. The link is not made to how socially responsible investments or philanthropy, such as towards child support organisations, should be made together with environmental investments. However, it is clear that investing in future generations and positively affecting their upbringing will have a significant effect on sustainable development of a country.

Concluding remarks

The editors understand that the contributions in this book do not provide comprehensive solutions, but mostly raise issues. This book aims to provide accessible information about the status of environmentally responsible and green investing, as well as socially responsible activities of businesses, in Russia.[19] It is an attempt to open up the dialogue and assist cooperation in this area and so lift the shroud of mystery that so often accompanies Russia.

On a more general note, the editors wish to encourage the private sector, whilst generating and increasing their own profits, to further participate in the global fight against poverty and to support the continuing worldwide commitment to sustainable development. Furthermore, given that UN Secretary-General Ban Ki-moon has warned that 'global military spending now tops $1 trillion per year, while funding for develop-

---

[19] See also WWF Russia publications '*Basic Principles for Environmentally and Socially Responsible Policy of Mining Companies*, at <www.wwf.ru/data/pub/energy/mining_eng.pdf> (September 2009) for information about sustainable business and the mining sector; and '*Promoting Responsible Finance in Russia*', at <www.wwf.ru/data/pub/business/sustainable_financing_eng.pdf> (May 2009).

ment remains woefully low by comparison',[20] more public and private investments in (sustainable) development, including funds that promote green and socially responsible business activity, are required to enable global peace. The editors hope that this book will encourage readers to engage in or promote more environmental finance transactions, as well as participate in environmentally sensitive and socially responsible business practices in Russia.

[20] 9 November 2009, see <http://www.un.org/apps/news/story.asp?NewsID=32891&Cr=disarmament&Cr1=>. In contrast, the UN newswire reported on 11 November 2009 that '…as the last preliminary negotiations before Copenhagen wrapped up in the Spanish city of Barcelona, Yvo de Boer, Executive Secretary of the UN Framework Convention on Climate Change (UNFCCC), said developed countries would need to provide at least $10 billion to enable developing countries to immediately develop low-emission growth and adaptation strategies and to build internal capacity,' see at <http://www.un.org/apps/news/story.asp?NewsID=32895&Cr=climate+change&Cr1=>.

Part I

Overview and First Comments

Chapter 1
THE RUSSIAN ECONOMY IN PERSPECTIVE

Philip Hanson[1]

Russia has been developing rapidly. Between 1999 and 2007, the economy grew at an annual average rate of 7%.[2] Most observers expect rapid growth to continue, if probably at a somewhat reduced rate. Still – its 2006 per capita gross domestic product ('GDP') was about a third of that of the UK.[3] There are, however, risks ahead. Some are the product of restrictive economic policies. Environmentally, also, there are many challenges to overcome. In this chapter, the growth record will be briefly described, policy issues be discussed and some of the environmental concerns and opportunities are noted that will be discussed more fully by other authors.

THE GROWTH RECORD

Russian economic growth since 1998 has meant real improvement in material circumstances for real people. Average wages and consumer spending, adjusted for inflation, have risen strongly. The proportion of people officially estimated to be living below the 'minimum subsistence' level fell from 25% in 2002 to 14% in 2007.[4]
Rather more debated are the sources and sustainability of Russia's economic growth. Part of the increase in production and income has been a recovery to levels just before the fall of communism. That recovery element in recent growth is now more or less played out.

1989 – late 1990s

Between 1989 and 1998 recorded Russian output fell by more than 40%.[5] The central planning and comprehensive state ownership that were fundamental components of

[1] Professor, Associate Fellow, Chatham House Russia and Eurasia Programme, The Royal Institute of International Affairs, UK. Email: *HansonPhl@aol.com*

[2] From Rosstat official data as cited by Troika Dialog, *Russia Economic Monthly*, July 2008.

[3] See World Bank, *World Development Indicators Database 2007*. This measure is not at 'purchasing power parity': that is, not simply comparing countries at the exchange rate but with adjustment for differences in price levels.

[4] Bank of Finland, *BOFIT Weekly*, 22 February 2008.

[5] For 1989-1991, see P. Hanson, *The Rise and Fall of the Soviet Economy*, London: Longman 2003, chapter 8. For 1991-1998, see Evgenii Yasin, *Rossiiskaya ekonomika*, Moscow: Higher School of Economics, 2002, chapter 16.

W.Th. Douma and F.M. Mucklow, eds., Environmental Finance and Socially Responsible Business in Russia
© 2010, T·M·C·ASSER PRESS, The Hague, The Netherlands and the Authors

the old Soviet system were done away with, but Russia was slower than, for example, Poland, the Czech Republic and Hungary in putting in place the policies and organisational reforms needed to establish a functioning market capitalist economy.

Nonetheless, the fundamental changes were made, albeit slowly. By the late 1990s, most production was in the hands of profit-seeking private firms. Consumer demand, not state planning, had the main influence on what was produced or imported. Commercial banks were operating. Money gave people and firms command over resources – in Soviet times only the dictates of the plan mattered, and money functioned as a unit of account but not as a reliable means of exchange and store of value.

The financial crisis of 1998 forced the Russian authorities to abandon an unrealistically high exchange rate for the rouble (of the order of RUB 6 to the US dollar). The rapid fall to RUB 20 or even more to the dollar suddenly made Russian production of many hitherto-imported goods viable. With imports priced out of the Russian market and real Russian firms in place to take advantage of the situation, output began to recover. Soon after, in 1999, the price of oil rose, adding another stimulus to Russian economic activity.

In this way, recovery started. Now that economic activity is roughly back to where it was in 1989-1991 – and above that level so far as consumer spending is concerned – the increased use of under-utilised, 'spare' capacity, as an easy source of growth, has disappeared.

1990s and beyond

After 1999, further rises in the world oil price fuelled the strong growth that followed. The Russian economy grew rapidly, not directly because oil prices rose, but as an indirect result of that rise. The increases in oil export earnings fed the state budget, through increased tax yields, fed company profits and also fed earnings. That in turn generated spending. Therefore, demand for Russian goods and services rose.

One effect of the huge rise in oil prices has been that Russia has been running very large trade surpluses. The country has tended at the same time to run a modest deficit in other current items – trade in services and cross-border flows of interest, profits, dividends and wage payments – but the net result has been a very large and persistent surplus in the current account of the balance of payments. In most recent years, a chunk of that inflow has been offset by a net outflow of capital – including capital flight – but a large amount has flowed into Russia's foreign exchange reserves.

This has tended to push up the real (that is, inflation-adjusted) exchange rate of the rouble, particularly, at times, against a weak US dollar. This appreciation of the rouble has been eroding the big competitive advantage that Russian producers received from the 1998 devaluation. If companies modernise, improve productivity and reduce costs, they can counter an appreciating exchange rate, but that is hard to do rapidly. Therefore, the Central Bank of Russia has, until 2008, pursued a policy of moderating the rise of the rouble by buying dollars (and, to a lesser extent, euros). Doing that meant issuing more rouble cash to do the buying, and that in turn has fed inflation. Consumer prices by late 2008 were rising at about 14% a year, but the prospects were for some deceleration. The increased money supply was less inflationary than might have been feared, since economic growth increased the demand for money, and growing confidence in the rouble led to an increased willingness to hold it, but it was still a problem.

Tight control on budgetary spending, assisted by the creation of a stabilization fund based on oil revenues, helped to limit these inflationary pressures. Consumer-price inflation was gradually reduced from 2000 through 2006. But state spending was increased in 2007, ahead of parliamentary and presidential elections; that, exacerbated by worldwide upward pressure on food prices, pushed the inflation rate back up in 2007 and early 2008. Consumer prices were rising at about 12% a year, after having been slowed to 9% in 2006.

The Putin-era record of economic growth and current-account surpluses, tracked against the price of Russian (Urals) oil is illustrated in Chart 1 (see below). Broadly speaking, accelerations in GDP have accompanied increases in the oil price, though the link is not perfect.

Chart 1. Russian economic growth, 1998-2007, balance of payments and the oil price. (year-on-year % change in GDP, current-account balance as % GDP, and average annual price of Urals oil in $ per barrel [right-hand scale]).

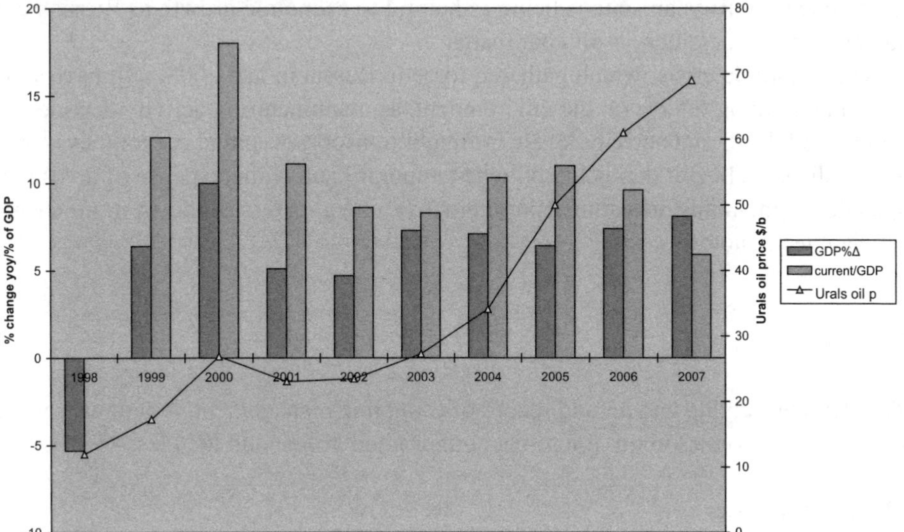

Sources: Rosstat; Central Bank of Russia; Troika Dialog.

Summary

It follows that Russian economic growth is vulnerable to a fall in the oil price. It slowed dramatically in late 2008, with industrial output turning down in December, following a rapid decline in the price of oil. Furthermore, Russia is already subject to some constraints from rouble appreciation and the disappearance of spare capacity inherited from Soviet times.[6] In addition, the working-age population has recently started to fall. The total population has been falling for a number of years, but this fall in the labour force is very recent. It adds another slowing influence. So does the global financial crisis which started in 2007. Russian banks and non-bank companies have to repay, and pay interest on, foreign loans to a total of about USD122 billion in 2008 and a further USD160 billion in 2009.[7] In most cases, they will have to pay more for fresh borrowing, and that will tend to curb investment at home in Russia.

Future tendencies

There are some favourable influences for the future. The growth of a Russian middle class continues apace and that is bringing a heady growth in demand for all sorts of middle-class accoutrements – more and better cars, new or enlarged apartments and houses, holidays abroad, and so on – as well as awareness of environmental concerns. The eager spending of newly-well-heeled Russians is matched by business confidence, exhibited in rising investment and a growth of business mergers and takeovers – in contrast to earlier post-Soviet patterns of business behaviour: taking one's short-term profit and salting the proceeds away offshore as soon as possible.

Whether the Russian state is being as helpful to economic growth as Russian consumers and businessmen, is another matter.

The economic crisis, which gathered force in Russia in late 2008, will have short-term ameliorating effects on the environment, as manufacturing activity declines, car purchases fall off and activity levels in freight transport drop and car journeys (probably) fall, as well. But this is likely to be temporary and at the expense of investment spending – including investment in alternative energy sources and in more energy-efficient technology.

Policies

The Putin leadership had an odd record of economic management. A summary would be: macro-economics good; micro-economics good at first and then less so.

Macro-economics

Throughout the Putin presidency (2000-2008), with a bit of a lapse in 2007, macro-economic management was sound in the best conservative sense. Faced with a flood of

[6] To be damaging, the price fall would have to be sustained. If oil halved in price one day and doubled back up the next, the fall would hardly matter. How long is 'sustained'? This is hard to say, but more than three months would be a fair guess. Merrill Lynch economists calculated that such a fall to $50 a barrel would slow Russian growth to a little over 5% a year (*Vedomosti*, 3 March 2008).

[7] Data from the Central Bank of Russia: <http://www.cbr.ru/statistics/print.aspx?file=[illegible] statistics/debt_maturity_08.htm&pid=svs&sid=vdRF_fisp>, accessed 6 January 2009.

petro-dollars, the Russian authorities did not spend them. The moneys accumulated in foreign exchange reserves and the stabilization fund. Industrialists and politicians called for more public spending: better pensions, better pay for public servants, more spending on health, education, social housing, roads and infrastructure generally. These were, indeed, good causes, and there was some real increase in spending on them, but the spending would most likely have not been very effective or caused inflation or – most likely – both. This was resisted, with a bit of weakening in 2007.

Micro-economics

On the micro-economic side – product and factor markets, regulation and competition – there was a period of liberal reform from the beginning of Putin's rule until midway through 2003. Legislation was passed that allowed a real market in land to develop. The tax burden on households and businesses was reduced. Bureaucratic impediments to the development of small businesses were reduced.

For at least one large company, such as Yukos, 2003 presented a different picture. Yukos was the leading Russian oil company. It was private, and like other large, privatised Russian companies, had a history that was untransparent in places. But it was becoming increasingly transparent, and had become, in the world of Russian business, a leader in improving accounting and managerial transparency – whereas, its chief executive and other senior executives were arrested on alleged charges of tax fraud.

This is not the place to probe more deeply into what happened and why. The important point here is that the Yukos affair appeared to have launched a shift in state policy towards some larger companies. Yukos itself went bankrupt after the courts froze its assets in face of large claims for allegedly unpaid back-taxes; thereafter, it was acquired by the state oil company Rosneft through a series of auctions.

State acquisitions and other sector developments

The Yukos affair was followed by further state acquisitions, most but not all of them in the oil industry. The state-controlled Gazprom acquired another large private oil company, Sibneft, in a deal with Sibneft's (reportedly) main owner, Roman Abramovich. Gazprom also acquired a controlling stake in the Shell-led Sakhalin-2 offshore project and is in the process of acquiring control of the big East Siberian gasfield at Kovykta from TNK-BP, a Russia-British joint venture.

These latter acquisitions were managed with less apparent use of state muscle than in the case of Yukos, but the purchases were preceded by the use of administrative pressure over taxes or natural-resource-licence conditions. Somewhat similar pressures were used in several more instances, including VSMPO-Avisma, the world's largest titanium producer – now in state control.

A somewhat separate development, with similar implications, has occurred in engineering industries associated directly or indirectly with defence production. State holding companies have been set up to control shipbuilding, aircraft manufacture, atomic energy, nanotechnology and an array of defence-related companies now being grouped into the holding company Rostekhnologii.

One result is that the share of Russian oil output coming from state-controlled firms has risen from about one-fifth in 2004 to about two-fifths currently. Another result is that the share of the private sector in generating GDP, as calculated by the European Bank for Reconstruction and Development (EBRD), has fallen from 70 to 65% – an unusual change for a country moving towards a market-economy.

Property rights and rule of law

More important than such statistics is the effect on the security of property rights and the rule of law in Russia. In the World Bank's Ease of Doing Business Survey for 2008, Russia overall ranked 106th out of 178 countries – a low ranking for a country of its 'upper-middle-income' development level.[8] There is also evidence from both Transparency International and the World Bank-EBRD Business Environment and Enterprise Performance Survey that in recent years corruption in Russia has worsened.

Foreign investment

Nonetheless, foreign business interest in investing in Russia, knocked back for a while by the Yukos affair, recovered and grew strongly, for four basic reasons. The Russian economy is big. It has been growing rapidly. It has large, underdeveloped oil and gas reserves. And the strength of the public finances has meant that the country's sovereign credit rating has been raised to very respectable levels. Russia's resources of people and skills are creating further challenges. The working-age population is declining and skill shortages are widely reported. It is claimed by many educational specialists that, despite the undoubted proliferation of business-related training, in management, accounting and the like, the core of the Russian educational system has been weakened over the past two decades.

The Russian leadership has moved to restrict foreign investment in some 42 'strategic' areas. These are mostly (in terms of numbers of headings) defence-related lines of production, but they also include oil, gas and metals, which are in economic aggregate much more important, and major media. The law on foreigners' access to strategic industries came into effect in May 2008. It lays down certain ceilings on the percentage stake a foreign company may normally have in companies in the industries listed. These are 50% in most of the activities (25% for a foreign state concern) and 10% in natural-resource industries (5% for a foreign state firm). A foreign investor can seek a larger stake as an exception, and such proposals will be vetted by the Russian government. Pre-existing deals that exceed the prescribed limits are supposed to be sacrosanct. The first two applications for more than threshold-level acquisitions were in gold prospecting and aerospace. Both were approved. Early signs are, therefore, that the Russian authorities' approach may be flexible and pragmatic.

---

[8] <www.doingbusiness.org/ExploreEconomies/?economyid=159>, accessed 21 March 2008.

Exactly how this law will work out in practice remains to be seen. It may serve some useful purpose at least in clarifying the rules of the game for foreign investors, though even that is not certain. The general idea seems to be to keep foreign companies from controlling any assets that are either especially attractive (most natural resources) or are linked (including tenuously) to arms production.

Summary

To sum up on policy: the Russian leadership has in recent times been reassuringly orthodox in managing taxing and spending. The value of strong public finances is evidently understood in the Kremlin. As Mr. Marc Franco once stated when quoting President Medvedev (see Chapter 12 (*Douma*)), in openness and market liberalism, the Russian state has 'regressed'.

Meanwhile, after 2004 the growth of Russian oil production – rapid in the immediately preceding years – slowed to a crawl. The growth of the volume of oil exports did likewise, though the scale of the export slowdown was masked by the meteoric rise in oil prices, ensuring that oil export earnings continued to rise steeply. This slowdown in the industry is the result of several factors, in which heavier state control and taxation loom large. That leaves Russia all the more vulnerable to a fall in the oil price: export volume growth has ceased to be buoyant and cannot offset even a modest price fall.

Environmental issues

At present, the Russian economy is a mixed bag. It is internationally competitive only in natural-resource-based industries. The service sector and some consumer-goods production are booming. Energy production problems lie ahead, particularly with a threat of shortfalls in gas production, given export contracts and domestic demand. However, President Dmitry Medvedev sees the link between Russia's future competitiveness and the environment. During his pre-election days at the Meeting of the Security Council held in January of 2008, he said, '…In the coming years the quality of the environment will be a key factor in the competitiveness of each Russian region and that of the entire country, not to mention the significant impact these factors have on the demographic situation and the health of the nation.'

The post-Soviet fall in output reduced some highly polluting heavy industry, which is unlikely to revive. The recovery has seen a shift towards service industries and consumer goods. So much – environmentally speaking – the better. On the other hand, the boom in car ownership and the gross congestion in the larger cities have meant a significant increase in vehicle emissions and air pollution. Some environmentally orientated companies are starting to emerge and thrive. However, it appears the 'environmental sector' is only just starting to develop.

Energy efficiency

In energy terms, though, Russia remains, as it was under communism, a lavish over-user of energy. Table 1 shows this.

Table 1. Energy usage in selected countries in 2005 (tons of oil equivalent per $1000 of GDP in year-2000 purchasing-power-parity dollars)

Russia	0.47	Canada	0.27
All OECD	0.18	World	0.21

Source: IEA, *Key World Energy Statistics* 2007.

Canada is included in Table 1 for climatic comparability with Russia. Part of the massive social adjustment Russia has undergone since the fall of communism is the reduction of settlements in the Far North as people have moved away from them. This is part of a monumentally painful set of changes, but it is in several senses beneficial. Many of these settlements were created in the Soviet period as forced labour settlements, and they were located to exploit natural resources in conditions and by methods that would not have been used in a democratic country with a market economy. But even after 14 years of this emigration away from such remote, frozen places, Russia is using energy far less efficiently than Canada. There is scope for more investment in energy-efficient electricity generation in Russia now that power stations, other than nuclear and hydro, have been 'unbundled' from the former state electricity monopoly, UES, and privatised to strategic investors who, in principle, would have the resources and the incentives to upgrade these capacities. However, Gazprom is the leading new 'private' investor in the sector, which in some circles causes concern, and the economic crisis is leading to delays in these investment programmes.

Part of the reason for the low energy efficiency of the Russian economy is the low prices of gas and electricity to domestic users, both residential and commercial. The incentive to economise on energy has been artificially blunted. The State Tariff Service calculated in late 2007 that if domestic gas prices were raised to those of gas exports to Europe, net of transport cost and export duty, they would have to rise threefold. This change is supposed to come by 2011 for commercial users and a year or so later for households (with special offsetting payments for poor families). It will most probably be resisted and the leadership may duck out of this commitment. Somewhat similar uncertainties attach to electricity pricing.

If the domestic price rises are carried through, there should be large savings in energy usage. Even if they are, however, Russian policymakers anticipate a need to substitute other fuels for natural gas in electricity generation – which relies heavily on gas at present. The Energy Strategy for 2020 and a draft Energy Strategy for 2030 have the same plan: to substitute coal and nuclear for gas, on a large scale, in electricity generation in order to free gas for export commitments and other domestic uses.

The paradox here is that the most polluting form (unless you believe in clean coal) of electricity production, namely coal, comes from the one industry in the Russian energy sector that has largely escaped state control – coalmining. Coal production is therefore in a tolerably buoyant state. The environmental implications are not encouraging.

Environmental policies and investments

The following chapters will describe in more detail where Russia is heading in context of its environmental laws, policies and in promoting environmental investments and opportunities in Russia. Core policy and investment areas may be in climate change mitigating Kyoto Protocol projects (joint implementation), solar energy technology, energy efficiency, environmental services or water. If biofuels develop, this may bring to life more novel opportunities of investment in agriculture and in conversion of old factories. As rings true for most sectors in Russia – the success and development of the 'environmental sector' will largely be determined by overall economic trends in Russia.

Conclusions

The Russian economy has been thriving. It has however a number of frailties, many of which may be attributable to strong state control, that may well slow growth over the next few years. So far as the environment is concerned, post-communist change has alleviated some forms of environmental damage but exacerbated others. Probably the single policy act that could do most to cut wasteful energy use in Russia would be to remove controls on electricity and gas prices or at least to raise those prices very substantially. That would be good both for economic efficiency in the narrow sense and for the environment.

Chapter 2

ENVIRONMENTALLY AND SOCIALLY RESPONSIBLE BUSINESS IN RUSSIA: WHY BOTHER?

Godfrey Cromwell[1]

This chapter provides a business perspective relating to the debate on environmental issues in Russia. It addresses questions such as is 'green business' possible? Is it understood in Russia? What are the prospects? And why bother with investment other than for short-term financial return?

A TRADITION OF PLENTY

Russian business attitudes to the environment are shaped by the culture and traditions of the country. Russian citizens are proud of the vast sweep of their country and its seemingly limitless resources. Anyone who has taken a train across even a relatively small part of the Russian Federation will – as the seemingly endless stretches of forest go past the window for hour after hour – know the feeling. Similarly, many Russians consider land and other natural features as almost limitless resources in a way that Western visitors – whose every square yard of land has been calibrated and its ownership precisely defined (and not infrequently fought over) – find hard to relate to.

Many pre-communist values reflected a 'wild frontier' approach to Russia's natural heritage as a huge 'out there' ready for taming, conquest and exploitation. The communist period retained these attitudes, with the natural resources as a basis for galvanising development – the ill-fated 'Virgin Lands' agricultural programme reflects the mentality. The era of production (rather than efficiency) targets encouraged often hugely wasteful and environmentally destructive exploitation of natural resources, while the dash for industrialisation under Stalin paid scant regard to environmental impacts. At the same time, the limitations on personal liberty and the lack of environmental controls or information prevented any real opposition to environmentally damaging production facilities to develop or be heard.

NEW WAYS?

What has changed with the greater opening of the country and the economy to outside investors? In many cases old attitudes remain and enterprises – due to a combination of

---

[1] At the time of writing, Executive Director, Russo-British Chamber of Commerce, <www.rbcc.com>.

W.Th. Douma and F.M. Mucklow, eds., Environmental Finance and Socially Responsible Business in Russia
© 2010, T·M·C·ASSER PRESS, The Hague, The Netherlands and the Authors

lack of redirection and lack of resources – continue in the same old way. Declines in control, or the ability to buy it off, in some areas have enabled some unscrupulous investors and Russians (often jointly) to exploit resources for quick profits in ways that would not be acceptable elsewhere.

On the positive side, the greater openness and ability to share information enabled by political changes in the last fifteen years have enabled new levels of environmental awareness-raising and expression. Similarly, Western business practices (at least those represented by some global and longer-term investors answerable to Western shareholders and media) have a key role to play in decreasing environmental impact of business activities.

Nevertheless, old habits die hard. It is a personal observation of 15 years working in Russia that many Russians still find it hard to take seriously the concept of the natural environment as a finite resource under pressure. The Kyoto Protocol is often seen – including among senior energy figures I have spoken to – as a 'Western-inspired nonsense' of little practical value and which, although it has now actually been ratified (as was promised in 2002) by Russia, was signed purely as a key to WTO accession. The 'financial mechanisms' implanted in Kyoto Protocol may be of interest longer term, but seem to have little appeal at present.

What is it like doing business in Russia?

The Russo-British Chamber of Commerce (RBCC) is regularly asked this question by our non-Russian members. Sometimes Russia seems like Western Europe, while at other times it seems very different indeed. Working successfully in Russia is often about navigating between these two extremes. It is certainly a mistake to think you are in a West European business environment, but equally there are many overlaps between Russian and European cultures that encourage a feeling of trust and understanding. Russians often enjoy to explain how 'Russia is different' in almost any situation (but particularly in the face of criticism). In reality, Russia is not as different as it has been said by some, but equally it is certainly not the same!

Space does not allow a detailed analysis of the particular requirements of Russia in terms of market entry, legislation and regulations, fiscal requirements, corporate structures, credit arrangements, etc. A good number of publications are already available on this subject and RBCC is happy to assist in all these areas and more.

Nevertheless, it might be helpful to touch on a few generic issues:

- *Building relationships*. Personal trust and credibility are important everywhere, but perhaps particularly so in Russia, where the rule of law is not yet always to be relied upon for resolving all disputes and where personal relationships underpin many business ventures and understandings. This means that being clear about what you want and ensuring that you are fully understood are essential. Equally, it is of paramount importance to ensure that your business partners are reliable and that investors understand both their immediate and their longer term agenda. Adequate research

is key to preparing a basis of a successful business. The RBCC consistently encourages Western investors to come to Russia and work at getting to know potential partners and building relationships before embarking on business activity.
- *Bureaucracy and corruption.* Yes, there is plenty of both, but as the ever-engorging European Commisson amply demonstrates, Western Europe is cheerfully, if perhaps somewhat absent-mindedly, encouraging both these evils much closer to home. However good an investor's personal contacts and relationships, it is essential to know the rules and play by them. Foreign companies regularly have to spend substantial time and resources to ensure that due diligence and all bureaucratic hurdles are cleared in full.
- *Risk.* Russia is an 'emerging market' and comes with the characteristics and risks associated with that status (see also Chapter 1). The rating agencies vary in their rating of Russia as an investment destination, but there are clear examples of businesses – with foreign, local and combined investment and management – that have achieved great success in the Russian Federation. In operational terms there are a range of companies in Russia and elsewhere that offer risk advice and mitigation services, legal advice and insurance cover tailored to Russia. RBCC is happy to take enquiries concerning experienced providers of these services.

Whether or not an investor is interested in environmental projects or engaging in socially responsible business, the above core issues apply universally.

Recipe for success?

Readers will not be surprised to learn there is no magic formula to achieve business success in Russia. Many of the same requirements apply in any business environment, e.g., information, planning, organisation, management, quality control. Perhaps what marks Russia out is that it can require additional levels of persistence and patience, of understanding of how the local system works, of risk management and of lateral thinking to achieve results. This makes Russia a very stimulating and exciting place to work. The apparent similarities and yet deeper differences referred to above can also make it a psychologically exhausting place for non-Russians to work.

What about green investment and socially responsible business in Russia?

For many investors and entrepreneurs, green and social credentials are becoming increasingly important. These attributes can add value to products, to services, and to companies if, for example, they improve customer or employee attraction/retention or enable companies to operate successfully within emerging environmental regulations (and thereby reduce the risk of state enforcement or civil actions). This is a welcome development, but should not be seen as a separate niche activity – it is part of successful mainstream investment strategy. In Russia, as anywhere else, such elements of investment are, thus, part of a whole strategy. A company evaluates financial return along with social and environmental effects as part of its full appraisal of the long term business potential of working in a sector or geographical area, including one such as Russia.

Future path

There is a green movement in Russia (photographs of Greenpeace activists being arrested make regular appearances in the newspapers). I personally have worked with Russian environmental NGOs and brought them together for face-to-face discussions with Russian parliamentarians and legislators to discuss improvement of Russian environmental legislation.

The generation now coming into their own will differ greatly from their forebears – environmental issues and advocacy are lively youth issues. Thus, it is likely that in the future, investors will increasingly need to adopt environmental and socially responsible operations. There is much still to do!

Is green investment good business?

This is the fundamental question that must be unequivocally answered and backed up with comprehensive proof by advocates of 'green' or 'socially responsible' investing. At present such initiatives face (and not only in Russia) a range of challenges.

First, there is the simple question of definition. What is a green investment or socially responsible business and how does one quantify and compare alternative investments?

Second, environmental issues tend to be associated with extra costs (clean-up, preventative technologies, etc.) rather than profits. Replanting trees, avoiding soil erosion, cleaning up rivers, installing pollution filters, recycling and a host of other environmentally beneficial activities are all extra cost – which comes straight off the bottom line. How are these to be justified?

Equally, some might argue that keeping alive unprofitable, heavily-indebted old Soviet-era enterprises (in provincial Russia quite often the only local employers and providers of everything from jobs to roads, from social fabric to healthcare) is a 'socially responsible' investment. No-one expects a rush of investors into such enterprises, however.

So is my conclusion gloomy? Not at all, for the following reasons:

a) Western enterprises, and increasingly Russian enterprises too, recognise that basic environmental controls are part of sensible health and safety practices.
b) Exactly because such practices cost money, there will always be those who seek to cut corners. But in a world that is increasingly litigious, where director's liabilities can be drastic, and where ever accelerating information dissemination means that a bad reputation can be around the globe in minutes, the disincentives to 'save' on such controls are increasing.

c) Employers, particularly those to whom attracting and retaining high-quality staff is a priority, also recognize that an attractive and healthy environment, as well adopting a socially responsible business strategy is a factor in staff motivation. While this can be only a localised issue, when coupled with employees' who wish to be associated with companies seen to be environmentally sympathetic, it can be a powerful factor.
d) Environmental business is also a business in itself. Worldwide, a number of companies have emerged in the renewable energy market, as well as to clean up, prevent or mitigate environmental damage. While the market for these exists or expands, investing in environmentally beneficial activities will remain attractive as a direct business proposition.
e) Long-term investors recognise (to sanitise an old expression) that 'you do not pollute where you eat'. In the short-term, environmentally sound investments and socially responsible business behaviour impact on company image and effectiveness. Longer term it reflects increasingly a global feeling that the world is a small place in which the actions of each affects the lives of all. Should investors be tempted to forget this, there are always the local and international pressure groups, along with their shareholders – and perhaps above all their own children – to remind them.
f) From the business perspective, sustainability is a word that has been used almost to the point of extinction (or at least meaninglessness). Nevertheless, at a time when the planet itself seems to be slipping out of control, the need for sustainability of the environment and economic development has never been more apparent or more pressing.

So where does this take us?

Incentives for environmental investment and socially responsible business cannot be left to personal conscience alone. If society wants something enough, it introduces and uses the sticks and carrots to achieve the desired outcomes. The rise of environmental awareness worldwide has been followed by legislation to put society's wishes into statute. The growing feeling that businesses, at least big businesses, are (or should be) about more than maximising profits, and should recognise overtly their social impact, will also encourage new legislation.

The challenge in many cases is to ensure the rigorous, but even-handed enforcement of legislation. This is also vital to create a level playing field for all businesses. Equally, however, environmental claims should be audited to discourage token gestures and empty rhetoric.

On the 'carrot' side, smart businesses recognise that a good environmental profile is (at least currently) a characteristic of being seen as a modern and leading business. International political and social acceptability are increasingly tied up with demonstrable environmental and social credentials.

Finally, environmental programmes funded by a business can be a demonstration of its success – after all, maintaining a beautiful garden has been sign of wealth and prestige since the very earliest civilisations.

Chapter 3

THE IMPORTANCE OF SOCIALLY RESPONSIBLE INVESTING TO PROMOTE SUSTAINABLE INVESTMENTS IN RUSSIA

Tatiana V. Guseva[1]

This chapter discusses the needs and opportunities with respect to the promotion of socially responsible investing ('SRI') in Russia. It considers such issues as (i) the impact of the concept of sustainable development in Russia; (ii) environmental and social responsibility of business; (iii) safety concerns; (iv) the historical background within which SRI has emerged in Russia; (v) Russian perspectives of SRI; and (vi) how foreign investors approach SRI in Russia.

BACKGROUND

Internationally, SRI is understood as an investment process that considers the social and environmental consequences of investments, both positive and negative, within the context of rigorous financial analysis.[2] Furthermore, one can say that SRI is a process of identifying and investing in companies that implement the standards of Corporate Social Responsibility ('CSR'). SRI is sometimes described as social investing, ethical investing, mission-based investing, but in most cases SRI reflects an investment approach that integrates social and environmental concerns into investment decisions.

Since SRI is closely linked to CSR, the author addresses the latter from the standpoint of the World Business Council for Sustainable Development ('WBCSD'), namely:

> 'Corporate Social Responsibility is the continuing commitment by business to behave ethically and contribute to economic development while improving the quality of life of the workforce and their families as well as of the local community and society at large.'[3]

Therefore, it is possible to conclude that CSR is about conducting business in a socially and environmentally responsible way, complemented by investment in communities for solid business case reasons. This approach is recognised in Europe, and many

---

[1] Professor, D. Mendeleyev University of Chemical Technology of Russia, Moscow, Russia; *tguseva@muctr.edu.ru*.

[2] Social Investment Forum <http://www.socialinvest.org/areas/research/trends/>.

[3] Lord Holme and Richard Watts, *Making Good Business Sense* (World Business Council for Sustainable Development publication, 2003).

W.Th. Douma and F.M. Mucklow, eds., Environmental Finance and Socially Responsible Business in Russia
© 2010, T·M·C·ASSER PRESS, The Hague, The Netherlands and the Authors

experts believe SRI is sustainable on the basis that social responsibility becomes an integral part of the wealth creation process, which, if managed properly, should enhance the competitiveness of business and maximise the value of wealth creation to society.

Sustainable development in Russia

Both SRI and CSR have been developing actively (if not emerged) as a result of the recognition and wider dissemination of the concept of sustainable development as put forward in 1987. Sustainable development is defined as 'development that meets the needs of the present without compromising the ability of future generations to meet their own needs'.[4]

The role of business in sustainable development is sometimes undervalued, but Agenda 21 clearly emphasises:

> 'Entrepreneurship is one of the most important driving forces for innovations, increasing market efficiencies and responding to challenges and opportunities… Responsible entrepreneurship can play a major role in improving the efficiency of resource use, reducing risks and hazards, minimizing wastes and safeguarding environmental qualities.'[5]

In Russia, the concept of sustainable development was initially only accepted by environmentalists and for a long time remained misunderstood by the general public. Even now, 'sustainability' is frequently addressed as a synonym of 'stability'[6] while the word 'development' is very often omitted. Environmentalists, who are familiar with the theory of ecosystem sustainability, appreciate the much more comprehensive meaning of sustainable development, but often tend to focus on environmental issues, rather than on the balance of economic, social and environmental development priorities. Consequently, sustainable development related issues have not been seen as a priority by core Russian research groups – and, whilst economists have stayed away from it, local and regional governments sometimes have used the concept to gain the votes of those who sought economic stability.

NGOs and some educational establishments have showed a strong interest in raising awareness of the concept of sustainable development, but they appear to have had little success. This was reflected in the year 2002 when Russia prepared an 'Environmental Doctrine' to be presented at the Johannesburg Summit, rather than a sustainable development related policy document.

---

[4] World Commission on Environment and Development, *Our Common Future* (Oxford 1987). See also <http://www.un.org/esa/sustdev/index.html>.

[5] Agenda 21. The United Nations Programme of Action from UN Conference on Environment and Development. Rio de Janeiro, 1992. See also <http://www.un.org/esa/sustdev/documents/agenda21/index.htm>.

[6] Partially, this happens due to translation difficulties; sustainable is translated into Russian by words that are close to 'steady', 'firm', 'stable', etc.

Environmental responsibility of business

In any case, in Russia, the environmental responsibility of business is mostly seen through the prism of expensive 'end-of-pipe' approaches: it is believed that companies are unlikely to select environmentally friendly (or environmentally acceptable) solutions if there are ways in which to run the business, which do not meet environmental requirements. It is necessary to point out that this position is not wholly wrong; moreover, it is easy to substantiate it by using examples in nearly any economic sector in Russia.

Environmental problems in Russia were not seriously considered before the 1980's, whereas social issues have often been addressed in context of lost socialist values and achievements. For example, today, a certain nostalgic feeling amongst Russians is fed by memories of the 1970's and 1980's when Russian industries were responsible for the development and operation of the so-called 'social infrastructure' (including kindergartens, summer camps, resorts, sport grounds and cultural centres). Under this system, the majority of the population used 'free' (not paid directly by the customers, but deducted – without any declaration – from their salaries) social and municipal services. Nowadays, municipal reform along with reduced 'free' social services, cause a wide-spread public perception that today's entrepreneurs take very little care of their employees and ignore communities living in cities and settlements where companies are located. Moreover, it is believed that companies which avoid taxes or which do not pay due sums to local and regional budgets, so prevent the development of municipal services and slow down municipal reform in Russia. Again, in Russia there are many cases that prove that certain industries follow such an approach.

Safety

The third factor, which needs to be considered while discussing the importance of SRI and CSR in Russia, is the issue of safety. 'Safety', as such, deals both with environmental and social issues, and quite often it is difficult to differentiate between the two.

Since the 1990's, information has increasingly been made available to the general public regarding major accidents that occur in industry, transport and the municipal services, which have caused the death of people and severe environmental pollution. On the one hand, this increase in reported accidents is a result of the growing transparency of Russian society, and on the other hand, due to the increase in the actual number of accidents.[7] It is stated that the reasons for these accidents is the 'human factor' (i.e., human non-compliance with regards to safety requirements), technical problems and poor technical conditions of the given industries themselves, and, especially, with respect to safety equipment.[8]

---

[7] Report 'On the state of industrial safety of hazardous enterprises, rational use of mineral resources and protection of entrails in the Russian Federation in 2003' (Federal Mining and Technological Surveillance Service 2004). Newer versions are not available.

[8] Idem.

Consequently, it is often understood by the public that industrial installations are operated without the required investments in technological and technical improvements, and that employees are not trained appropriately (due to the lack of concern and investments in the 'human factor'). This point of view is often shared with respect to the regional authorities and the municipalities, which are at times regarded as preventing industrial developments in order to minimise the risk caused to society and to the state of the environment.

Historical background

As mentioned above, the concept of sustainable development, SRI and CSR emerged in Russia at the point when Russia was going through a difficult period of economic and social restructuring. At the same time, there were (first) significant foreign investments; the transformation of some national oil, gas and non-ferrous metal businesses into transnational corporations; rapid deterioration of some other sectors; and the availability of adequate environmental and social information.

Against this historical backdrop, it is not surprising that a number of stakeholders tend to assume that environmentally and socially irresponsible investment policies and management, which were typical of the 1990's, caused negative social changes and environmental problems and that they persist within the conditions of the recent economic decline.

Naturally, in reality, the reasons for environmental and social problems are far more complicated and rooted in both historical and current peculiarities of national policy and governance. These, in return, reflect both internal and international problems, even though it is impossible to exclude the issue of irresponsibility from the whole spectrum of reasons.

Russian perspectives on SRI

In 2004, the Russian Union of Industrialists and Entrepreneurs (RUIE),[9] published their 'Social Charter', which became a document reflecting the understanding of social and environmental responsibility by Russian entrepreneurs. According to this document, the

> 'Social mission of business lies in the achievement of sustainable development of independent and responsible companies. This development complies with long-term economic priorities of business, contributes towards social peace, safety and wellbeing of citizens, environmental protection, and provision of human rights.'

---

[9] For information regarding RUIE, see <http://www.rsppr.ru/ and, generally: http://www.corp-gov.org/links/link.php3>.

In this document, the term 'sustainable development of companies' is based on the Russian understanding of the term 'sustainable' and has not much in common with the concept of sustainable development as developed under international environmental law. At the same time, all components of socially responsible business are there: economic, environmental and social issues are addressed and it is hoped that it includes that the long-term economic priorities of business do not contradict long-term national development priorities.

At the beginning of 2005, the Russian Ministry for Economic Development and Trade prepared a draft 'Mid-term Programme of Social Development of Russia in 2005-2008',[10] which, among many other issues, contained a requirement to draft a new Federal Act 'On Compulsory Social Reporting of Natural Monopolies, State Owned Enterprises and Enterprises with Large Contribution of State Shares' (although this Act had to be passed in 2006 its appropriateness is still being discussed). Therefore, although not actually concerned with SRI *per se*, not only has the Russian government, but also Russian entrepreneurs, begun to show some concern for SRI issues. The former has done so by developing longer and mid-term plans, programmes and legislation. The latter do this by means of preparing and adopting voluntary codes and disseminating open reports (including those complying with the Global Reporting Initiative[11] requirements). Moreover, since the year 2001, the Association of Russian Managers[12] has been promoting research and running various public events (conferences, round tables, discussions, etc.) devoted to the social responsibility of business.

During 2004-2008, the amount of 'open' corporate reports prepared and published by Russian companies has been growing – and the quality of such reports has been improving. Russian NGOs have been encouraging entrepreneurs to disclose sustainable development related information. For example, WWF Russia[13] runs a longer term project which ranks 'open' corporate reports and discusses their shortcomings, while the Agency for Social Information conducts numerous conferences and workshops (so called dialogues) aimed at clarifying positions of various stakeholders and/or at presenting open corporate reports.

Foreign investors and SRI

Foreign investors may wish to bear in mind that stakeholders often pay more attention to foreign investors than to Russian investors. There are various historical and cultural reasons for this and environmental and social considerations have a serious role to play.

Firstly, it is assumed by many stakeholders (or the general public) that often out-of-date technological solutions are utilised and installed at Russian-European joint-ven-

[10] See <http://www.niisp.ru/News/Events/art34>.

[11] *Sustainability Reporting Guidelines* (Global Reporting Initiative 2002). See also <http://www.globalreporting.org/guidelines/2002/gri_2002_guidelines.pdf>.

[12] See <http://www.amr.ru/>.

[13] See <http://www.wwf.ru>.

ture enterprises. It is, therefore, supposed that such out-of-date installations present a high risk of causing accidents and severe environmental impacts. It is unlikely that one can immediately find cases that evidence this position. However, the prevailing stakeholder perceptions and second of all, Environmental Impact Assessment ('EIA') and Environmental Review ('ER') procedures of very reasonable projects were often organised formally, and the necessary information was neither interpreted appropriately, nor disseminated widely. This led to the major change in Russian legislation according to which an Environmental Review became an insignificant part of so called Construction Review. In fact, EIA procedure remains, therefore, the only preventive instrument of environmental regulation, which is used in Russia. Under these conditions foreign investors play a very positive role by ensuring that EIA procedures are conducted in accordance with internationally recognized approaches. Leading consulting companies are often called to run quality assessment of EIA procedures and EIA Reports ('EIRs') and, thus, in some ways substitute the former state Environmental Review Commissions.

From a social responsibility standpoint, foreign investors are sometimes criticised based on the public assumption that new or re-structured enterprises tend to employ younger specialists trained abroad and in some cases foreign employee – thus, neglecting middle-aged local labour, or that foreign investors offer unfair employment conditions (and that they discharge employees on grounds of staff reduction). In this respect, the older generation is more concerned about these issues than younger people, arguing that foreign investors are not interested in paying pensions and/or taxes in local and regional budgets that are used, for example, to pay salaries to teachers and medical doctors.

Considering the above, the role that can be played by SRI to ameliorate a foreign investor's image and to rectify public opinion becomes not just important, but crucial. There is an urgent need for really transparent SRI and investment, which involves stakeholders. At the same time, it is necessary to enhance the profile of CSR in Russia, generally. One can argue that the former goes hand in hand with the latter, but this is not quite so. Public opinion is shaped also by the dissemination of solid reasoning and the provision of such reasoning needs time and resources. CSR will need to become a well-understood concept, and genuine SRI and CSR practices ought to substitute philanthropic exercises, which are peripheral to the main business, but which are increasingly explored by larger companies not only in Russia, but elsewhere, too. It is necessary to emphasise again that SRI becomes sustainable when social and environmental responsibility forms an integral part of business, helping to enhance its competitiveness and to maximise the value of wealth creation to society.

Concluding remarks

From a general perspective, SRI can act as a strong factor in promoting economic development and can also contribute towards the solution of environmental and social challenges.

Economically, SRI can serve as a powerful instrument to balance regional differences, and to support the revamping of the processing sectors of industry, municipal services and agriculture. Socially, SRI can create new employment and livelihood opportunities (and thereby contribute also to the development of special and higher education). It can increase the revenues of local and regional budgets (and, thus, help to support the so called 'budget sphere' entities – e.g., schools, hospitals, sport grounds) and, by exploring high social accountability standards, form the basis of good social practices. Environmentally, SRI can strengthen the role of sustainable approaches in environmental and natural resource management by means of thorough implementation of EIA practices and – again – by way of adopting voluntary approaches that go beyond compliance, e.g., the wider dissemination of Environmental Management Systems,[14] cleaner production practices, and compliance with ethical codes in nature resource management.

The above-described areas, where SRI can lead to significant improvements, determine its importance for Russia. The roles and responsibilities of different stakeholders in Russia are very similar (if not identical) to those recommended by Agenda 21:

> 'A stable policy regime enables and encourages business and industry to operate responsibly and efficiently and to implement longer-term policies... Governments should encourage the establishment and operations of sustainably managed enterprises. The mix would include regulatory measures, economic incentives and streamlining of administrative procedures to assure maximum efficiency in dealing with applications for approval in order to facilitate investment decisions, advice and assistance with information, infrastructural support and stewardship responsibilities.'[15]

It is too early to say that the practice of sustainably managed businesses is being encouraged to the full in Russia. This is possibly due to the incomplete understanding in Russia of the meaning of SRI. To conclude, it is possible to say that although there are constraints caused by peculiar Russian policy concerning the development of small and medium sized enterprises and taxes, in general, the policy and regulatory regime in Russia can be described as rather stable. Nevertheless, the below quote sets out a reasonable description of the responsible business, which needs to be promoted in Russia:

> 'Through more efficient production processes, preventive strategies, cleaner production technologies and procedures throughout the product life cycle, hence minimizing or avoiding wastes, the policies and operations of business and industry, including transnational corporations, can play a major role in reducing impacts on resource use and the environment ... Business and industry, including transnational corporations, should recognise environmental management as among the highest corporate priorities and as a key determinant to sustainable development... Leaders in business and industry, including transnational corporations, are increasingly taking voluntary initiatives, promoting and

---

[14] See chapter 16 (*Dayman*) for more detail.

[15] Agenda 21. The United Nations Programme of Action from UN Conference on Environment and Development. Rio de Janeiro. 1992. Chapter 30.1 See also <http://www.un.org/esa/sustdev/documents/agenda21/index.htm>.

implementing self-regulations and greater responsibilities in ensuring their activities have minimal impacts on human health and the environment.'[16]

Finally, as with any process based on the collective activities of various stakeholders, SRI in Russia needs to be developed transparently, based on lessons learnt internationally and by thoroughly considering national priorities and specific economic, social, and environmental conditions of transition from an older totalitarian regime to a more democratically and sustainably managed country.

[16] Ibid., Chapters 30.2 and 30.3. See also <http://www.un.org/esa/sustdev/documents/agenda21/index.htm>.

Part II

Legal and Policy Perspectives

Chapter 4

RUSSIAN ENVIRONMENTAL LAW – AN OVERVIEW FOR BUSINESSES

Daria N. Ratsiborinskaya[1]

Russia has carried out major environmental policy reforms during its transition period since the early 1990s, including with respect to global issues such as climate change, loss of biodiversity and ozone layer depletion. In view of these changes, this chapter provides a brief overview of current (and forthcoming) Russian environmental law as applicable to businesses operating in Russia. By touching upon the main difficulties that investors may face, e.g., environmental charges and fines, implementation of the Environmental Impact Assessment ('EIA') procedure, this chapter may assist businesses to seek out and share experiences of best practices, provide innovative policy advice, as well as meet the challenges in the implementation and enforcement of Russian legislation and policies.

In addition, this chapter analyses and investigates (i) economic and administrative enforcement instruments, (ii) institutional arrangements, (iii) the Russian system of environmental compliance control, and (iv) new strategies which have been developed to protect the environment in Russia, ranging from the 'end-of-pipe' approach to pollution prevention strategies (through new technical regulation, best available techniques and 'cleaner production' methodology).

Environmental problems in Russia

In many respects, Russia is a major net contributor to European and global environmental quality. 65% of Russia's territory has not been affected by economic activity and huge geographical areas remain virtually undisturbed by man. Russia contains 20% of global water resources and 22% of global forests.[2] Thus, the conservation of these immense environmental assets is a key concern.

Although both the government and businesses are increasingly taking steps to protect the environment, Russia continues to face immense environmental challenges. According to Russian official sources, the environmental situation typical of densely

[1] Staff lecturer at Erasmus University Rotterdam (section of European law) and Associate Researcher at the Asser Institute. Email : *ratsiborinskaya@frg.eur.nl.* Statements made in this chapter do not constitute or purport to constitute legal advice.

[2] OECD, Environmental Performance Reviews: Russian Federation (Paris, OECD 1999) p. 3.

W.Th. Douma and F.M. Mucklow, eds., Environmental Finance and Socially Responsible Business in Russia
© 2010, T·M·C·Asser Press, The Hague, The Netherlands and the Authors

populated urban and industrial areas (constituting 10-15% of the country's territory) is 'alarming'. For example:[3]

- municipal water supply is unreliable in many cities, with threats of water pollution, including severe contamination by heavy metals. One-third of all water pipes and 17% of sewage pipes urgently need replacement, and waste is exceeding the capacity of sewage pipes by 60%;
- systems of industrial and domestic waste handling are inadequate in Moscow and other cities. 1.8 billion tons of toxic waste have accumulated and this increases annually by 108 million tons;
- industrial and vehicle emissions cause severe air pollution in many cities;
- chemical and oil pollution of soil is characteristic to the Northern Caucasus, Kurgan, Irkutsk, the Cheliabinsk regions, the Volga basin and Western Siberia. The condition of soil (and land, generally) is unsatisfactory and critical in these regions due to soil erosion, decline in humus content, desertification and flooding, salination and nitrification, and pollution by pesticides, heavy metals and radionuclides;[4]
- the health situation in Russia is also considered a matter of deep concern. Diseases and poisoning from heavy metals and other toxic materials are a significant factor in the decline in life expectancy, which for men is now said to be only 58 years.[5] On current trends, the population in Russia is expected to decline from 140 million today, to 135 million by 2015 and 100 million by 2050;[6]
- the potential for energy savings in Russia is vast. For example, the potential for energy-efficiency gains is estimated at 400m toe (compared with annual natural gas production of 490m toe);[7]
- Oil companies in Russia spill some 20 million tonnes of oil each year (5% of total extraction).[8] In comparison, the entire spillage of crude oil into Alaskan waters as a result of the Exxon Valdez spillage in 1989 would be less than a day's spillage in the Russian countryside.[9] It is noted that Russian environmental law provides that the

---

[3] Ministry of Natural Resources, Государственный доклад 'О Состоянии и Охране Окружающей Среды Российской Федерации в 2003 году' [State Report 'On Condition and Protection of Environment of Russian Federation in 2003'] (2005) p. 83.

[4] Country Profile of Russian Federation to the UN Commission on Sustainable Development, 1997.

[5] Commission of European Communities Country Strategy Paper 2002-2006, National Indicative Programme 2002-2003, Russian Federation (27 December 2001).

[6] Russian Ministry of Natural Resources, Государственный доклад 'О Состоянии и Охране Окружающей Среды Российской Федерации в 2003 году' [State Report 'On Condition and Protection of Environment of Russian Federation in 2003'] (2005) p. 83.

[7] Organization for Economic Cooperation and Development, Environmental Performance Reviews: Russian Federation (Paris, OECD 1999) p. 6.

[8] Communication from the Commission on EU-Russia environmental cooperation of 17 December 2001 [COM (2001)772 final – Not published in the *Official Journal*], p. 4.

[9] Oleg Mitvol, Deputy Director of the Russian federal service managing the oversight of natural resources (Rosprirodnadzdor) said in a recent interview that 'spills of oil and other oil products take place almost every two weeks' [in Russia], RBC Daily Russian news agency reported, see as reported at <http://www.bellona.no/en/energy/39771.html>. Yet, according to the Bellona Foundation, Russian officials do not have the statistics available that would help evaluate the scale of the oil leakage and spillage problems in Russia.

company which causes an oil spillage must undertake all works relating to the clean-up of an ecological (pollution) incident.

Many of the above problems have a transboundary or global dimension. For example, Russia accounts for 7% of world carbon dioxide emissions for 1% of gross domestic product ('GDP').[10] Other, more regional dimensions of Russian environmental issues relate to the need to reduce pollution in the Baltic, Caspian and Black Seas, which requires joint action by Russia and its neighbours. Furthermore, control of air pollution, chemicals and hazardous waste, including spent nuclear fuel and radioactive waste, also require close co-operation with Russia's neighbours and other nations and stand to offer important mutual benefits.[11]

Russian economy and the environment in context

In the past five years, the Russian economy has grown at an annual rate of above 7%, except in 2005, when the Russian GDP expanded by 6.4%.[12] In 2006, Russia's GDP grew by 7.4%,[13] whereas according to Russia's State Statistics Service, Russia's GDP grew 8.1% in 2007. Industrial output increased by some 17% in the same period.[14] This economic growth has been accompanied by increased environmental degradation. Thus, with the recent global economic decline since late 2008, pollution levels have declined, but these are expected to increase again as the economy recovers and starts to expand again.

Pressures on the State budget have, however, led to constraints on funding for environmental policy, institutions and enforcement. Sovereign guarantees for environmental loans have been unavailable for several years.[15]

The rate of public and private investment in environmental projects is critically low. Massive technological and management modernisation is needed – including in environmental technology, much of which will need to come from abroad. Thus, there are many opportunities for (technically orientated and other) international businesses interested in investing in environmental projects in Russia, or projects, which contain 'environmental' elements. Although it is growing, foreign direct investment ('FDI') in Russia remains rather low compared to other emerging economies. For example, FDI

---

[10] Communication from the Commission on EU-Russia environmental cooperation of 17 December 2001 [COM (2001)772 final – Not published in the *Official Journal*], p. 4.

[11] For example, a recent study estimated that the benefits to non-EU countries (notably Russia, Belarus and Ukraine) from actions by the EU candidate countries to meet the requirements of EU directives would be at least EURO 9.5 billion a year, see ECOTEC et al., Benefits of Compliance with EU environmental acquis for the candidate countries (2001).

[12] See also chapter 1 (*Hanson*), for an overview of the Russia economy.

[13] Source: Deputy Economics Minister Andrei Belousov, *RIA News Agency*, 21 February 2008. <http://en.rian.ru/russia/20080221/99794450.html>.

[14] See also chapter 1 (*Hanson*), for more detail on the Russian economy.

[15] Communication from the Commission on EU-Russia environmental cooperation of 17 December 2001 [COM (2001)772 final – Not published in the *Official Journal*].

in Russia between 1991-1999 amounted to about Euro 23 billion (compared to Euro 36 billion in Poland or Euro 340 billion in China).[16]

Although the economic situation in Russia has not enhanced the improvement of Russia's environmental performance, legal certainty, as well as the implementation (and enforcement) of environmental legislation still needs to be optimised. Nevertheless, provided that the recent improvement of the general macro- and micro-economic conditions (including the current reform of natural monopolies) continues, there will be growing opportunities for businesses, which are interested in environmental investments in areas such as:[17]

- water and other municipal services
- energy efficiency
- reductions of GHG emissions

In addition, there are likely to be increasing opportunities as a result of new economic instruments and due to the continued harmonisation and approximation of environmental standards, norms and legislation.

Russian institutions and judicial system

For businesses to fully understand Russian environmental law, being familiar with Russian institutions (relevant to environmental issues) and with the Russian judicial system is key. The implementation and enforcement of Russian environmental laws (and, of course, other laws relating to businesses' activities) are strongly determined by the responsible institutions, the Russian legal system, and, more generally, the rule of law. The below provides a very broad-brush overview of both the institutional and judicial system currently prevailing in Russia.

Institutions

It is well documented that the Russian Federation is a presidential republic, which was founded in 1991 following the collapse of the USSR. According to the Russian Constitution, adopted by way of referendum on 12 December 1993, the Russian Federation

---

[16] Communication from the Commission on EU-Russia environmental cooperation of 17 December 2001 [COM (2001)772 final – Not published in the *Official Journal*] It is noted that the EU is by far Russia's main trading partner and accounts for about one third of Russia's imports and exports. This figure could increase up to 50% after the recent EU enlargement. Russia has a large trade surplus with the EU. Russian energy supplies are particularly important for both Russia and the EU. 53% of Russia's oil exports and 62% of natural gas exports go to the EU. This represents 16% of EU oil and 19% of EU gas consumption. This expanding trading relationship offers incentives for harmonising environmental standards.

[17] Analytical data according to RosBusinessConsulting, <www.rbc.ru> (4 February 2005). It is noted that a recent study estimated that the benefits to non-EU countries (notably Russia, Belarus and Ukraine) from actions by the EU candidate countries to meet the requirements of EU directives would be at least Euro 9.5 billion *per annum*.

contains 89 subjects (or 'Regions'), including 21 republics, 49 areas, 6 territories, 11 autonomous Regions, and also cities of federal significance – Moscow and Saint Petersburg. Constituents of the Federation are devised into 7 federal districts which have no legislative authorities, but administer the Regions. Constituents of the Federation are subdivided into districts and municipalities.

The essential role of the federal government is to develop federal policy, strategy, legislation and the legal base to be introduced as directly applicable within the entire territory of the Russian Federation.[18] Federal authorities must exercise direct regulation and control only in cases when the main interregional and international questions need resolving, for example, relating to the construction of a transnational pipeline, transboundary waste shipment, and so forth.

The following barriers at times hamper the effective distribution of governmental responsibilities:

- similar functions are carried out at different levels that results in duplication of functions;
- there is no clarity regarding the distribution of responsibilities, which can result in unpredictable 'freedom of interpretation';
- constituents of the Russian Federation are accorded tasks which exceed their responsibilities.

Environmental institutions in Russia

The Ministry of Natural Resources and Ecology ('MNR')[19] is the key governmental authority responsible for environmental protection and natural resources. According to a governmental resolution of July 2004,[20] the responsibilities and functions of the MNR include:[21]

- state policy formulation and normative and legal regulation in the sphere of the study, renewal, and conservation of natural resources, including management of the State subsoil stock and forestry;
- the use and conservation of the inventory of water resources; the use, conservation, and protection of the stock of wooded forests and reproduction;
- operation and safety of multipurpose reservoirs and water-resources systems, protecting and other hydraulic structures (except navigation hydraulic facilities);
- the use of wildlife resources and their habitat (except wildlife resources assigned to hunting resources);

[18] According to Art. 71 of the Constitution, the establishment of federal policy principles and federal programmes on the environmental development of the Russian Federation, defence and safety, as well as issues surrounding the generation of hazardous substances are under the exclusive jurisdiction of the Russian Federation.

[19] See <http://www.mnr.gov.ru/part/?pid=398>.

[20] Resolution of the Government of the Russian Federation No. 370, dated 22 July 2004.

[21] Ibid., p. 1.

- specially protected natural areas, as well as in the sphere of environmental conservation (except the sphere of ecological supervision).

The MNR has the right to issue environmental permits and to bring draft environmental laws and regulations before the State Duma (which is the second Chamber of Russian Parliament). MNR comprises 5 federal agencies: (i) the Federal Nature Management Supervision Service, (ii) the Federal Subsoil Use Agency, (iii) the Federal Service on Ecological, Technological, and Atomic supervision, (iv) the Federal Water Resources Agency and (v) the Federal Service for Hydrometeorology and Environmental Monitoring. The structure of MNR includes:

- *at the federal level*: departments ('UPRs')
- *at the regional level*: central departments of MNR ('GUPRs')
- *at the municipal level*: authorized representatives of MNR branches.

Other federal institutions, which are responsible for environmental regulation, are:

- *Ministry of Public Health and Social Development*,[22] develops norms relating to health protection based on four levels of toxicological danger of waste. This ministry is also responsible for the protection of labour and hygienic issues.
- *Ministry of Industry and Trade*,[23] which comprises, amongst others, the Federal Agency on Energy and the Federal Agency on Technical Regulation and Meteorology.[24]
- *Ministry of Agriculture*,[25] which consists of the Federal Veterinary Supervision and the Federal Agency on Forestry.[26]
- *Ministry of Finance* ('Minfin').[27]
- *Ministry of Economic Development of Russian Federation*[28] ('MED'), comprising amongst other agencies, a Federal Agency for Management of Special Economic Zones[29] and a Russian Federal Property Fund.[30]
- *Ministry of Energy*.[31]
- *Federal Service on Ecological, Technological and Atomic Surveillance* (which is accountable to the MNR).[32]
- The Regional prosecution departments specialized in environmental protection make part of the *Public Prosecution Department*[33] and are also involved in enforcing environmental law.

22 See <http://www.minzdravsoc.ru>.
23 See <http://www.minprom.gov.ru>.
24 See <http://www.minprom.gov.ru>.
25 See <http://www.mcx.ru>.
26 See <http://www.rosleshoz.gov.ru>.
27 See <http://www.minfin.ru>.
28 See <http://www.economy.gov.ru>.
29 See <http://eng.www.rosoez.ru>.
30 See <http://www.fpf.ru/eng/>.
31 See <http://minenergo.com/>. (Note, website address is subject to change).
32 See <http://www.gosnadzor.ru>.
33 See <http://eng.genproc.gov.ru/>.

After the abolition of environmental funds (in 2000),[34] Minfin took on the responsibility of distributing revenues (to stakeholders involved in nature-use and nature protection) earned from environmental taxes imposed by the Federal Treasury. Minfin comprises, amongst others, the Federal Tax Service, which is responsible for the collection of environmental taxes. As the structure of an 'environmental tax' is not yet fully established, the scheme which was formerly used for the collection of environmental payments is still employed in most of the Regions.

When discussing environmental regulation, it is crucial to highlight the importance of the main law-making institution in Russia: the *State Duma*. The Duma is a parliamentary body and the lower chamber of the Federal Assembly (Parliament). It comprises, amongst others, the Committee on Natural Resources and Ecology, which is mainly concerned with lawmaking. Unfortunately, cooperation between the Duma and the executive authorities is very often low-effective or even on the brink of conflict. From a business point of view, this decreases legal certainty, as well as makes environmental lobbying a very complicated enterprise.

The essential role of federal government is to develop federal policy, strategy, legislation and the legal base to be introduced and directly applicable within the entire territory of the Russian Federation. Federal authorities must exercise direct regulation and control only in cases when main interregional and international questions need solving, for example, relating to the construction of a transnational pipeline or, transboundary waste shipment.

Regional level

The procedures relating to the distribution of power between the federal and regional authorities is currently being developed. In this respect, the Specialised Commission for Power Distribution,[35] which develops proposals for distinct distribution of responsibilities, was established under the Presidential Administration.

While the MNR plays the central role in context of Russian environmental law, the determination of environmental policy, control and coordination functions over its territorial branches, State policy at the regional level is carried out by the relevant Regional Duma and Regional Administration. Their responsibilities as regards environmental regulation are set out in regional legislation and are based on the Constitution.[36] According to Article 76 paragraph 5 of the Constitution, laws and other normative acts of the subjects (i.e., the Regions) of the Russian Federation cannot contradict federal laws. In case of contradiction, federal law is deemed to prevail.

---

[34] Federal and regional budgetary resources deemed to be spent on environmental needs were allocated in so-called environmental funds, which, until 2000, were part of the budget structure in every region of the Russian Federation.

[35] *Государственная комиссия по вопросам разграничения полномочий между уровнями государственного управления.*

[36] Para. 2, Art. 78 of the Constitution of the Russian Federation reads: 'Federal executive authorities under agreement with executive authorities of the Russian Federation constituents can delegate them a part of their responsibilities if it does not contradict the Constitution of the Russian Federation and federal laws.'

Municipal regulation

Each municipality within the Regions is free to choose the means by which to implement environmental regulations. Municipalities are responsible for waste management on their territory. The key environmental authority at this level is the administration of the relevant municipality. This gives some room for manoeuvre for national and foreign investors.

Judicial and arbitration practices

The high-level judicial bodies, such as the Supreme Court (*Верховный Суд*) or the higher arbitration court,[37] participate in the environmental regulatory process by issuing instructions and 'informational letters' (*информационные письма*), which interpret legislative provisions, and by way of enacting decrees. In certain cases, informational letters summarize the judicial practice, which then create judicial precedents.[38] Such informational letters have advisory force only and are addressed to the lower courts. In Russian legal theory, some authors argue that due to their advisory character, such decrees may not be recognized as sources of law.[39] In any event, when describing the Russian legal system, it is not permitted to omit information on judicial activities in the environmental area, as such information provides interpretation of legislative provisions, helps avoid contradictions amongst such provisions and fills in the legislative gaps. Despite the advisory character of the above-mentioned instructions, the lower courts tend to follow the regulatory decisions of higher courts.

With regard to environmental law, the Plenum of the Supreme Court issued a Decree on the Practice of Judicial Enforcement of the Legislation on Environmental Liabilities.[40] This decree aims to ensure the correct and uniform application of environmental liabilities legislation. The document also points at the high level of threat to society emanating from environmental offences, and instructs the courts to make a deep and a comprehensive study of a given committed offence. In particular, the decree requires courts to provide explanations on the serious consequences of the offence, as well as on what is to be understood by 'moral damage' caused by the offence.

The Supreme Court also provides periodical analyses of judicial practice in the settlement of environmental disputes and in taking liability measures against environmental offences. Thereafter, it publishes surveys in the 'Bulletin of the Supreme Court of the Russian Federation', a publication intended mainly for lawyers.

Procurators' offices (*Prokuratura*) are empowered to generally supervise compliance with legislation. They play an important role in combating environmental of-

[37] I.e., which are the commercial courts (not to be confused with arbitration tribunals).

[38] It is noted that the Russian Federation has a civil law system, which means that officially courts are not legally bound to apply judicial precedents.

[39] O. Kolbasov and I. Krasnova, *Russian Federation*, International Encyclopaedia of Laws (London, Kluwer Law International 2003) p. 42.

[40] Dated 5 November 1998, *О практике применения судами законодательства об ответственности за экологические правонарушения.*

fences. There are currently 66 specialized procurators' offices, which were set up some 20 years ago and which have by now gained good experience in uncovering environmental offences and punishing offenders by applying existing procedures.

Any proceedings which relate to economic disputes, which have environmental consequences are evaluated, coordinated and analyzed by the Higher Arbitration Court of the Russian Federation.[41] The Higher Arbitration Court regularly issues decisions instructing regional arbitration courts on questions of the adequate settlement of claims relating to compensation for property damage inflicted due to the violation of environmental legislation, as committed by legal persons (including organizations, institutions and enterprises). Such decisions are often issued as 'letters' of the Higher Arbitration Court.

Appeals

Enterprises can appeal the government bodies' decisions in any case when they believe their interests are affected. For example, the *Law on Ecological Expertise* provides for a possibility to challenge the results of an expertise (Art. 18 Para. 8).

All cases are reviewed by a court of general jurisdiction, except for cases which have an economic interest at stake. Thus, arbitrary courts can review, amongst other matters, industries' appeals against licence/permit refusals. Currently, the Law on Arbitrary Procedures is being drafted by the State Duma, which may bring further changes to the appeals procedure. Neighbouring enterprises and other government authorities can also be involved in the process in different ways: as co-plaintiffs, co-defendants or third parties (if recognized by the court).

Private parties can challenge the government bodies' decisions on licence / permit issue or refusal if such decisions constitute a breach of citizens' constitutional rights. A remarkable recent case to illustrate this is the case of *Nadezhda Fadeyeva* who lodged an application against the Russian Government with the European Court of Human Rights ('ECHR') and won the case.[42]

Non-governmental organization and the general public can issue a plea aimed at the protection of common environmental interest(s).[43]

---

[41] Art. 127 of the Constitution reads: 'The Higher Arbitration Court of the Russian Federation shall be the supreme judicial body for settling economic disputes and other cases examined by courts of arbitration, shall carry out judicial supervision over their activities according to federal law-envisaged procedural forms and provide explanations on the issues of court proceedings.'

[42] Application No. 55723/00, *Fadeyeva* v. *Russia;* ECHR 9 June 2005. Based on the Article 8 of the European Convention Human Rights, the ECHR ruled that governments are legally responsible for preventing serious damage to their citizens' health caused by pollution from industrial installations, even when they are privately owned and run. Consequently, the ECHR held that the Russian Federation was guilty of violating the human rights of Ms. Fadeyeva, resident of the town of Cherepovets, north-east of Moscow. Ms. Fadeyeva had complained that her health had been harmed by long-term exposure to pollution from a Severstal steel plant in the town. The court said the state had failed to protect Ms. Fadeyeva by either resettling her away from the plant or reducing its pollution levels. It ordered the Russian government to pay her €6,000 in compensation and to ensure it resolved her situation. The ruling is binding and if Russia fails to comply, it could ultimately be expelled from the Council of Europe, which administers the Court. The case marked the first time the ECHR had held a state responsible for damage caused by a private company.

[43] See Arts. 10 and 12 of the Environmental Protection Law (as defined below).

It should be noted that 'group actions' are not available in Russia. Claims may be brought by groups of claimants, but all claimants must be identified at the beginning of the trial. At the discretion of a judge, such independent claims may be united in a single trial.

Penal or exemplary damages are not available under Russian law, as only material damage and lost profit incurred by the injured party are subject to compensation.

All major cases are published in information bulletins of Supreme Arbitrary Court and Supreme Court of Russian Federation.

General Russian environmental law

Environmental law is an independent branch of Russian law consisting of legislative acts and executive by-laws aimed at the preservation of the natural environment, the rational use of natural resources, and the sustainable development of nature, man and society. The main legislative rules are found in the federal framework legislative acts. These general laws are supplemented by Presidential and Government decrees, as well as ministerial regulatory acts that specify certain legislative provisions and provide concrete implementation measures. The body of general Russian environmental law is rooted in Soviet legislation, which aimed to address mounting environmental problems in the USSR. Partly as a result, contemporary Russian environmental law, for example, focuses on placing limitations on discharges of hazardous substances or harmful/polluting emissions, or imposes permitting systems on operations, which have negative environmental impacts.[44]

The protection of different environmental elements – such as soil, water, air, protected areas, biodiversity – as well as the protection of the environment against certain impacts – e.g., chemicals, waste, radiation – is regulated additionally by specific legal acts, which are based on general principles and rules, taking into consideration the specific qualities of these elements.

In general, businesses operating and investing in Russia will need to assess what kind of impacts their operations and investments may have on the environment – and ensure compliance with the relevant regulatory system(s), not only relating to laws relevant to investments in Russia, but also relating to the general (and more specific) Russian environmental laws, as may apply to their particular operations and investments.

The below provides an overview of the various key environmental laws and regulations which are likely to apply generally in context of business activities in Russia.

Federal environmental legislation – an overview

The Constitution of the Russian Federation provides the legal basis for Russian environmental regulation, as follows:

---

[44] O. Kolbasov and I. Krasnova, *Russian Federation,* International Encyclopaedia of [illegible] (London, Kluwer Law International 2003), p. 49; and see chapter 7 (*Nystén-Haarala*).

> *'every citizen has [the] right to [a] favorable environment, true information on the state of [the] environment and reimbursement for damage to health or property caused by [a] breach to environmental legislation'* (Art. 42).

Moreover, Article 58 of the Constitution provides that 'every citizen is obliged to protect nature and [the] environment'. This obligation also applies to companies. It is important to note that international agreements (e.g., multilateral treaties) constitute an important factor in further shaping national environmental policy and legislation. According to Russian legislation, provisions of international agreements have supremacy over the existing provisions of Federal legislation.[45]

In the past decade, around 30 federal laws addressing various environmental issues have been adopted. These laws can be divided into three general categories and are as set out below:

a) *Integrated federal legislation*
In this context, the term 'integrated legislation' means legal norms of general application. Most often an integrated law is of federal importance and it establishes the general legal regime of the regulated object, while specific conditions of such regime may be stated in sectoral (environmental) legislation.

- The Law 'On Environmental Protection' (*Об охране окружающей среды*, 2002) (the Environmental Protection Law)
- The Law 'On Ecological Expertise' (*Об экологической экспертизе*, 1995)
- The Law 'On Sanitary and Epidemiological Well-Being of the Population' (*О санитарно-эпидемиологическом багополучии населения*, 2001)
- The Law 'On Specially Protected Areas' (*Об особо охраняемых природных территориях*, 1995)
- The Law 'On the Protection of Lake Baikal' (*Об охране озера Байкал*, 1998)
- The Law 'On Air Protection' (*Об охране атмосферного воздуха*, 1999)
- The Law 'On Waste of Production and Consumption' (*Об отходах производства и потребления*, 1998)
- The Law 'On Hydrometeorological Service' (*О гидрометеорологической службе*, 1998)

b) *Federal legislation relating to natural resources*
- Land Code (*Земельный Кодекс*, 2001)
- Forestry Code (*Лесной Кодекс*, 2006)
- Water Code (*Водный Кодекс*, 2006)
- The Law 'On Subsoil' (*О недрах*, 1992)
- The Law 'On Wildlife' (*О животном мире*, 1995)

c) *Federal legislation relating to environmental safety*
The notion of 'environmental safety' is understood as 'life conditions, protected from industrial and natural disasters'. Various federal laws, as listed below, establish a frame-

[45] Constitution of Russian Federation, Art. 15 Para. 4.

work or provide detailed rules regarding the operation of industrial facilities, their environmental obligations and state control measures:

- The Law 'On Use of Nuclear Energy' (*Об использовании атомной энергии*, 1995)
- The Law 'On Radioactive Safety of the Population' (*О радиационной безопасности населения*, 1995)
- The Law 'On Destruction of Chemical Weapons' (*Об уничтожении химического оружия*, 1997)
- The Law 'On State Regulation of Gene Modifying Activities' (*О государственном регулировании в области генноинженерной деятельности*, 1998)
- The Law 'On Protection of Population and Areas from Natural and Industrial Emergencies' (*О защите населения и территорий от чрезвычайных ситуаций природного и техногенного характера*, 1994)
- The Law 'On Industrial Safety of Dangerous Production Facilities' (*О промышленной безопасности опасных производств и объектов*, 1997)
- The Law 'On Technical Regulation' (*О техническом регулировании*, 2002).

The fundamental principles of Russian environmental law are set out in the Environmental Protection Law and public agencies are expected to apply these when enforcing environmental law. The Environmental Protection Law provides the basis for environmental regulation in Russia. It includes principles such as the *polluter pays principle, the principle of potential environmental danger, full compensation for damage caused to the environment, principle of environmental impact assessment*. In a separate chapter, the Environmental Protection Law establishes general requirements for various economic activities, including the choice of location, construction and operation of various types of facilities, and regulates matters connected with the import of radioactive waste. It also determines the general features of the legal regime relating to specially protected areas.

In October 2007, an attempt to codify some integrated environmental legislation was undertaken which resulted in a draft of an *Ecological Code*.[46] If codified, it would result in rendering the following Federal Laws void: The Environmental Protection Law, the 'Law On Ecological Expertise', the 'Law On Air Protection', the 'Law On Waste of Production and Consumption'.

Strategic approach: 'End-of-Pipe'

Until this day, the main strategic approach in Russia as regards environmental protection is the 'end-of-pipe' approach. The assessment, evaluation, monitoring and control of the actual industrial pollution characterize this approach. Within this approach, each industrial plant is required to obtain special permits with respect their air emissions,

---

[46] The Russian text of the Code can be found at the website of the Ministry of Natural Resources <www.gov.mnr.ru>.

water discharges, and waste disposal. Such permits set the 'Emission Limits Values' ('ELV'), which are calculated to meet ambient standards (i.e., maximum allowable concentrations), but usually permit pollution levels according to the actual emissions of the industrial facility. At the same time, the 'end-of-pipe' approach includes the polluter pays principle, which provides that the polluting entity pays for the environmental pollution it has caused (e.g., in the form of ecological fees, fines, and compensation payments for environmental damage). The 'end-of-pipe' approach includes the putting-in-place of environmental compliance control mechanisms for working facilities, as well as of administrative fees and penalties for non-compliance.

Thus, the 'end-of pipe' approach does not incorporate pollution prevention measures and provides neither stimulation to apply best available technologies nor encourage environmental protection.

Environmental impact assessments

All installations in Russia are subject to permitting, regardless of their size, nature or environmental impact. Under Russian law, all draft legislation and decrees that could have environmental impacts must themselves be submitted for an environmental '*expertiza*'.

Prior to 1985, there was no formal EIA process in the USSR. Environmental impacts of selected major activities (e.g., projects and plans) were appraised by various sectoral expert committees or *expertizas*. Attempts to introduce EIA in the USSR and, then, in Russia, have led to the development of two distinctive regulatory concepts which are in practice today:

- the concept of the State Environmental Expertise ('SEE' or *gosudarstvennaya ekologisheskaya expertiza*); and
- the concept of the Assessment of Environmental Impacts ('OVOS' or *otcenka vozdeistvija na okruzhajushuju sredu*).

In 1989, the Supreme Soviet (Parliament) issued a law which made the initiation of projects which had not passed the SEE unlawful. The SEE process continues to be regulated by the Ministry of Ecology's SEE Regulations (1993). These regulations require all developers to submit the project, plan or programme documentation to the Federal or Provincial SEE Departments (SEEDs). SEEDs checks whether each activity meets the environmental requirements. The SEED may hire individually licensed experts or expert teams in order to assist the SEED to carry out the SEE. The developer is required to cover the cost of the SEE. The outcome of the SEE process is the so-called 'SEE Resolution',[47] which may be positive or negative. For example, if the SEE Resolution relating to the relevant project, plan or programme deems that a particular activity thereunder is 'negative', then the relevant activity cannot be implemented and the relevant developer must modify the project (plan or programme) and resubmit it for a further SEE.

---

[47] In Russian: *Reshenie Gosudarstvennoi Ecologicheskoii Expertizi.*

The concept of OVOS originates from a more process-oriented attempt to translate the Western understanding of what constitutes EIA into the Russian context. It focuses on trying to establish an adequate process for identifying likely environmental impacts, consulting with concerned parties and balancing environmental impacts against social and economic considerations. No Soviet or Russian legislative act (i.e., passed by the Parliament), mentions the concept of OVOS. OVOS was, however, addressed by a number of regulatory guidelines and instructions (i.e., adopted by the Ministries). The Order of the Minister of Environment No. 222 of 18 July 1994 introduced the 'OVOS Regulations'. This was done, amongst other reasons, in order to comply with the requirements of the *1991 UNECE Convention on Environmental Impact Assessment in a Transboundary Context* (the 'Espoo Convention')[48] to which Russia is a party. Currently the OVOS procedure is set out in the *Regulation on the Assessment of Environmental Impacts*.[49]

Unlike the SEE process, the OVOS process is required only for projects listed within the OVOS Regulations, which is slightly broader than the Espoo Convention list. The provincial and federal authorities are given the discretion to initiate an OVOS for other types of projects. The OVOS Regulations require (i) the developer to prepare an environmental impact statement ('EIS') and (ii) the competent authorities to initiate the consultation and public participation process. These Regulations were followed by the 1992 Guidelines of the Ministry of Ecology[50] (this ministry is now replaced by the MNR). These Guidelines detail the process for submitting the draft (and final) EIS to the competent authority within the project development cycle and provide what information is to be supplied with respect to the EIS.

Although there has been significant progress in introducing elements of EIA within Russian law, many loopholes still exist. The required process for a SEE has poor screening requirements, does not provide for the evaluation of complex and uncertain environmental impacts and has only very limited consultation provisions. Whereas, the OVOS process is not well integrated into the project development cycle and its requirements are not supported by sufficient provisions for research, training and development of guidance. Therefore, it has only limited practical application. Perhaps, most of the current difficulties relating to OVOS and SEE could be more effectively dealt with if both processes were more closely integrated. It is also quite apparent that both OVOS and SEE need a more sound legislative basis.

---

[48] See <http://www.unece.org/env/eia/welcome.html>, see also the 2003 Kiev Protocol on Strategic Environmental Assessment. Russia signed the Espoo Convention on 6 June 1991 but has not yet ratified this convention, see <http://www.unece.org/env/eia/convratif.html>.

[49] Dated 16 May 2000, No. 372 (*Об утверждении Положения об оценке воздействия намечаемой хозяйственной и иной деятельности на окружающую среду в Российской Федерации*).

[50] Issued 1 January 1992 by the Ministry of Ecology of the Russian Soviet Federative Republic (*Руководство по проведению оценки воздействия на окружающую среду (ОВОС) при выборе площадки, разработке ТЭО и проектов строительства (реконструкции, расширения и технического перевооружения) хозяйственных объектов и комплексов*). The text of the document can be found at <http://oboc.narod.ru/ovos1.htm>.

According to a study by Peterson in 1997 (produced under the auspices of the National Council for Eurasian and East European Research),[51] foreign firms operating in the Russian market, especially those operating in the pollution-intensive sectors such as the oil, logging, and consumer goods sectors, employ more efficient equipment and technology and generally use more 'environmentally friendly' practices than Russian firms. For example:[52]

- An international consortium developing oil and gas off Sakhalin Island is using the latest equipment and practices. Elsewhere, the Polar Lights joint venture between CONOCO and its Russian partners received an environmental achievement award.
- The numerous oil development projects by Western companies currently underway in the Caspian Sea region pose fewer environmental threats than comparable Russian and Soviet projects undertaken over the last 40 years.
- The Russian Industrial Consortium for Protecting the Environment, a group of international packagers, is attempting to improve waste management and recycling schemes to lessen the impact of disposable packaging in Moscow and St. Petersburg.

These examples demonstrate how businesses are increasingly adhering to best environmental practices regardless of the legal environment.

Environmental permits

Under the current framework of existing Russian law, the stages of obtaining an environmental permit are as follows:

1. The relevant company must arrange (and pay for) an inventory of its emission and discharge sources, draft emission limit values, draft discharge limit values and draft waste disposal limits. These calculations of the various 'limits' are usually made by specialised consultancies on behalf of a company.
2. The draft limits are submitted to the MNR Regional Direction for approval within one month. *Sanepidnadzor* (the Federal Centre on Hygiene en Epidemiology[53]), must agree with these limits. If the limits are not accepted and the various draft limits is refused as incorrect, grounds must be provided.
3. Once approved, the company's draft emission and waste disposal limits are valid for 5 years, whereas its draft discharge limits are valid for 3 years, provided there is no change in the company's production processes. Temporary limits are valid for one year.

---

[51] D.J. Peterson, Environmental Activism in the New Russia: Prospects for the Coming Decade, Report to the National Council for Eurasian and East European Research (Washington DC, National Council for Eurasian and East European Research 1997).

[52] National Intelligence Council, *The Environmental Outlook in Russia* (January, 1999).

[53] See <http://www.fcgsen.ru/>.

4. Permits for a company's emission, discharge and waste disposal limits are developed and approved by the Federal Nature Management Supervision Service of MNR (*Федеральной службой по надзору в сфере природопользования*), responsible, amongst other matters, for ecological '*expertise*' and the permitting. Thus, based on these draft emission and waste disposal limits, the company applies to the Federal Nature Management Supervision Service for a permit for its emission, discharge and waste disposal limits.
5. Permits are issued for 1 year and can be extended, based on a so-called 'technical report' on extension of limits.
6. The Federal Nature Management Supervision Service reserves the right, if necessary, (i) to carry out an inspection in order to reach its decision on approval of the limits, (ii) to organize an inspection of the installation, or (iii) to request information about inspections performed by the State Control Division of the territorial body.

As mentioned, both during the EIA and permitting stages, investors and operators are encouraged to use specific consultancies to perform the numerous necessary studies and calculations.

There is no fee when a permit is issued. Nevertheless, the preparation of ELVs' volumes and other preliminary data collection and expertise requires time and money. According to the enterprises interviewed in pilot regions, it usually takes around 30 days to collect and prepare the information necessary for a permit application. Thus, payments associated with permit applications may vary from tens to hundred thousands of roubles.

It is difficult to estimate the cost of a project approval, as the preparation of the EIA, including calculations of emissions and limits, requires obtaining baseline information from different authorities. The cost of obtaining such baseline information varies from one authority to another. Furthermore, the cost of an EIA differs according to the type of industry, complexity of the project and required level of expertise.

Under general Russian legislation, it is possible to appeal any administrative decision. In practice, this possibility is rarely used as individuals and companies try to avoid conflict with the authorities. However, this has happened recently in other areas and it is likely that companies may start to appeal decisions on environmental permitting in the near future.

If a requested permit (licence) is not granted, an applicant is entitled to challenge the public agency's decision in court within three months of the date of the relevant decision. Similar procedures exist with respect to appealing against the conditions contained in the environmental permit.

Environmental regulators have some enforcement powers regarding violations of permit conditions. If the environmental procedures described in the project/design documentation are not observed or other environmental violations are committed during the implementation of an industrial project, civil, administrative or criminal liability measures may be imposed (see below for more detail). If the conditions stipulated by other types of permit/licence are violated (e.g., subsoil use licence, hazardous waste licence), these licences can be withdrawn in particular circumstances and grounds, which will result in the termination of the person's right to undertake the activities listed in the licence. If, regardless, such activities are continued in the absence of the necessary licence, such activity can be treated as illegal entrepreneurial activity, which might entail the compulsory liquidation of a company.

Quality of goods, production processes

The federal Law 'On Technical Regulation' (the 'Technical Regulation Law') has attracted much attention in various circles, especially government and industry. It entered into force on 1 July 2003 and regulates the quality of goods, processes of their production, operations, transportation, as well as their sale and treatment (utilization). The objective of the Technical Regulation Law is to guarantee the safety of consumers, including environmental safety, as well as the safety of all processes and goods relating to human life and health. This objective is expected to be achieved through the establishment of a special procedure for law-making and changing technical requirements.

The Technical Regulation Law applies to two types of regulatory acts: *technical regulation laws* and *standards*. It prescribes that technical regulation laws are to be adopted according to a special procedure as set out in detail in Article 9. Further to the Technical Regulation Law, the federal executive bodies (ministries) may also adopt 'optional regulatory acts'. This means that the Russian government may temporarily adopt technical regulations that will remain in force until a respective technical regulation law is adopted.[54] Under the Technical Regulation Law, the President may also adopt decrees in cases of emergencies and such decree will become invalid as soon as a technical regulation law on this issue is adopted. Since 1 July 2003, laws that establish requirements relating to goods, production processes, use, storage, transportation, and utilization of goods have been implemented in a mandatory way only in context of protecting human life and health, as well as the environment. Technical regulation laws should be adopted within seven years after entry into force of this law. An active process of law-making is currently under way and many sectoral technical regulation laws are already in place. However, in some areas temporary laws still exist.

The impact of the Technical Regulation Law on the environmental obligations of businesses has yet to be assessed. However, it is understood that it will result in considerable operational changes being required, such as relating to standards and emission values, monitoring requirements, as well as certification requirements. Current environmental permit and project description documents for industry may become invalid in the future, unless specific regulations relating to these are developed. Even the validity of certain existing laws, such as the *Law On Industrial Safety of Dangerous Industrial Facilities* (dated 21 July 1997) may be questioned, at least in 5 years' time, if it is not replaced. However, an interpretation of the Technical Regulation Law is still underway.

---

[54] A technical regulation (law) [*tehnicheskii reglament* – in Russian] is a piece of sectoral environmental legislation or a legal norm, introduced during the development of integrated federal legislation, namely, the Federal Technical Regulation Law.

Liabilities

Breach of environmental law or permits can give rise to civil, administrative or criminal liability under Russian law.

In the Environmental Protection Law, there is no complete list of environmental violations. Such violations are dispersed among a number of legislative acts. The *Russian Federation Criminal Code and Code of Administrative Offences* determine in detail what types of violations and punishments exist.

Civil liability might result either in the need for full compensation for damage caused to the environment or performance of remediation. Chapter 17 of the Civil Code[55] specifically deals with land ownership issues, matters related to leasing, as well as the use of natural resources and civil liability (compensation for damage) relating to natural resources. Court practice evidences that remediation should replace monetary compensation if the responsible party is willing and capable to undertake such remediation (or to finance it), or if remediation is more effective than financial compensation. Under Russian law, the amount of financial compensation is estimated using applicable calculation methodologies and, in their absence, on the basis of the actual expenses incurred during/necessary for remediation of the damaged environment, including lost profits (if any).

Administrative liability may result in the imposition of fines (in comparatively insignificant amounts) or suspension of a facility's operation for up to 90 days. In certain rare cases, closure of a facility can be required. Imposing a fine or suspension of operations does not remove the obligation to eliminate a violation. Thus, if a particular violation is not eliminated, fines/suspension may be imposed repeatedly. (See the *Code on Administrative Violations*,[56] which was adopted in 2002).

Other laws contain similar provision, for example the 1995 *Law on the Continental Shelf of the Russian Federation*[57] covers all issues connected to activities within the continental shelf, including the use of minerals and living resources.

The *Law on Social Protection of Citizens Subjected to Radiation Caused by the Chernobyl Accident*[58] provides comprehensive regulation relating to Chernobyl victims' social protection and establishes the legal regime relating to areas polluted by radiation.

The *Law on State Secrets*[59] provides a list of matters which are considered to be state secrets, including various information concerning natural resources and certain activities related to the use of the environment – and provides a restricted regime of access to such information.

---

55 *Grazhdansky Kodeks*, adopted in 1994 (Part I), 1995 (Part II) and 2001 (Part III).

56 *Administrativny Kodeks*, for details see Baker & McKenzie report 'New Code of Administrative Violations of the Russian Federation: Increased Responsibility of Commercial Organizations and Their Managers', 2 July 2002: <http://www.bisnis.doc.gov/bisnis/bisdoc/020710AdmCode.htm>.

57 *O kontinentalnom shelfe Rossiskoi Federatsii.*

58 *О социальной защите граждан, подвергшихся радиации в следствие катастрофы на Чернобыльской АЭС*, 1992.

59 *O gosudarstvennoi taine,* 2005.

Environmental control agencies are empowered to suspend the operation of enterprises and organizations which systematically violate environmental law provisions relating to pollution and other types of environmental deterioration. A decision to suspend or close such operations is taken only in cases where an order or a prescription of administrative inspectors is not complied with. Such measures are not often taken in practice, due to the complicated procedural barriers (related to tax and bankruptcy regulation, as well as relating to the adverse economic consequences).

The administrative tools which are used in Russia to assist environmental enforcement and compliance include the following:

- Environmental pollution permitting, including the evaluation of Environmental Quality Standards and Emissions;
- Limit/Levels based of existing 'Maximum Allowable Concentrations' with respect to harmful components;
- Environmental impact assessment (ecological expertise);
- Ecological compliance control.

The economic instruments for environmental protection currently in force are as follows:

- *Ecological fees*: Special obligatory payments for air emissions, water discharges, and waste disposal. This includes the individually calculated payments agreed with ELVs specified in permits, and 5-multipletime payments when a plant exceeds its permissible levels of emissions or operates without an appropriate permit.
- *Environmental fines*: Administrative fines (penalties) for non-compliance. The effectiveness of these fines from an environmental protection perspective is limited due to their low rates; it is more convenient for polluters to pay low administrative fines than implement all the environmental requirements and invest in pollution prevention and control.

All the revenue collected from the abovementioned fees and fines are accumulated in the federal budget of Russia and in the appropriate regional budgets. A part of such funds collected is spent in accordance with the aimed budget planned for environmental protection activities and to executive authorities carrying out the ecological compliance control.

In 1996, a new Criminal Code[60] was adopted which lists at Chapter 26 (*Environmental Crimes*) criminal offences relating to the illegal and 'socially dangerous' uses of natural resources. *Criminal liability* may result in criminal fines varying on average from (the Rouble equivalent of) Euro 1,150 to 14,500, corrective labour, imprisonment from 3 month to 8 years, prohibition to hold certain official posts or to engage in certain activities. Punishments for individuals and officials differ. It should be noted that compa-

[60] *Ugolovniy Kodeks*, last amended on 8 December 2003.

nies and legal persons cannot be punished under the criminal procedure, only individuals can face criminal charges under Russian law.

Environmental crimes are still rarely enforced in practice. One of the reasons is that enforcement bodies do not have strong capacity to find and prove criminal deeds or to establish causal links in connection with environmental crimes. Illegal hunting, water pollution and illegal forest cutting are amongst the most typical environmental crimes.

Administrative fines and other less strict administrative liabilities are imposed in cases when an enterprise, for example, exceeds its permissible emission levels, and they, thus, precede the suspension of operations or the closing of the enterprise. It is noted that fines for non-compliance with environmental regulations are at times disguised as bribes to corrupt authorities.[61]

Environmental insurance

Russian law provides for obligatory and voluntary environmental insurance. For example, environmental insurance is required with respect to hazardous industrial facilities (e.g., oil and gas development and transportation, most industrial manufacturing processes). In general, however, the environmental insurance market in Russia is in its very early stages of development and does not play a significant role. The information on environmental insurance claims is not open to the public in Russia, so it is very hard to obtain any figures on environmental insurance.

New (environmental) legislation underway

Intensive legislative activities relating to environmental protection are still on-going. At present, the State Duma is preparing several draft laws, e.g., the *Federal Law on Payments for Negative Impacts on the Environment*,[62] the laws on hunting and fishing, as well as the *Law on the Protection of Marine Biological Resources*.[63] Several drafts of the new *Law on Subsoil Resources* have been prepared by various governmental bodies and are currently under negotiation in the State Duma.

The Russian government's draft law on 'Procedures for making foreign investment in Russian commercial organizations of strategic importance for national security of the Russian Federation' (No. 455348-4) was submitted to the Duma on 17 July 2007, and approved in a first reading on 14 September 2007 – it entered into force in May 2008 (the 'Strategic Investments Law').[64] Yet it seems that more amendments are needed

[61] For example, pressure politics helped Ikea win its battle against certain bribery-inclined local authorities in Moscow. Ikea was temporarily stopped from opening its shopping mall in Chimki outside Moscow in the beginning of 2005 on the basis of a gas leak threat from the Ikea building. However, Ikea manager for Russia, Lennart Dahlgren, claimed the real reason was that Ikea refused to pay bribes to the local authorities (source *The Moscow Times*, 27 April 2005).

[62] *О плате за негативное воздействие на окружающую среду.* The draft Federal law regulates the relationships between state executive organs and legal and private persons on the basis of the polluter pays principle.

[63] *Об охране морских биологических ресурсов.*

[64] Федеральный Закон «*О порядке осуществления в Российской Федерации иностранных инвестиций в коммерческие организации, имеющие стратегическое значение для*

to regulate this sensitive issue. Thus, the Federal Antimonopoly Service ('FAS') together with MED, the Federal Security Service ('FSB') and the Commission on State Secrets are preparing some adjustments to be introduced in early 2009.[65]

The new Strategic Investments Law requires that any transactions between 'foreign investors' and 'Russian commercial organizations' (later Russian company or companies) must be subject to preliminary government approval if the transaction meets two criteria simultaneously:

- The Russian company is engaged in *activity of strategic importance to national security* as defined by the law; and
- The foreign investor's transaction will result in direct or indirect *control of a Russian strategic company* as defined by law.

The transaction may include equity acquisition or other agreements resulting in foreign control of a Russian strategic company. This law defines 'control' as the capacity of a foreign investor (or a group of persons) to determine directly or indirectly decisions made by a commercial organization of strategic importance. (Art. 2, Section 4 of the law).

The Strategic Investment Law covers both types of companies and identifies 39 types of activity considered of strategic importance to national security. In addition to defence, energy, aircraft and aerospace industries, and critical infrastructure – all of which have long been deemed as strategic – the list also includes activities related to 'the use of pathogens of infection diseases' and activities that fall under the jurisdiction of the Federal Service for Hydrometeorology and Environmental Monitoring. In addition, new amendments proposed by the Presidential Administration and endorsed by the State Duma Committee for Construction and Land add certain natural resources, radio and television companies, publishing and printing companies, telecommunications companies, and Internet service providers, among others, to the list.

New steps towards cleaner production

It is noted that the Ministerial Declaration adopted at the meeting of European environmental ministers in Sophia, 1995, stipulated that 'cleaner production' means the continuous implementation of integrated environmental strategies for production and processes, directed to decrease harmful effects on humans and the environment.[66] By comparison, 'cleaner production' methodology in Russia is considered the method by

национальной безопасности Российской Федерации»; English version to be found at <http://www.fas.gov.ru/english/legislation/21182.shtml>.

[65] The FAS is responsible for contacts with foreign investors and has a good English-language portal. See <http://www.fas.gov.ru/english/>.

[66] Declaration by the Ministers of Environment of the Region of the United Nations Economic Commission for Europe (UNECE), 25 October 1995. See <http://www.unece.org/env/wgso/pre-kiev.declar/Sofia.E.pdf>.

which to shift the perspective from 'end-of-pipe' regulation to preventive strategies and effective mechanism for strengthening environmental compliance and enforcement.

The development of the concept of 'cleaner production' in Russia began in 1994 thanks to activities of the Russian-Norwegian Cleaner Production Center.[67] This Center began the 'cleaner production programme', which consists of three components: (1) the cleaner production training programme (including cleaner production theory); (2) financial engineering and investment projects for cleaner production; and (3) preparing enterprises to develop EMS and ISO 14000 certification.[68]

The main principles of the 'cleaner production programme' are:

- Education 'from engineer to engineer'
- Identifying the primary sources of environmental problems of an enterprise
- Pollution prevention through projects aimed at decreasing air emission, water discharges, waste generation and energy saving

The first steps to implement this programme were focused on training programmes at the industrial plants in Northwest Russia. During the 10 years of the activities of the Russian-Norwegian Cleaner Production Center in Russia, more than 1,600 engineers from 600 enterprises were trained. The cleaner production programme led both to financial benefits (for each $1 invested into the 'cleaner production' programme resulted in a $2-5 profit[69]) and resulted in positive environmental impacts, such as pollution prevention and energy saving at enterprises.

The results of the cleaner production programme introduced in the Northwest of Russia prompted the need to spread the cleaner production methodology across Russia as a means by which to prevent pollution. The Federal Service on Ecological, Technological, and Atomic supervision of MNR is the competent authority for the elaboration of legislative acts in the field of environmental pollution prevention, and this agency is now taking the appropriate steps for the development of such cleaner production laws and policies.[70]

Considering its positive results, businesses interested in improving their environmental impacts may wish to investigate ways in which the cleaner production programme was implemented in Northwest Russia.

---

67 See <http://www.ruscp.ru/tsygeng.htm and http://www.ruscp.ru/>.

68 See chapter 16 (*Dayman*), for a discussion of EMS in Russia.

69 A.Tsygankov, L.Yanchik, On the development of the Cleaner Production Program in Russia (2004) p. 4.

70 The International Conference 'Cleaner Production as a way to Sustainable Development' took place in December 2004 in Russia. The presentations made at this conference displayed that due to the mutual interests and common efforts of governmental bodies, non-governmental organizations and enterprises, in Russia the 'cleaner production' path may become the universal way of achieving a cleaner future through pollution prevention.

Conclusions

This brief overview of Russian environmental law and practice provides a means by which to analyze the current status of environmental protection in Russia, including the gaps and challenges which may be faced when undertaking environmentally sound investments. On this basis, and in the author's view, both national and foreign businesses which are interested in undertaking environmental investments and operating in a socially responsible manner in Russia are recommended to encourage the following:

- Strengthening of the legal and regulatory framework;
- Intergovernmental cooperation, especially as regards permitting and inspection activities at the federal, regional or territorial level;
- The building of an effective compliance and enforcement system for existing environmental laws and regulations;
- More active public participation, by specifying whose responsibility it is to provide environmental information;
- Investments in environmental management and control;
- Strengthening of monitoring systems and the adoption of better environmental indicators;
- Better technical, scientific and policy analyses.

The results of the implementation of the current system for environmental protection in Russia based upon the 'end-of-pipe' approach shows that there still remains space for further improvement. The preferred manner to achieve such improvement appears to be the development of policy promoting environmental compliance, including the provision for a broader application of preventive strategies, economic incentives and targeted financial support actions.

The Technical Regulation Law regarding the establishment of common technical regulations for ecological safety is expected to be the foundation of the new system of Russian environmental law, stipulating the norms for an improved system of environmental enforcement and compliance, based on pollution prevention principles. Furthermore, the draft federal *Law on Payments for Negative Impacts on the Environment* is being prepared, which is hoped to be very beneficial to foreign investors. The draft law stipulates new principles for economic incentives and targeted financial support activities. For example, according to the draft, industrial plants using 'best available technologies' in their operations will be relieved of paying ecological fees.

In January 2008, in Cheliabinsk, the then First Vice-Premier Dmitry Medvedev (currently the President of Russia) suggested to bring in changes into environmental legislation related to industry.[71] According to him, polluting industry should either change their profile or be closed. Medvedev underlined that solving environmental problems becomes one of the Russia priorities. He pointed out that the system whereby limits and norms of environmental impact are individually set for each given factory is

[71] See materials of the press-conference: <http://chelyabinsk.rfn.ru/rnews.html?id=77335>.

extremely outdated, and a coherent system that covers a region, or several regions, or the whole country should be introduced instead.

President Medvedev has commented[72] that it is of equal importance to strengthen liability in certain circumstances and to tighten the sanctions for negative industrial impact on the environment. He has remarked[73] that a number of Russian companies have already taken measures to improve the environmental situation at their factories. Thus, currently, a new environmental policy based on a preventive strategy is being elaborated in Russia. This promises to provide a framework within which environmental investments will be encouraged. In addition, the application of cleaner production methodologies are increasingly shifting the current Russian law focus on 'end-of-pipe' regulation, towards more preventive strategies.

---

[72] Ibid.
[73] Ibid.

Chapter 5
RUSSIAN NGO'S AND PUBLIC PARTICIPATION – LEGAL AND PRACTICAL PERSPECTIVES

Olga A. Razbash[1]

This chapter aims to provide a brief overview of Russian law and practice relating to the principle of public participation in Russia. It also focuses on the way in which Non-Governmental Organizations ('NGOs') engage with Russian businesses and describes the steps that have been taken to progress public participation in Russia. In particular, the activities of the Russian Regional Environment Center ('RREC') are reviewed in context of how RREC can interact and support businesses to engage in green investments and socially responsible business practices.

SUSTAINABLE DEVELOPMENT AND PUBLIC PARTICIPATION IN RUSSIA

Whilst attracting large amounts of FDI in recent years, Russia continues to face the challenge of pursuing development and an investment pattern which balances economic with socio-environmental needs – in other words, sustainable development.

The concept of sustainable development has its roots in international law – and encompasses a number of principles, including the principle of public participation. Thus, before focusing on this particular aspect of sustainable development, the below very briefly touches upon the integration of the concept of sustainable development within Russian law.

Sustainable development in Russia

Under international law, the concept of 'sustainable development' was most prominently discussed in the Brundtland Report,[2] wherein it is defined as:

> 'development that meets the needs of the present without compromising the ability of future generations to meet their own needs'.

[1] Environmental Law and Public Awareness Expert, Russian Regional Environment Center ('RREC'), Moscow, Russia; Email: *razbash@rusrec.ru.*

[2] See Report of the World Commission on Environment and Development: *Our Common Future* (1987), Transmitted to the General Assembly as an Annex to document A/42/427 - Development and International Co-operation: Environment see at <http://www.un-documents.net/wced-ocf.htm>.

W.Th. Douma and F.M. Mucklow, eds., Environmental Finance and Socially Responsible Business in Russia
© 2010, T·M·C·ASSER PRESS, The Hague, The Netherlands and the Authors

This concept has also been elaborated upon in numerous international declarations, as well as regional and multilateral treaties – which Russia, too, has signed and ratified.

The official commitment of the Russian Federation (the 'RF') to protect the environment and to implement the principles of sustainable development can be found in numerous documents – legal acts, policies and declarations – adopted and articulated during the last decade: in the RF Environmental Doctrine, the RF Concept of Sustainable Development, the RF Policy in the area of Bio- and Chemical Safety, and in numerous technical regulations on environmental and biosafety issues. Indeed, the speeches of President Dmitriy Medvedev at the session of the Council of Safety in January 2008, Prime-Minister Vladimir Putin in July 2009 at the special governmental meeting on environmental safety in Irkutsk, and Minister of natural resources and environment Yurii Trutnev at the session of the upper chamber of the RF Parliament in September 2009 as well as the recent change of the name, structure and competence of the Ministry of Natural Resources – now named the Ministry of Natural Resources and Environment ('MNR')[3] – promise that Russia is increasingly committed to following a path of sustainable development.

Evidence of Russia's growing commitment to sustainable development can also be found in the practical arrangements of the Russian Government. It uses best international environmental standards and is creating a legal precedent by applying provisions of the UN ECE 1991 Convention on Environmental Impact Assessment in a Transboundary Context (hereinafter the Espoo Convention)[4] to the large scale transboundary 'Nord Stream'[5] gas pipe-line project.

International framework – public participation

As mentioned above, the principle of 'public participation' is part of the concept of sustainable development – and is thereby embedded within the legal and policy framework of the Russian Federation. Numerous international (legally binding and non-binding) documents, signed and ratified by the Russian Federation, express the importance of public participation and the need to institutionalize such practices within the private and public sector in order to enable a nation's sustainable development.

In particular, Principle 10 of the *1992 Rio Declaration on Environment and Development*[6] signed by more than 100 heads of State worldwide, including the Russian Federation, established that:

---

[3] It is noted that there are still strong anticipations that a separate environmental protection governmental body will be established in Russia – in order to divide the competence of the use of natural resources and environmental control and supervision.

[4] See <http://www.unece.org/env/eia/> for further details. Not ratified by the Russian Federation.

[5] See <http://www.nord-stream.com/en/>; Nord Stream is a gas pipeline which is to link Russia and the European Union via the Baltic Sea.

[6] See <http://www.unep.org/Documents.multilingual/Default.asp?DocumentID=78&ArticleID=1163>. The Rio Declaration sets out that some basic components must be present in all participatory democracies, e.g., public participation to allow government actions to become transparent and to assist avoiding corrupt practices.

> 'Environmental issues are best handled with the participation of all concerned citizens, at the relevant level. At the national level, each individual shall have appropriate access to information concerning the environment that is held by public authorities, including information on hazardous materials and activities in their communities, and the opportunity to participate in decision-making processes. States shall facilitate and encourage public awareness and participation by making information widely available. Effective access to judicial and administrative proceedings, including redress and remedy, shall be provided.'

Under international law (especially under the Aarhus Convention, as defined below), 'public participation in decision-making on environmental matters' is understood to mean wide, adequate public[7] involvement in the process of decisions, which have or could have an impact on the environment at the very early stage 'when all alternatives are still open'.

In addition, the definition and meaning of public participation has been established and elaborated upon in various multi-lateral treaties and non-binding instruments, such as:

- the 1991 Espoo Convention
- 1998 UN ECE Convention on Access to Information, Public Participation in Decision-making and Access to Justice in Environmental Matters (the Aarhus Convention):[8] The Aarhus Convention elaborates Principle 10 of the Rio Declaration, which '*stresses the need for citizen's participation in environmental issues and for access to information on the environment held by public authorities. As such it is the most ambitious venture in the area of environmental democracy so far undertaken under the auspices of the United Nations*' (Kofi A. Annan, former Secretary-General of the United Nations)
- 1992 United Nations Convention on Biodiversity (the CBD)
- 2000 Cartagena Protocol on Biosafety to the CBD (the Biosafety Protocol)
- 1992 United Nations Framework Convention on Climate Change (the UNFCCC)
- 1997 Kyoto Protocol to the UNFCCC (the Kyoto Protocol)
- 1992 Convention on the Protection and Use of Transboundary Watercourses and International Lakes[9] (the Water Convention)
- 1999 Protocol on Drinking Water and Health to the Water Convention (the Protocol on Drinking Water and Health)
- 1971 Convention on Wetlands of International Importance especially as Waterfowl Habitat (the Ramsar Convention)[10]

Public participation in Russia

In Russia, the issue of public participation is debated in context of the interaction of all sectors of Russian society – and is applied to discussions amongst Russian decision

---

[7] In this context public means 'one or two individuals or legal persons'.

[8] Adopted on 25 June 1998, entered into force on 30 October 2001; see <http://www.unece.org/env/pp/documents/cep43e.pdf>. Not signed or ratified by the Russian Federation.

[9] See <http://www.unece.org/env/water/text/text.htm>.

[10] See <http://www.ramsar.org/key_conv_e.htm>.

makers, ministries and agencies which are responsible for environmental issues, as well as environmental NGOs and business associations.

The public participation debate in Russia exists within the general understanding that it is often difficult to analyse or evaluate the benefits of a given project or business activity against the interests of the Russian Government, or of society at large.

The issues relating to public participation are also strongly linked to the principle of democracy. Public participation underpins the democratic process and energizes it continuously by creating a permanent connection between the governed and those who govern. Democracy can, thus, be understood to be a 'joint venture' between society, the public and the private sector in order to allow for more balanced decision-making, to enable better understanding of the concerns of a given society or group of people, and to allow all interested parties to cooperate towards possible and more viable solutions to difficult issues. The public participation debate in Russia, therefore, also touches upon the need to develop and encourage investment and business activity within a democratic system. It is argued that it is a way in which to provide the business community with clear transparent 'rules of the game', whilst citizens are provided with an institutional framework which allows for their effective participation in issues, which affect them.

Businesses and public participation

Thus, NGOs which promote the evolution of an integrated society and, in many cases, the protection of the environment, are keen that businesses are encouraged to embrace public concerns at the early stages of such businesses' projects and to involve interested parties, including the public, throughout the project.

Increasingly, businesses are beginning to realize that public participation is a key instrument with which timely and comprehensive risks assessments can be gleaned at the very early stages of a project. Furthermore, applying public participation as a business tool enables businesses to reduce various potential risks (e.g., environmental, economic, social, or image-related risks) and assists substantially in making a business venture successful in the long term. By engaging in public participation this also assists businesses (i) in making their projects more likely to qualify for IFO funding (e.g., from EBRD, the World Bank or IFC) and (ii) in increasing their competitiveness internationally.[11]

Public participation under Russian law

Under Russian law, public participation has been addressed relatively comprehensively. The following table sets out a summary of current Russian legislation which includes and refers to public participation:

[11] See chapter 12 (*Douma*).

Table 1. Examples of federal laws (public participation)

- Federal Law on 'On Environmental Protection' 2002 (the Environmental Protection Law)
- Federal Law on 'On Environmental Expertise' 1995, as amended in 2006 (the Environmental Expertise Law)
- Federal Law on 'On Nuclear Energy Usage' 1995
- Federal Law on 'On Sanitary-Epidemiological Wealthy of the Population' 1999
- Federal Law on 'On State Regulation in the Field of Genetic Engineering Activity' 1996
- Federal Law on 'On Specially Protected Natural Areas' 1995
- Federal Law on 'On Wild life Protection' 1995
- Land Code 2001 (the Land Code)
- Forest Code 2007 (the Forest Code)
- Water Code 2007 (the Water Code)
- City Construction Code 2004, as amended in 2006 (the City Construction Code'
- EIA Regulation (MNR RF) 2000 (No. 372) (the EIA Regulations)
- Regulation on Federal Service of Environmental, Technological and Nuclear Supervision 2004 (No. 401) as updated in May 2008
- Regulation on Federal Service of Nature Protection Supervision 2004 No. 400
- other federal ministerial regulations and instructions

The below table provides examples of regional laws which provide for public participation:

Table 2. Examples of regional legislation (public participation)

- Moscow City Law 'On Citizens Rights Protection during Realization of City Construction Decisions in Moscow' 1997
- Moscow City Law 'On Citizens Complaints' 1997/2000
- Moscow City Law 'On Consultative Poll of Citizens in a Moscow City district' 1997
- Regulation on the Moscow City Government Commission for consideration of citizens' complaints during construction in Moscow (approved by Moscow City Mayor's Decree of 19 February 2001 No. 133-PM and of 22 October 2002 No. 614-PM)
- Saint-Petersburg Law on Public Participation in City Construction Activities, 2006

Fundamentally, the right to public participation under Russian law is enshrined in the Russian Constitution, which guarantees that each human being has the right to live in a 'wealthy' environment and has a right to – timely, full and adequate – access to environmental information (Art. 42). In addition, Article 58 of the Russian Constitution establishes responsibility of each individual or legal entity to protect and improve the environment.

In order to guarantee these rights, various Russian laws provide for public participation and 'control' procedures in environmental decision making and in enterprises' activities.

Public participation/control rights

The principle of public participation is an integral part of the Environmental Protection Law[12] and other specific environmental laws, in particular relating to Environmental Impact Assessment ('EIA') and the State Environmental Review ('SER') procedures. There are also various economic incentives and other Russian regulations which enable and are aimed at promoting green investments[13] – these economic incentives and regulations, however, merely encourage and prescribe public participation in such green investments in an *indirect* manner.

Furthermore, under Article 1, paragraphs 1-4 of the Land Code, Russian citizens and NGOs have the right to participate in preparing decisions, which could have an impact on the state of lands during their usage and protection.

According to Russian law, governmental bodies, municipals and subjects of economic activities (i.e., businesses) are obliged to provide – organize, support and allow – for public participation in the manner and form established by law. The various Russian federal laws which provide for public participation in environmental decision-making include the Environmental Protection Law, the Environmental Expertise Law, the law '*On Nuclear Energy Usage*', the City Construction Code, the Land Code, the new Forest Code and the Water Code.

Thus, businesses which wish to initiate a project in Russia must disseminate information about its project plans in a timely manner; provide access to the technical resumé of the project (or in the case of a full-scale EIA – to project documentation on the environmental aspects of the project); in cooperation with the local authorities, arrange for public consultations, public hearings, collect all public comments, analyse them and integrate them into the project design materials as part of the EIA – in order to guarantee that 'the voice of the public is heard'.

Environmental decision-making

Russian law defines 'environmental decision-making' as meaning any decisions (relating to individual projects, normative acts, programmes, etc.) made by authorized federal governmental bodies, local municipals, special administrative bodies which have or could have an impact on the environment, public health and wealth.

A rather detailed description of the public participation procedure as a form of public environmental control can be found in the EIA Regulations.

Definition of 'public'

Russian law does not directly define the terms 'public' or 'public concern'. Rather, the term 'public' is understood to be very general in meaning, covering citizens, NGOs

---

[12] For example, Article 3 of the Environmental Protection Law considers public participation as one of the basic principles of environmental protection.

[13] See chapter 7 (*Nystén-Haarala*).

and other non-commercial associations whose environmental interests are touched upon by decisions and who show interest in environmental consequences of planned activities. In some cases, municipal and local self-governance bodies are considered to be 'public' (see Article 31 of the Land Code).

Moreover, Russia signed the Protocol on Drinking Water and Health and ratified it on 31 December 1999. In accordance with Part 4 Article 15 of the Russian Constitution, this international agreement became an integral part of Russian national legislation, including the definition used in this Protocol for 'public': 'public is one or more physical or legal persons', which is the same definition as applied in the Aarhus Convention.

Procedures and forms of public participation

Under Russian law, the public has the right to participate:

- at each stage of preparation, making and realization of environmental decisions;
- in discussions of environmental policies, programmes, plans (e.g., relating to territorial development, development of industries);
- in discussions of draft laws, regulations and non-normative acts with potential adverse impact on the environment;
- in EIAs and environmental 'expertise' activities (public or state);
- in meetings, referenda and other environmental protection measures; and
- in other forms of public participation.

Furthermore 'concerned' citizens and NGOs have the right:[14]

- to file requests to governmental bodies and local municipalities, as well as to other organizations and officials in order to obtain information on the state of the environment in their residential area and on environmental protection measures being taken;
- to file complaints, appeals and proposals on environmental protection and negative environmental impacts and to receive timely and adequate responses;
- to propose and initiate a public environmental 'expertise' – i.e., not an EIA (which applies to businesses); and
- to participate in the EIA process initiated by a business or other initiator of a 'relevant activity'.

For businesses operating in Russia this means that they will need to take the above into consideration when planning any projects in Russia – and ensure that public participation is part of the planning process.

[14] Articles 11 and 12 of the Environmental Protection Law.

EIAs, PUBLIC HEARINGS AND BUSINESS

In addition, Russian law requires[15] that when a project is being analyzed (or evaluated) that different scenarios of development of a business' project be formulated within a framework of discussions with all interested parties. Furthermore, Russian law requires that the indirect benefits of a given project, such as the additional public benefits and any increase in quality of life caused by the given project, should be taken into account, too. However, there is no accepted methodology in Russia to assess these 'indirect' benefits to the public.

Of key importance to investors and businesses are the EIA Regulations. Under these regulations, a proposed project investor (or initiator of a project) must undertake an EIA – and inform and involve the public in discussions of the plans relating to the proposed project (or of its future activity). The EIA should be carried out for those sections which must be submitted for purposes of a SER. According to the EIA Regulations, the investor should involve the public by way of public discussions, hearings or preliminary public consultations with the investor in cooperation with local municipalities at all stages of the EIA procedure.

In addition, NGOs and other non-commercial associations are authorized:

- to organize meetings, collect signatures, propose referenda and discussions of any project with environmental impacts;
- to carry out public hearings relating to the design and location of enterprises with possible adverse environmental impacts;
- to organize and carry out a public environmental 'expertise';
- to recommend that their representatives participate in a SER.

Public environmental 'control'

Russian law provides for the right of 'public control' under a number of legal acts.[16] For example, Article 72 of the Land Code provides that local self-governance bodies (municipalities) have the right to control the procedure of environmental decision-making by way of specially authorized governmental bodies and in compliance with citizens' rights and lawful interests. This provision under Russian land law correlates with Article 13 of the Environmental Protection Law, which sets out that governmental bodies and local municipalities and officials must provide assistance to citizens, NGOs and other non-commercial organizations in the realization of their environmental rights. General provisions exist under Russian law, which establish responsibility of officials who create obstacles for citizens and NGOs in their environmental protection activity.

Russian law also provides for public 'control' through the referendum process during decision-making on individual projects relating their potential adverse environ-

---

[15] E.g., Russian EIA Regulations 2000; Russian Federal Law on Environmental Expertise; Russian City Construction Code; the new Water Code and new Forest Codes.

[16] See also Articles 11, 12 and 13 of the Environmental Protection Law; Articles 31 and 72 of the Land Code, Articles 28, 30, and 31 of the federal City Construction Code; Articles 19, 20 of the Federal Law 'On Environmental Expertise', and the EIA Regulation.

mental impacts. Furthermore, business projects must receive governmental – local, regional or federal – approval in order to obtain a land plot for purposes of their projects. This is a further level of 'public control' with respect to business activities.

The public is entitled to participate at all stages of the preparation, adoption and realization of environmental decisions: in disputes on state environmental policy, different programmes, plans, in EIAs and SERs. In addition, any member of the public can independently carry out a 'public environmental expertise', participate in public hearings, referenda, and citizens assemblies. It is noted that any member of the public can initiate and register a 'public environmental expertise' for any type of project. Russian law does not differentiate between 'business' or 'public' projects, the only relevant criteria is the project's environmental impact.

Environmental Impact Assessments

Under Russian law, an EIA must be undertaken in order to prevent or minimize the economic or other negative impact of a given project on the environment or on society. Furthermore, an EIA must be undertaken with respect to activities which should be submitted also to 'state environmental expertise', as set out in the Environmental Expertise Law. These activities are as follows:

- draft Russian laws and regulations, the realization of which could lead to negative consequences on the environment;
- drafts of normative-technical and instructive-methodical documents;
- feasibility studies and substantiations and designs of construction, reconstruction and development of enterprises with potential negative environmental impacts, including on the territories and the environment of neighbouring countries.[17]

An 'initiator' of any one of the above activities is required to notify the public if any one of the above listed activities is taking place – and how the public will be involved in the EIA process relating to such activity. In addition, preliminary public consultations must be held in order to determine all interested stakeholders in a given project.

The EIA Regulations also prescribe[18] that an investor must consult with the public at the stage of scoping, screening and elaboration of the terms of reference relating to an EIA. These materials and the results of an EIA, should then be made available to the public.[19] The EIA Regulations also provide that the public is entitled to give its comments at all these stages.

At the final stage of an EIA, materials prepared and public comments received must be taken into account. Furthermore, the minutes of public hearings and information on public comments should be included into the final version of the EIA materials.

---

[17] In order to prevent and minimize (and monitor) substantial negative consequences on the environment in general and in the transboundary context, as regulated in the Espoo Convention (see Art. 1).

[18] See paragraph 3.1.2.

[19] These provisions correlate with the requirements of the Aarhus Convention.

Positive effects of public participation

Generally, public participation benefits are intangible in that they constitute and result in ethical, moral and normative improvements within society. Public participation is also considered to preserve and improve 'public goods' on the global and regional levels. Where businesses engage in public participation, this is also considered to diminish the risks and potential costs of a given project. Furthermore, public participation can positively impact on the 'public good' by improving public health through the phasing-out harmful industrial facilities.

Thus, developed public participation procedures in Russia based on internationally accepted standards and approaches are likely to lead to:

- further development of civil society in Russia;
- improvement of instruments of societal partnership, e.g., instruments of cooperation and coordination between civil society, business and local authorities;
- improvement of sectoral interactions and enable intersectoral consensus, which will assist in diminishing potential social tension;
- achieving a balance of interests of different societal groups and layers;
- improvement of the quality of the environment; and
- better observation of citizen's constitutional rights.

Public access to environmental information and public participation in the early stages of environmental decision-making processes is beneficial to all stakeholders:

- *For investors*: The benefits for investors include:
  (i) Solution of (potential) conflicts and better understanding of public needs and views enables investors to change their plans (if necessary) with respect to their projects at the early stages and so avoid negative consequences arising as a result of a project early on.
  (ii) Losses can be avoided at the development stage of a project.
  (iii) Reduction of image risks through use of public participation procedures can improve the attitudes of credit institutions towards an investor and so assist investors in obtaining loans or other financial support.

- *For governmental authorities*: The benefits for governmental authorities include:
  (i) Results of public participation enable governmental authorities to better understand public attitudes and anticipations with respect to certain projects or activities.
  (ii) Increased public awareness about environmental problems could redistribute responsibilities.
  (iii) Increase of public trust towards governmental decisions.

- *For citizens*: The benefits for citizens include:
  (i) Being informed about important environmental decisions relating to projects and activities;

(ii) Ability to influence decision making processes and to object to certain decisions. Thus, responsibility of decision makers will increase and make decisions more sustainable.
(ii) Awareness of decisions (relating to projects and activities) and their consequences will increase, enabling a better understanding of the value of the environment and meaning of decisions during decision-making processes.
(iii) Broader democratic representation and improved social partnership, thus, leading to the development of a more sustainable and inclusive society.

NGOs and public participation: Russian Regional Environment Center

The way in which NGOs engage with the business community to enable public participation can be exemplified by the work of the Russian Regional Environment Center ('RREC').[20] RREC was founded in 2000 (starting operations in 2001) and is funded by the European Commission and the RF Academy of Civil Governance on behalf of the Russian Government.

RREC is part of the network of Regional Environmental Centres ('REC') in the NIS. REC offices also exist in Moldova, Georgia – for the three South Caucasus Republics – as well as in Central Asia. Apart from managing grant programmes to NGOs, the RECs similar to the original Regional Environmental Centre for Central and Eastern Europe ('REC CEE'), are intended to:

- support and develop civil initiative based on best international practices;
- act as intermediaries between governments and the public;
- work with the private sector to enable sustainable development; and
- be a focal point for information and training.

RECs are independent from existing NGOs and from any other pressure groups. They are non-advocacy organisations and do not undertake any partisan activity.

Within this broader framework, RREC's mission is to:

> '[enable] co-operation on [the] international, national and regional levels among government bodies, local authorities, non-governmental organisations (associations), the business community and any other entities and individuals involved in environmental protection through *exchange of information and assistance.* The Centre also promotes broad public participation in the decision-making process relating to the protection of the environment.'

RREC's strategic goal is to provide relevant services and support to all groups of stakeholders (NGOs, governments, and the private sector) with respect to environmental issues – in particular, its goal is to:

---

[20] See <www.rusrec.ru>.

- promote the most efficient decisions relating to environmental problems which Russia faces;
- support and develop dialogue and co-operation among all stakeholders, legal entities and individuals in environmental protection and transition to sustainable development.
- contribute to the development of civil society in Russia;
- provide expertise to governmental bodies at the federal and regional levels relating to environmental and sustainable development issues;
- assist in the exchange and dissemination of information and best standards relating to the environment and sustainable development between governmental bodies and NGOs;
- strengthen the dialogue between Russia and the European Union in the area of environmental protection and sustainable development;
- disseminate the best practices in environmental protection and replicate positive results at the regional and federal levels;
- involve the business community in solving environmental problems;
- promote public participation in environmental decision-making;
- become the focal point in the area of implementation by the Russian Federation of its international environmental and sustainable development obligations;
- be the resource centre for supporting public initiatives in the field of environmental protection and sustainable development.

Furthermore, RREC is a member of the Renewable Energy and Energy Efficiency Partnership ('REEEP'),[21] and acts as the Russian Regional Secretariat for REEEP.

RREC and business support and cooperation

RREC representatives regularly participate in different thematic meetings at the Trade Industrial Chamber, the Russian Union of Industrialists and Entrepreneurs ('RUIE'), and attend public hearings on draft technical regulations in the RF Ministry of Industry (which is responsible for this process).

Experts of RREC have organized public participation processes for several businesses in Saint-Petersburg (road construction project), Chita Oblast (golden mining project) and Khabarovsk Krai (road construction project).

Recently, in May 2008, upon the request of EBRD, RREC together with the REC CEE organized a public consultation process relating to the new EBRD Environmental and Social Policy ('ESP') and the new Public Information Policy ('PIP') involving the Russian business community, NGOs, experts and governmental bodies. Representatives from Moldova and Ukraine also took part in this public consultation. It was a

---

[21] See <http://www.reeep.ru/>. REEEP is a public-private partnership and was launched by the United Kingdom along with other partners at the Johannesburg World Summit on Sustainable Development in August 2002. REEEP currently comprises 150 partners representing 29 governments, businesses and NGO's committed to accelerating the uptake of renewable energy and energy efficiency.

challenging experience. Public comments which were collected were comprehensively analysed and the final RREC report was presented to EBRD (as part of the REC CEE comprehensive analysis). Information on the process was widely disseminated among all interested stakeholders, including the Russian banking community.[22]

In addition, taking into account RREC's mission and goals, cooperation between RREC and the business community can be developed further through the following possible projects:

- Round tables, discussions on the issues of environmental safety, implementation of the All-Russian Environmental Doctrine with involvement of the business community, the Russian federal and regional governments and NGOs;
- Promotion of wide public dialogue on cooperation and interaction between public interest, NGOs and businesses, federal and regional legislatures and environmental law enforcement bodies;
- Development of joint scientific research in environmental areas;
- Environmental management training and sustainable development programmes; and
- Seeking possibilities for the improvement of relations between local authorities, the general population, environmental control bodies and environmental NGOs.

Furthermore, RREC facilitates the development of climate change policy of the Russian Federation, as follows:

- by supporting business initiatives to create and develop new institutions and procedures to implement the mechanisms of the Kyoto Protocol;
- through organization of inter-sectoral cooperation to prevent the adverse consequences of climate change;
- through training activities, conferences and consultations with the business community on the realization of industrial enterprise initiatives in the area of climate change.

Other areas of potential cooperation include:

(a) *Russian Federation accession to WTO*: RREC is able to cooperate with Russian businesses in connection with the issues relating to Russia's accession to the WTO, as follows:
    - through the preparation of recommendations for the Russian business community and various authorities as regards the minimization of environmental consequences of Russia's accession to the WTO;
    - interaction with environmentally oriented business groups (in forestry and other sectors) to define ways in which to harmonize further market competence goals with environmental protection goals;

[22] It is noted that there is now a special EBRD standard concerning 'financial intermediaries' which operate loans and credits granted by EBRD. See also chapter 12 (*Douma*) for further information on the EBRD in Russia.

- by supporting pilot projects which are environmentally sound; and
- by promoting and supporting environmental initiatives at the regional and local levels.

(b) *Promotion of environmentally friendly business activity*: RREC can assist businesses in making their operations more green or socially responsible, notably as follows:
- by stimulating development and introducing environmental management systems (EMS) an internationally approved instrument for economic growth;
- by disseminating information as regards 'eco-audit' principles at industrial enterprises;
- by supporting and promoting the image of enterprises to become leaders in environmental protection;
- by attracting the attention of international organizations, the international community and consumers to environmentally friendly Russian businesses;
- through assisting businesses in seeking additional opportunities to develop 'green' businesses opportunities with foreign partners;
- by assisting enterprises to attract 'green' investors;
- by assisting enterprises through the elaboration of the advantages and privileges enterprises can obtain from 'green' investments; and
- by training businesses with respect to their strategies of fundraising for 'green' projects and as regards participation in grant-competitions and applications at the international level.

(c) *Dissemination of information, assistance in creating partnerships*: RREC is able to assist with the dissemination of information and in creating partnerships, as follows:
- through its Russian-English language business page at the RREC website (<www.rusrec.ru>) which includes information on environmental technologies, environmental protection measures, environmental standards and problems, and information on enterprises and engineering consulting companies (specializing in the introduction of new environmentally oriented and energy efficient technologies);
- by providing the business community with interactive information on the environmental safety of investments;
- by carrying out educational and practical seminars for the business community.

Thus, there are many potential areas of cooperation between RREC and the Russian business community – or with foreign businesses operating in Russia. It is noted that during 2007-2008, there were several public consultations and information campaigns, wherein the RREC acted as a facilitator in the public participation process with business, including with respect to the Nord-Stream project.

CONCLUDING REMARKS

Russian NGOs – especially those involved in environmental protection and other relevant issues (such as RREC) – can assist and cooperate with Russian and foreign businesses interested in undertaking green investments in Russia and in operating in a socially responsible manner. Such NGOs constitute a valuable source of information (local and regional). They have knowledge of the local experts community, of both the relevant Russian legal framework and international standards on public participation procedures; and have capacity and experience in organizing public participation processes.

The 21st century is an era of globalization within which conflicts of interests are increasing. Thus, public participation procedures represent one of the most efficient way to try to diagnose, project and avoid or diminish such conflicts for the mutual benefit of all interested stakeholders. This is so, especially, in large cities with intensive construction activities. Thus, the growth of citizens' self-consciousness and competence of local authorities and communities, the development of property owners class in Russia should also be taken into account while planning any project.

Chapter 6
HARMONISATION OF RUSSIAN ENVIRONMENTAL STANDARDS WITH EU LEGISLATION

Dietrich Hahn[1] and Michael Begak[2]

It is no surprise that environmental standards and approaches to setting such standards differ from one country to another. Different historical developments, standards of living, governments, public participation, the influence of industrial lobbying groups – all these factors influence the formulation of environmental standards and related legislation. However, these differences present no reason for states not to cooperate with respect to nature protection, ecological safety, and solving global environmental problems facing humanity. Environmental standards should not act as a barrier to international cooperation, but should facilitate such cooperation, as well as promote free trade and investment.

The partial harmonisation of Russian environmental legislation with relevant EU legislation is regarded as necessary in order to support the development of a joint EU-Russia economic space and in order to support international commercial activities generally. Furthermore, it is considered to be important that Russian producers ensure that their production processes adhere to such harmonised environmental standards in a transparent manner (which is increasingly expected by end consumers of goods). As a basis to promote such harmonization, EU experiences and examples relating to industrial environmental management serve as an excellent source of information and provide useful experiences, which can be applied to the Russian Federation.

In this chapter, the EU-TACIS project on 'Harmonisation of environmental standards' ('HES II')[3] is described as an example of an attempt to harmonise Russian environmental standards with the environmental standards and approaches of the European Union ('EU').

HES II Project

European and world experience of industrial environmental management and administration shows that the most convenient way to reach a balance between industrial activ-

[1] Team Leader, Harmonisation of Environmental Standards (HES) II, Moscow, Russia; Email: *D.Hahn@ippc-russia.org or Dietrich.Hahn@web.de.*

[2] Key Expert, Harmonisation of Environmental Standards (HES) II, Moscow, Russia; Email: *mbegak@gmail.com.*

[3] See <http://www.ippc-russia.org>.

W.Th. Douma and F.M. Mucklow, eds., Environmental Finance and Socially Responsible Business in Russia
© 2010, T·M·C·Asser Press, The Hague, The Netherlands and the Authors

ity and citizens' needs, is the introduction and use of 'integrated permits' relating to enterprises' emissions, discharges and waste disposal – and based on technological standards using Best Available Techniques ('BAT'). The wide scope of such integrated permits allows, firstly, the simplification of the permitting procedure and, secondly, the minimization of negative environmental impact.

The first phase of the 'Harmonisation of Environmental Standards' project showed, amongst other results, the possibility and suitability of developing such an integrated permit system in the Russian Federation. The second phase of this project (i.e., the HES II project) is aimed at solving the practical issues of the realisation of an integrated permit system in Russia. The HES II project focuses on the implementation of the integrated permit system, which requires essential changes in all parts of Russian environmental law: from the rules on environmental impact assessment ('EIA') of projects to environmental monitoring of the operating enterprises. Hence, the HES II project has a rather wide scope.

The HES II project is a two-and-a-half year follow-up project to an Europe Aid cooperation activity that was carried out as Phase I in 2002-2004 ('HES I').[4] The basic recommendation of the HES I project was to introduce in the Russian Federation an integrated environmental permitting and controlling system for industrial installations harmonised along European rules, based on the concept of BAT and adopted in compliance with EC Directive 96/61/EC concerning integrated pollution prevention and control ('IPPC Directive'). This was suggested in order to enhance the implementation of the Russian Federal Law 'On Technical Regulations' ('Technical Regulations Law').

The goals of the HES II project are to clarify the possible fields of harmonisation of Russian environmental legislation with EU legislation and to make proposals for the implementation of such harmonisation. The HES II project covers the following 10 clusters, all of which relate to the introduction of a system of integrated ecological permits for industrial enterprises in Russia:

- Institutional Issues
- Environmental Impact Assessment
- Environmental Auditing
- Environmental Certification
- Environmental Monitoring
- BAT / BREF
- Environmental Permitting
- Environmental Pollution Fees (or charges)
- Environmental Quality Standards
- Environmental Insurance and Liability

---

[4] See also W.Th. Douma and A.E. Kellermann, eds., *Economic Aspects of Environmental Policy in Russia, Selected Papers and Seminars* (January 2004, Moscow), with support of A.V. Kharchenko, project funded by EU Tacis Programme, implemented by ERM lead Consortium; and see 'Improving Russia's environmental permitting regime for industry. Recommendations on harmonisation of Russia's environmental law and practice with that of the EU', Moscow, November 2003. TACIS Project 'Harmonisation of Environmental Legislation, Russia'. The report is available at the European Environmen[illegible]v website (<www.eel.nl>) at National pages, Russian Federation.

The below discusses in more detail the issues relating to each of these cluster areas covered by the HES II project.

INSTITUTIONAL ISSUES

The procedure of granting environmental permits to industrial enterprises operating in Russia is still complicated, contradictory and tiresome for enterprises. In order to make this permit procedure user-friendly, transparent and consistent, a number of institutional and administrative changes are needed.

Administrative reforms of the Russian government during 2000-2004 weakened the management system of nature protection in Russia. As regulative and control functions were dispersed between several bodies this resulted in partially overlapping functions and competences. During the last administrative reorganization of the Russian government in May 2008, all functions relating to environmental control were transferred to Rosprirodnadzor.[5] Whereas, the other environmental management bodies, i.e., Rostechnadzor[6] and Roshydromet[7] (responsible for environmental monitoring), were placed under the umbrella of the Russian Ministry of Natural Resources ('MNR', recently renamed the Ministry of Natural Resources and Ecology).[8] As a result of these reforms, the federal bodies of environmental control in the Regions of the Russian Federation are now lacking qualified and experienced personnel and are so unable to meet their commitments properly.

With respect to the harmonisation of institutional arrangements in Russia (relating to environmental protection), HES II research has, in particular, identified the need to put in place one single executive body responsible for environmental matters that would coordinate both domestic and international government activity across Russia's ministries and agencies.

Furthermore, HES II experts established that the concept of BATs applies more widely in Russia than just to environmental issues only. EU BAT Reference Documents ('BREFs') can become important reference elements for the Russian standardisation system. Institutionally, this would mean an additional role of the Federal Agency on Technical Regulation and Metrology ('Rostechregulirovanie')[9] in implementing BAT reference documents within the Russian technical regulation system. In order to introduce an EU-like technological exchange process in Russia, which creates BAT reference documents, this would require the creation of a special body – a BAT office – maybe similar to the EU IPPC Bureau which is operational for the EU Member States within the EU Joint Research Centre. HES II experts have held discussions on this issue with organisations such as the Russian Union of Industrialists and Entrepreneurs (RSPP), the Russian Chamber of Commerce and Industry ('TPP'), the Russian

5 See <http://control.mnr.gov.ru/>.
6 See <http://www.gosnadzor.ru/> .
7 See <http://www.meteorf.ru/default.aspx>.
8 See <http://www.mnr.gov.ru/>.
9 See <http://www.gost.ru/wps/portal/pages.en.Main>.

Scientific Research Centre of Ecological Safety ('RAS'), Rostechregulirovanie and others. Of course, the issue of BAT/BREFs cannot be resolved until a single Russian environmental executive body is set up and consolidated, including the putting in place of functional and sufficiently equipped structures in the Regions. Furthermore, real political will, which favours environmental responsibility, should be incorporated into Russia's development policy and strategy, with a view of caring for future generations.

Environment impact assessment

Environmental impact assessment ('EIA') procedures in Russia consist of an impact assessment ('OVOS') and a verification procedure, i.e., a state environmental review ('SER').[10] The Russian EIA procedures were well organised and commended in Europe and elsewhere before the RF Urban Code (the Urban Code)[11] was changed in December 2006 due to pressure from the construction and industrial lobby. These amendments to the Urban Code meant that fewer scenarios were required to undergo an EIA. However, further amendments made in May 2008 resulted in some improvements being achieved by instituting the new requirement that a SER be undertaken for facilities being constructed within 'natural areas of preferential protection'. Although this is a move in the right direction, progress of further developing EIA procedures in Russia is rather slow. However, it is encouraging that during his meeting with human rights activists on 15 April 2009, President Medvedev promised 'to charge the examination of the issue [of EIAs in Russia]'.[12]

HES II experts have finalised their work on the harmonisation of EIA procedures in Russia and propose that the following legal acts and changes are required within Russia's institutional structure:

- a single 'Environmental Review Department' should be established as part of an independent Russian federal agency authorised to deal with environmental protection, state environmental control and environmental reviews;
- an obligatory accreditation process for experts involved in environmental assessments and reviews should be introduced. Such accreditation should be aimed at ensuring that EIAs and SERs are conducted by competent specialists capable of assessing the economic impact of a given activity on the environment and public health – so ensuring compliance with applicable Russian standards, rules and regulations concerning environmental protection and safety.

Furthermore, and on the basis of EU and World Bank experience, HES II proposes to divide certain types of facilities and activities into four categories for purposes of environmental review: Categories A, B, C, and D.

---

[10] See also chapter 3 (*Guseva*) and chapter 4 (*Ratsiborinskaya*).

[11] Russian Federal Urban Code, No. 190-FZ of 29 December 2004 (last amended on 17 July 2009).

[12] See <http://www.kremlin.ru/transcripts/3767>.

Category A includes highly hazardous, technically complex and unique facilities or large-scale activities that have an obvious environmental impact and are included in similar EU lists (e.g., Annex 1 to Directive 2008/1/EC[13]) and are indisputably subject to the environmental review procedure. The category A list is closed and can only be amended if an order is passed by the Russian government.

Category B includes projects of facilities that have an obvious environmental impact, but which implementation does not result in any unpredictable effects, as proven by previous reviews or similar procedures. Depending on certain natural and industrial conditions, an EIA may be required for category B projects and project components that are likely to result in reduced or full environmental impact. The list of category B projects is also closed and can be amended in accordance with the same procedure as for category A. In some cases, categories A and B cover the same types of installations – however, whether or not the installation falls within category A or B is determined by a given threshold value. For example, combined heat and power plants ('CHPPs') and other combustion plants with electric capacity of at least 300 MW fall within category A – whereas, all other CHPPs with a lesser capacity fall within the B list. However, if a CHPP is coal- and (or) black oil-fuelled – it will fall within category A as its negative impact can be substantial, irrespective of its capacity.

In the case of categories A and B, the design documents of such a facility or activity should be subject to a mandatory EIA and then a SER. No funding should be permitted nor provided to such a facility or activity if the SER committee decides that the given facility or activity does not meet the requisite standards.

Facilities or activities falling with category B should be subject to a reduced EIA requirement at the stage when the party undertaking the activity or construction of a facility, declares its possible impact. Thereafter, the authorities responsible for the subsequent SER must decide either to:

- authorize the reduced EIA;[14]
- ban the implementation of the project or the activity; and/or
- order the carrying out of a comprehensive EIA and make a decision in accordance with its outcome.

The EIA requirement should be further simplified for facilities or activities falling with category C. Such a simplified EIA should be carried out when the developer (or originator of an activity) prepares its Statement of Environmental Impact ('SEI'). An originator of a given activity or developer of a facility is required by Russian law to confirm in the SEI that they comply with existing Russian environmental requirements or that

---

[13] See Council Directive 2008/1/EC of 15th January 2008 concerning integrated pollution prevention and control; and see <http://eur-lex.europa.eu/LexUriServ/LexUriServ.do?uri=OJ:L:2008:024:0008:0029:EN:PDF>.

[14] Such a reduced EIA may provide for the facility's or activity's impact on only one of the following natural media: air, water or soil.

the planned activity has no environmental impact. An example of a category C facility is where a facility owner declares (in the SEI) the level of the planned activity's environmental impact in its SEI and proposes impact mitigation actions with respect to such environmental impact.

Facilities and activities falling within category D should not be subject to an EIA requirement, but should merely be subject to the requirement of a state review of the given facility's or activity's design documents (in accordance with Article 49 of the Urban Code).[15] The following activities and facilities would fall within category D: projects with insignificant environmental impact and which incorporate requisite engineering solutions in order to reduce such impacts to the required levels, as well as all facilities and proposed activities related to housing and civilian construction.

Experts of the HES II project further propose that the authority determining whether a facility or activity falls within category B, C or D may delegate this task to a Russian constituent entity.

The proposed list of facilities and activities for each category has been determined[16] and the concept of a new EIA Regulation, which uses the above classification of facilities and activities has been proposed by HES II experts. This proposal recommends that two procedures, which have never been used in Russia before, be included in the assessment: screening and EIA scoping – both in order to save resources and to speed-up EIA procedures.

Environmental auditing

Environmental auditing ('EA') is a relatively new instrument in the field of Russian environment protection and is hardly mentioned in Russian environmental law. HES II experts considered the experiences made in connection with the EU Eco-Management and Audit Scheme ('EMAS')[17] as useful when applied to Russia. Especially so, with respect to EMAS' experiences with institutional and organisational arrangements of industry, as well as the EMAS practice of having companies (which are audited for self-management), financing institutions or investment companies, use EA reports.

It was not considered necessary to propose a new Russian law on EA, as HES II experts view EA as a market instrument and as Russia has no law on environment management systems ('EMS').[18] Nevertheless, it is noted that EMS systems are becoming popular in Russia and enterprises operating in Russia are increasingly obtain-

[15] Art. 49 of the Urban Code requires the review of project (design) documentation only. Russian law requires that project (design) documentation contains information on the environmental impact of the project, as perceived by the designers (and which is not an independent EIA). The state review commission (with one environmentalist) then estimates the adequacy of the entire documentation with respect to the applicable requirements, including as regards technical issues, construction safety, fire safety and environmental safety.

[16] See <http://www.ippc-russia.org/public/cluster03/CL3_Final_Report_EIA_short_version_EN.pdf>.

[17] See <http://ec.europa.eu/environment/emas/index_en.htm>.

[18] See chapter 16 (*Dayman*) for more detail on EMS in Russia.

ing relevant certificates voluntarily. The degree to which EA results can be used by Russian environmental authorities remains to be explored in more depth. The mandatory use of EA was raised as an issue within the HES II project, but it was agreed that EA cannot be forcefully established by the government since it is a voluntary market tool.

As Russian entities export many goods and services, the question was raised what other auditing systems should be reviewed within the HES II project. Some HES II experts voiced the opinion that no federal law is needed in order to regulate auditing. Other experts insisted that EA in Russia should retain its guise of being an internationally accepted and transparent management tool and that there is no need to adapt EA to Russian circumstances by way of adopting a federal law. In any event, HES II project experts recommend that:

- accreditation and development of EA procedures in Russia should follow ISO standards;
- auditors should be able to compete freely on the market; and
- a system should be set up, which includes the putting in place of a national accreditation body – ideally recognised by other countries' accreditation organisations. It is noted that in Russia, EA standards are set in line with the ISO 14000 series standards.[19]

Environmental certification

Environmental labelling (or eco-labelling) and certification has been widely and successfully applied in the EU since the early 1990s. In particular, companies promoting and selling environmentally friendly products have used this tool in order to inform their customers and partners of the environmental aspects and characteristics of a given product, including its development, production, utilization and disposal processes.[20]

HES II findings show that in Russia eco-labelling is primarily being applied in the food sector. Eco-labelling is rare in the industrial sector despite the ISO 14020 series standards having been translated into Russian and adopted as Russian standards. Furthermore, these findings showed that there is a need for a consistent legislative concept of environment-friendly (acceptable) products for Russia as a whole, as well as for consistent use of applicable legal definitions (definitions range from 'environmental labelling', 'eco-labelling', 'environmental label' to 'eco-label'). HES II experts established that there is still a need to apply consistent criteria throughout Russia in order to distinguish between environmentally sound and 'ordinary' products. There are only few examples of actual use of eco-labelling in Russia (with the food sector being the exception), which means that there is low consumer awareness and, consequently, a

[19] See also chapter 16 (*Dayman*).

[20] The EU 'Eco-Flower' logo is the most well known and promoted within the EU, while there are many others in many EU states.

lack of eagerness and preparedness of Russian consumers to pay more for environmentally sound products.

In accordance with its findings, the HES II project recommendations with respect to environmental certification and eco-labelling in Russia include the need to:

- establish and develop a Russian National Environmental Certification and Labelling System and registering this system on the international level;
- further develop a consistent legislative concept of environmental certification and labelling within Russia;
- introduce regulatory restriction in order to prevent unjustified 'self-declarations' by producers who state that their products are 'environmentally friendly', when they are not;
- adopt international and national standards of eco-labelling;
- implement and adapt EU experiences in environmental certification and labelling of goods and services to Russian circumstances;
- create a list of producers, manufacturers and products that have obtained eco-labels (Environmental Producers) in order to raise customers' awareness of eco-labelling;
- create a demand for environmentally labelled products through wide-scale promotional campaigns (i.e., through federal level social advertising) and information campaigns of Environmental Producers;
- introduce 'green' public procurement practices in Russia;[21]
- promote international partnerships as regards environmental certification and labelling; and
- formulate Russia's position regarding environmental labelling for WTO negotiations, specifically with the Committee on Trade and Environment ('CTE') within the Agreement on Technical Barriers to Trade.[22]

Environmental monitoring

Russian industrial ecologists treat environmental monitoring ('EM') as a type of 'environmental control procedure' (within the framework of the permitting process, which includes 'self-control' of industrial enterprises) and 'state environmental control'. In Russian environmental practice, however, the terms 'environmental control' and 'environmental monitoring' have different meanings. Russian environmental bodies have

[21] See also chapter 10 (*Khmeleva*) for more detail on green public procurement in Russia.

[22] However, after 16 years of negotiations, Russia's negotiation position for WTO membership is currently unclear due to recent statements that Russia would pursue WTO membership jointly with Belarus and Kazakhstan as a customs union (see Reuters, 18 June 2009, 'WTO in confusion after Russia customs union plan', at <http://www.reuters.com/article/reutersEdge/idUSTRE55H18920090618>; see also WTO website, 13 August 2009, 'Medvedkov will head WTO negotiations of Customs Union (RIA Novosti, 12 August 2009)', at <http://www.wto.ru/en/news.asp?msg_id=24904>; and 24 August 2009, 'WTO members want more clarity on the Customs Union Accession to the WTO (RIA Novosti, August 22, 2009)', at <http://www.wto.ru/en/news.asp?msg_id=24931>. Most recently, at the time of writing, indications were made that Russia may, however, aim to become a WTO member separately, not within the customs union.

broad experience of 'state environmental control', but the procedures relating to enterprises' self-control (or monitoring) are, sometimes, excessive and overly complicated for enterprises. This often causes confusion and makes it difficult for enterprises to adhere to the regulations on environmental 'self-control' or monitoring.

Although Russian law requires that Russian enterprises must carry out 'industrial environmental self-control' (or monitoring) in accordance with the provisions approved at the state level, there is no normative document, which provides the exact requirements of such industrial environmental self-control. However, in his speech at the 5th Baikal Economic Forum,[23] Minister Yury Trutnev stated that the Russian system of enterprises' environmental self-control and reporting to the relevant controlling bodies will be modified in a radical way. Minister Trutnev stated that the revised system of environmental self-control is to be based upon and implemented only with the help of instrumental methods. HES II experts are of the view that this approach is possible only if the number of controlling parameters is drastically reduced and the use of surrogate or marker parameters is introduced.

HES II project work has further established that Russian industrial circles express an interest in the environmental monitoring approaches as described in the 'EU Reference Document on the General Principles of Monitoring' (the 'EU Monitoring Document') (this document was translated into Russian and made publicly available via the HES II project website[24] and is used during training and pilot exercises). Russian industrialists consider this document to be a useful manual, which sets out how to organize and implement industrial environmental 'self-control'. Thus, the EU Monitoring Document could be helpful in order to clarify what is meant by 'industrial self-control' or 'monitoring' under Russian law and provide useful guidance with respect to the envisaged reform mentioned by Minister Trutnev.

BAT / BREF

The BAT/BREF system constitutes a relatively new concept for the Russian Federation and appears to be an attractive concept which, in particular, industrialists could consider. BREF documentation is in high demand in Russia and much interest was shown in this subject at an international conference held in April 2008.[25]

Within the EU context, BREFs are used as the main instrument for integrated environmental permitting. The main purpose of using BREF documentation is to ensure that industrialists and relevant permitting authorities have a similar understanding of

---

[23] The Baikal Economic Forum took place on 9 September 2008 and was devoted to the complex developmental programmes for Eastern Siberian and Far East regions and their integration into the world's economic space. See <http://en.baikalforum.ru>.

[24] See HES II project homepage <http://www.ippc-russia.org/content.html>.

[25] See 'Best available techniques ('BAT') as a tool to reduce negative environmental impacts and improve energy efficiency of Russian enterprises' International Conference, arranged by HES II with the additional support of the British Foreign and Commonwealth Office within the Global Opportunities Fund ('GOF'), 3-4 April 2008, Moscow, Hotel 'Saljut'.

production processes, associated environmental problems and of the applicable BATs for any given industrial sector. BREFs are not legally binding, but provide guidance only. Thus, BREF documents do not prescribe any particular technology, but provide essential references (and parameters) for the main permit conditions of a given activity, including guidance on applicable emission limit values. The BAT and relevant permit conditions for an installation are then decided upon by the relevant permitting authority on a case-by-case basis.

The Russian Federation has declared strong political interest in elaborating and implementing a BAT-based environmental permitting system. In addition, there is interest in using BAT guidelines for EA and environmental certification activities in Russia.

HES II experts consider that aside from BREFs serving as a key tool for the reform of the Russian environmental permitting system, BREFs could be used:

- for guidance and as a point of reference for environmentally and economically sound investments, as well as for production planning and design;
- for benchmarking of existing installations (e.g., in the framework of environmental audits);
- for educating and training engineers and environmental specialists (this applies especially to 'horizontal' BREFs[26]);
- as an element of the new Russian standardisation system; and
- as a basis for discussion to adjust Russian environmental quality standards used in environmental permitting.

The most preferred method of implementing BREF-documents adjusted to Russian conditions (within the Russian system of technical regulation) is to prepare BREF-documents in such a manner that they can be a part of the Russian national standardisation system. This means that codes of practice or national standards should be elaborated and adopted based on BREF-documents, in line with the procedures established under the Technical Regulations Law. Furthermore, upon analysing the Technical Regulations Law, HES II experts concluded that BAT guideline documents can be best implemented within the Russian system of technical regulation as follows: the 'vertical' BREF-documents[27] can be best implemented in the Russian system of technical regulation as 'codes of practice', whereas 'horizontal' BREFs are best adopted as 'national standards'.

Environmental permitting

While informing the deputies of the State Duma about the envisaged measures of the Russian government as regards the improvement of the environment in the Russian

---

[26] 'Horizontal BREFs' are BREFs developed for separate branches of industry or agriculture.

[27] 'Vertical BREFs' are BREFs which are applicable to all or several industrial or agricultural branches.

Federation, Minister Trutnev has made some important statements.[28] He expressed the opinion of the MNR on the necessity of simplifying the Russian procedure of industrial permits, by making it more transparent and by decreasing its 'corruption capacity'. The ministerial agenda details the use of the BAT approach for the industrial permitting procedure and the 'declarative permitting' procedure[29] for enterprises with small negative environmental impact.

HES II experts are of the view that the 'declarative' approach to permitting could be combined with an 'integrative' approach to permitting – and that the application of the 'declarative' approach in the Russian permitting system could be restricted by way of thresholds based, for example, on the capacity and types of activity. This approach was used by HES II experts in the EIA cluster, although the lists of facilities and activities established that there could be a basis for differentiating permitting procedures.

The procedural provisions of the IPPC Directive seem especially applicable to big industrial enterprises – and so makes harmonisation of Russian law with the IPPC Directive in this field seem required and feasible. HES II research shows that specific attention should be paid to the economic attractiveness of integrated environmental permitting based upon BAT and the corresponding industrial technical standards and norms.

Other areas of harmonisation that can be considered are: the type and size of installations for which a new integrated permitting system should be applied; public participation within the permitting system; the 'one window' concept;[30] the role of environmental quality standards and the establishment of key environmental indicators. Reform of the Russian permitting procedure might require the redefinition of the permitting authorities' responsibilities and the ways in which they participate in the process. HES II experts are working on formulating the procedure which should apply to environmental permits, but uncertainty about the future executive body in the environmental field is hindering progress in this area.

Environmental pollution charges

The issue of pollution charges (or fees) in Russia is closely linked with the introduction of BAT and permitting procedures. Enterprises receiving integrated environmental permits should potentially enjoy some exemptions from certain charges and fees. Environmental taxation can be used as one of the most effective tools in speeding-up the necessary transition of an enterprise operating in an environmentally harmful manner (e.g., using fossil fuels as its energy source) to making its operations environmentally

[28] See the report of Minister Trutnev at the 'Governmental hour' in the State Duma on 3 December 2008, at <http://www.ecosinform.ru/userfiles/file/1_64_ec_12_08.pdf> (in Russian).

[29] 'Declarative permitting' is a simplified permitting procedure, whereby an enterprise with small negative impact submits (to the relevant environmental authority) its declaration of its negative environmental impact in the form of an environmental auditor's conclusion.

[30] The 'one window' or 'one-stop shop' concept means that the permit applicant deals with one permitting agency and does not have to have its permit application endorsed by other numerous instances.

sound (e.g., using renewable sources of energy or 'climate-neutral' energy sources) and becoming energy efficient. For example, in Finland and Sweden, energy taxation of fossil fuels forms the main instrument to promote renewable energy in the short term.

In Russia, the share of pollution charges in prime production costs is relatively small. The share of Russia's budget revenue (federal or regional) from pollution charges is negligible – nor does Russia use pollution charges as an economic regulatory tool to motivate Russian enterprises to reduce their emissions. Thus, harmonisation of Russian law with the EU's approach to pollution charges appears to be important in this field.

HES II experts' work on this topic includes the preparation of recommendations relating to long- and short-term approaches of improving Russian environmental law and developing some specific suggestions in reference to the draft federal law 'On Charges for Negative Environmental Impact'. For these purposes, HES II understands that valuable lessons can be taken from the EU experience with redistributable charges.[31] Furthermore, in EU countries, fines for accidental or fugitive pollution and excessive emissions are usually high and levied by courts upon submissions received from environmental authorities. However, in the medium-term, HES II experts are sceptical about Russian law being successfully harmonised with EU legislation on environmental taxes.

HES II experts believe that, in future, penalties for the violation of environmental permit conditions should replace environmental pollution charges as regulative instruments. It should, however, be emphasised that the full-scale development of an integrated environmental permits system in Russia is impossible without changing the entire chain of environmental instruments, i.e., EIA, environmental charges and penalties, environmental control, audit and certification.

Environmental Quality Standards

The issue of developing Environment Quality Standards ('EQS') in the framework of the HES II project is merely focused on the issue of the need for industrial permits granting and further environmental monitoring (control).

Ambient air quality standards

It is clear from preliminary comparative analysis of standards for ambient air that for some substances (nitrogen dioxide (NO_2), sulphur dioxide (SO_2), ozone (O_3), benzo[a]pyrene) the standards are the same or almost the same in the EU as in Russia. The World Health Organisation standards for NO_2 and SO_2 coincide with those existing in Russia. However, Russian ambient air standards do not include various pollutants, which the EU considers as 'priority harmful pollutants', i.e., fine particles and

[31] Redistributable charges are charges, which are accumulated in special funds and distributed to enterprises that are deemed to have the best environmental performance.

ground-level ozone.[32] HES II experts, thus, propose that PM_{10} and $PM_{2,5}$ should be included in Russian ambient air standards.[33]

Russian law considers the above mentioned standards to be sanitary-hygienic standards. The assumption under Russian law is that environmental standards for ambient air are set – however, in practice, no actual ambient air standards exist within the Russian territory.

The new EU Directive on Ambient Air Quality and Cleaner Air for Europe[34] envisages standards for the protection of vegetation. Analysis of these standards and their application to Russian conditions is another area, which could be harmonised. The HES II project is in process of reviewing this within the EQS cluster.

Sanitary protection zones

The next possible field for standards harmonisation in Russia is in the area of sanitary protection zones ('SPZ'). Although EU law does not define the term of 'sanitary protection zones', some European countries do have similar concepts to the Russian SPZs. Germany, for example, has issued a 'distance decree' at the *Länder* level, which creates 'Distant Zones' and is designed for use in spatial planning procedures. Distant Zones and the associated permitted land-uses merely determine the value of that particular piece of land, and no taxes or fees are levied on enterprises situated in Distant Zones.

By comparison, in Russia, the SPZ regulations contradict principles of market economy: an enterprise owner pays land tax in order to own land within the territory of an SPZ, but is restricted in its use of the land within the SPZ.

Water quality standards

Harmonising Russian water quality standards with EU standards is more complicated. For some substances, especially for heavy metals, Russian standards for fishery basins are very strict. The emission limit values ('ELV') based upon such standards are, thus, often unrealizable. For many years, Russian enterprises worked with permits based on temporary ELVs agreed within procedures which were potentially prone to corruption. The setting of these temporary ELVs was a compromise between the very strict legal requirements and an enterprise's ability to meet the required standards.

HES II experts consider that in order to harmonise Russian and EU water quality standards, the following steps should be taken:

- setting of reasonable water quality standards; and

---

[32] These substances are a significant threat to human health in that they are said to shorten average life expectancy by almost one year and threaten the healthy development of children.

[33] The notation PM_{10} is used to describe particles of 10 micrometers or less and $PM_{2.5}$ represents particles less than 2.5 micrometers in aerodynamic diameter.

[34] 2008/50/EC of 21 May 2008.

- revising the methodology of the determination of 'impact standards' and searching for new approaches for ELV determination (for example, acute and chronic toxicity of the effluent together with restrictions for the most harmful substances).

In his speech at the Baikal Economic Forum, Minister Trutnev declared that the Russian permitting system for water discharges will be changed drastically. He stated that the practice of temporary ELV setting will be cancelled. This follows HES II experts' recommendations and will lead inevitably to the revision of water basins quality standards.

Environmental insurance and liability

Within the EU, environmental insurance has developed from being a compliance-driven product to being a tool, which supports environmental risk management strategies and which protects shareholders from the financial consequences of their company's environmental liabilities. Environmental insurance now plays a major role in a variety of European business transactions and financial engineering situations. In mergers and acquisitions, it is used to protect the acquiring company against the risks of unanticipated cost overruns associated with known clean-up obligations, as well as the discovery of unknown, pre-merger pollution conditions.

The environmental insurance market continues to develop in member countries of the Commonwealth of Independent States ('CIS'). The Inter-Parliamentary Assembly of Member Nations of the CIS[35] adopted the model law 'On Ecological Insurance' in 2000 and adopted a new version in November 2003. Furthermore, the Republic of Azerbaijan passed the law 'On Environmental Insurance' in 2002 (which came into force in 2004) and, in 2005, the Republic of Kazakhstan adopted the law 'On Mandatory Environmental Insurance'.

In Russia, several legislative acts, such as the federal law 'On Environmental Protection',[36] establish the basis for the development of environmental insurance. However, no specific law on environmental insurance has been passed yet. An Expert Committee on Insurance was set-up under the former Rostechnadzor.[37] This committee works as a permanent consulting body for working out the proposals on the improvement of the system of insurance of dangerous industrial objects.

HES II experts' opinion is that before any further steps are taken to design an environmental insurance law in Russia, first basic materials (and, if needed, proper guidelines) should be prepared in the following areas:

- environmental liability
- environmental risks calculation
- environmental damage valuation

[35] See <http://www.iacis.ru/html/index-eng.php>.

[36] See also chapters 4 (*Ratsiborinskaya*) and 7 (*Nystén-Haarala*).

[37] See <http://www.gosnadzor.ru/slugba/sovet_strahovanie.html>.

In particular, strong emphasis should be placed on comprehensively regulating the area of environmental liability in Russia before considering any further issues connected to developing Russian environmental insurance laws. It is noted that efforts by HES II experts might lead in the near future to the adoption by the Inter-Parliamentary Assembly of Member Nations of the CIS of a revised model law on environmental insurance based upon EU liability schemes.

Conclusion

The harmonisation of Russian environmental standards with EU law is an ongoing process. We are not so optimistic to believe that the results of the HES II project will drastically change the existing system of Russian environmental law governing industrial environmental management. However, we see today that there is an increased demand within Russian society to change relations that currently exist between industry and the environment.

HES II work shows that, in particular, the BAT approach is an accepted way of reaching a balance between the interests of industry and economic development – and those of nature protection. By establishing a BAT system in Russia, this opens-up the rare possibility for industrialists and environmentalists to start a new era of cooperation and to find a common language. Thus, the HES II project is creating the preconditions for developing a positive dialogue between industrialists and environmentalists. Further efforts will, however, be required by the EU Commission and the Russian government (and executive bodies) within the official sub-groups of the EU-Russia Environmental Dialogue[38] in order to foster further understanding of and responsibility for the common EU-Russian environmental space.

[38] See <http://ec.europa.eu/environment/enlarg/russianis_en.htm>.

Chapter 7
MECHANISMS TO PROMOTE GREEN BUSINESS IN RUSSIA[1]

Soili Nystén-Haarala[2]

INTRODUCTION

Promoting green business is an important challenge for Russia, considering, in particular, the magnitude of the need for technological restructuring after the collapse of the Soviet Union. Today, pollution remains a problem and it has become one of the most important target areas of international aid to Russia. Thus, not only is there a strong need for the development of a 'green' business sector and green business practices in Russia, but there are great opportunities for businesses that wish to undertake environmental investments or promote green business practices.

The general Russian legal environment is a challenge for green or, for that matter, any kind of business operations. There are currently no significant domestic incentives, such as tax reductions or low-cost loans available for 'green' businesses.[3] Furthermore, Russian environmental law is constantly changing and it is punitive by nature, as well as based on strict regulations, inspections and fines.[4] Thus, the mechanisms to promote green business in Russia are still in their infancy.

Nevertheless, there are non-regulatory mechanisms available to companies and investors in order to promote green business and green business practices. Although

[1] This chapter has been funded by the Programme 'Russia in Flux' (2004-2007) of the Academy of Finland (<www.aka.fi/russia>). The author lead the research project 'Governance of Renewable Natural Resources in Northwest Russia' (No. 203864) within the framework of the programme. The project focuses on participation of different stakeholder groups in governance of forest and fisheries resources at local, regional and federal levels. The author has conducted interviews with managers of paper and pulp mills, wood trading and harvesting companies, saw mills, fishing enterprises, as well as representatives of regional administration and municipalities in the Leningrad region (March 2004), the Karelian Republic (July and September 2004), the Arkhangelsk region (February and November 2005) and the Murmansk Region (November 2005). The joint publication of the project, by Soili Nystén-Haarala, ed., *The Changing Governance of Renewable Natural Resources in Northwest Russia* (Ashgate, Farnham 2009).

[2] Professor of Civil (Private) Law University of Joensuu, Finland; Email: *soili.nysten-haarala@uef.fi*.

[3] However, see chapter 15 (*Maksimenko and Gorkina*) as to the loans available in context of the Russian National Pollution Abatement Facility.

[4] The constant changes in Russian environmental law are due to the following reasons: (i) the current debate relating to federalism, whereby the federal central government is now introducing changes, which reduce the power of the regions and municipalities (with respect to decision-making relating to the use of natural resources); (ii) due to the fight over private property rights. Enterprises have mostly been privatized, but natural resources still remain State property; and (iii) the failure of formal institutions and priority of informal relations in Russian society. Law and legal institutions are numerous, but they do not always work in a predictable manner. Therefore, it is important that larger businesses maintain good relations with the State and municipal authorities.

W.Th. Douma and F.M. Mucklow, eds., Environmental Finance and Socially Responsible Business in Russia
© 2010, T·M·C·ASSER PRESS, The Hague, The Netherlands and the Authors

Russian civil society is still relatively weak, NGOs co-operate with businesses through, for example, voluntary certification of products and processes.[5] This type of cooperation with NGOs can allow businesses and investors to gain positive publicity and support from such consumers and citizens who are interested in environmental issues.

This chapter aims to provide a brief overview of the available mechanisms to promote green business in Russia. In order to do so, the general Russian legal framework is introduced, the procedures relating to obtaining licences and permits (to use natural resources) are described, and further discussion sets out how companies can use the existing legislation, official State programmes and new non-governmental mechanisms, such as certification, to implement and promote green business in Russia.

The need for green business

Towards the end of the 1980s, *glasnost* enabled the open discussion and criticism of the pollution of the Russian environment during the Soviet era. It was one of the main themes of discussion within the new emerging Russian civil society.

However, it can be claimed that, in general, today's attitudes in Russia have not significantly shifted to being more environmentally friendly and, thus, as having moved away from such attitudes which prevailed in Soviet times. Russia has been struggling with the inevitable social and economic difficulties generated by economic transition. Initially, people were less interested in the pollution of the environment and more concerned with retaining their industrial jobs.[6] As the Russian economy moves towards a free market economy, ordinary people simply do not appear to have the time to consider civil society issues, as supporting their families takes all their energy. Economic growth increased consumption, but did not turn the interest of Russian citizens back to environmental problems. With the increasing budget revenues brought along with the economic growth, the Government preferred to pay attention to still extensive social and economic problems before environmental ones.

The full technological modernization of Russian industry still lies ahead and the Russian people are wary of the fact that modern and more environmentally friendly technology can potentially lead to increased unemployment, despite the overall benefits it brings to society (such as relating to health and safety). Corporate social responsibility ('CSR') is more important both for companies and the government at the moment. The interest of Russian companies in environmental responsibility is growing[7] , in particular, with respect to exporting companies who have realized the potential competitive advantage, which investing in environmental friendly production could offer.

---

[5] See chapters 5 (*Razbash*) and 21 (*Lawry and Rogers*), on details how the government and NGOs are cooperating with businesses to promote socially responsible business and environmental projects.

[6] J. DeBardeleben and K. Heuckroth, 'Public Attitudes and Ecological Modernization in Russia', in Ilmo Massa and Veli-Pekka Tynkkynen, eds., *The Struggle for Russian Environmental Policy* (Helsinki, Kikimora Publications 2001), Series B 17.2001, pp. 49-76.

[7] See chapters 19 (*Kostin*) and 20 (*Dayman*) on trends of CSR in Russia.

The restructuring of industry was delayed on the basis that post-Soviet 'insider privatization'[8] left most of the former Soviet enterprises in the hands of their former managers and employees, who did not have the necessary means or know-how to put in place new technology or production processes.[9]

However, the financial crises of 1998 and the strengthening of domestic production that followed the crisis, made economic growth possible for industries other than just the oil industry, the latter of which has long flourished due to exports and high oil prices. As a result, some enterprises were able to invest in new technology, which, as a side effect, also resulted in environmentally friendlier production. Nevertheless, there remain substantial amounts of industrial enterprises which still require restructuring, and which, inevitably, is linked to the need for further green business and investment in Russia. Thus, the scope for investing in and promoting 'green' technology, and green business in general, still remains significant, especially in view of replacing ageing industrial technology.

From an environmental perspective there is a need for green business, but the attitudes of citizens, politicians and state officials towards such green business are rather mixed. Those who start green businesses, especially foreign investors, may encounter people being suspicious of green business and environmental investments. Foreign investors are, in any event, viewed by the public with suspicion and there is a fear that foreign investors use Russian natural resources at the expense of the Russian people.[10] These attitudes will hopefully change in the next few decades to come. However, the campaign, which former President Putin started from the beginning of his presidency, to strengthen state power in relation to business, as well as his campaign to weaken the influence of international financing of NGO's appears to have resulted in channelling national sentiments of the people towards suspicion for the aims of foreigners and foreign countries. Russian domestic industry is also using these sentiments to prevent competition from abroad. For the moment, the effect of such attitudes should not be underestimated or disregarded by 'green' investors and businesses, be they foreign or

---

[8] The original plan for privatization was the mass privatization and sale of enterprises to investors, who could then have restructured them. However, in order to fulfil this plan, the managers of the state enterprises had to be persuaded to support the privatization processes. Thus, an alternative plan was launched to transfer 51% of the shares to the directors and employees of the state enterprises before the auctions and mass privatization occurred. As a result, the managers often held more than 70% of the shares in the privatized state enterprise. This meant that insiders privatized state enterprises for their own benefit. See S. Nystén-Haarala, *Russian Law in Transition. Law and Institutional Change* (Helsinki, Kikimora Publications 2001) p. 148.

[9] The takeover by the managers would not have been possible without the support of the employees, who were promised that everything would continue as in previous times and they could keep their jobs, see Nystén-Haarala, *supra* n. 8, at p. 147.

[10] See A. Kits, S. Novitskov and A.G. Stroganov, *Paradoksy rossiskogo biznesa. Ili snova ob investitsijah* (Kompania ASTSh, Moskva 2001). The authors of the book are former members of the Russian Duma. They complain that the attitudes of state authorities towards foreign investors are often negative and cause damage to the Russian economy. The interviews of the Governance project, *supra* n. 1, however, suggest that employees of foreign owned companies start to trust their new employer as soon as they become convinced that the company will stay in Russia and do business and not just squeeze out profits and leave. Nystén-Haarala, *supra* n. 1.

domestic. Thus, running any larger business operation in Russia will require good public relations[11] in order to lessen suspicions and it will also require informed negotiations with state authorities.

Enterprises are frequently requested to assist with alleviating the financial challenges faced by the municipalities and regions. Thus, environmental and socially responsible investments can often be marketed to the state and municipal authorities as means by which such challenges can be addressed in cooperation with businesses. Furthermore, paying attention to 'ecological plans' (see more detail below) may also be valuable as business development opportunities. However, corruption remains a big issue in Russia and (foreign) investors are advised to be aware of this and, in particular, to avoid that good business relations with the authorities do not develop into corrupt relationships.

General legal framework – overview and key issues

Russian environmental law does not provide for any particular mechanisms to promote 'green' business. However, it does provide a framework of laws, which businesses and investors (be they environmental or not) will need to take into consideration in their operations. For these purposes, the below sets out the general structure of Russian environmental law.

Russian environmental legislation has been changing rapidly since *perestroika* and the fall of the Soviet Union. It should be noted that Soviet environmental law was administrative by nature. Furthermore, under the Soviet system, the state owned all the natural resources and means of production, and Soviet environmental law only regulated how state property was to be used and protected. The emphasis was placed on natural resources and the regulating laws were classified according to the types of the resource: forest law, water law and subsoil resources law.[12] In essence, all these laws were part of Soviet land law, but on the basis that there were specialized laws and decrees, land law was treated as regulating other forms of land use (industrial land, agricultural land etc.). The protection of the environment was treated as a separate issue and the Soviet laws on nature protection were (and today continue to be) very strict. There are large protected areas (*zapovedniki*), in which, despite the weakened inspection and existing corruption practices, there still exist huge areas of wild nature and ancient forests.[13]

General structure of environmental legislation

Russia's transition towards a free market economy has resulted in various new requirements under Russian environmental law. For example, there have been attempts to

[11] See also chapter 4 (*Ratsiborinskaya*) on a review of Russian environmental law and chapter 14 (*Small, Barton and Lanman*) regarding the issue of building relationships and communicating key messages.

[12] I.A. Ikonickaya,'Zemel'noe pravo Rossiiskoi Federacii', *Iurist* (Moscow, 2002).

[13] V.-M. Tynkkynen and I. Massa, 'The Russian Environment – a Source of Risk or the Lungs of the Planet?', in Massa and Tynkkynen eds., *supra* n. 6.

integrate (with difficulty) into Russian environmental law the principles of international environmental law, such as the *polluter pays principle*, as well as to improve channels for participation of citizens. Despite these changes, in general, Russian environmental law has remained much the same since Soviet times.[14]

The most important environmental legislative acts are federal legislative acts and they are as follows (Acts of Parliament):[15]

- Constitution of the Russian Federation 1993;[16]
- Law on the Protection of the Environment 2002;[17]
- Law on Environmental Expertise 1995;[18]
- Water Code 2006;[19]
- Law on Subsoil Resources 1992;[20]
- Land Code 2001;[21]
- Law on Air Protection 1999;[22]
- Forest Code 2006;[23]
- Law on Fauna 1995;[24]
- Law on Payment for Land 1991;[25]
- Law on Licensing Various Functions 2001;[26]
- Law on Protection of Juridical Persons and Entrepreneurs in Applying State Control 2001.[27]

---

[14] Nystén-Haarala, *supra* n. 8, p. 177.

[15] See also chapters 4 (*Ratsiborinskaya*) and 5 (*Razbash*).

[16] Dated 12 December 1993 (the 'Russian Constitution').

[17] Dated 10 January 2002, as amended in June 2007 (the 'Law on the Protection of the Environment').

[18] *O ekologicheskom ekspertize*, dated 23 November 1995, as amended in 1998 (hereinafter referred to as 'Law on Environmental Expertise').

[19] Dated 3 June 2006 (the 'Water Code').

[20] 21 February 1992, amended significantly on 5 March 1995 and slightly on 2 January 2001, 2002 and 2004. The amendment of 2002 diverted the payments to the federal budget (the 'Law on Subsoil Resources'); a new Subsoil Law was approved by the Government (which was submitted to the Parliament (*Duma*) already in 2005) but has not yet been accepted; The most recent amendments to this law pertain to the limiting of rights of foreign investors and tightening up of environmental protection.

[21] Dated 25 October 2001 (the 'Land Code').

[22] Dated 4 May 1999, as amended in 2004.

[23] Dated 4 December 2006 (the 'Forest Code').

[24] Dated 24 April 1995, as amended in December 2006.

[25] Dated 11 October 1991, as amended in June 2007. The amendment in fact refers back to the Land Code (Art. 65) determining the grounds for payment The Tax Code, Part II (Art. 31) defines land tax among municipal taxes. The grounds for minimum payments on leasing the forest are defined in the Forest Code. The minimum payment level is regulated in governmental decrees. General principles on the use of subsoil resources (licenses) are regulated in the Law on Subsoil Resources and regulated in more detailed Government Decrees.

[26] *O litsenzirovanii otdel'nyh vidov deyatel'nosti*, dated 8 August 2001, as amended 2002, 2003, 2004, 2005, 2006 and 2007 (the 'Law on Licensing'). The law provides a framework for admitting licenses. The more detailed rules on specific licences are regulated by decrees, e.g., the Decree on the Procedure for Licensing the Use of Subsoil Resources (dated 15 July 1992, amended in June 2007) A model contract for licensing has been passed as a Government Decree on 11 April 2006).

[27] Dated 8 August 2001, as amended 2003 and 2004.

The rules on liability for damage to the environment are set out in the Civil Code (Part I, 1995) and in the Code on Administrative Offences.[28] Decrees relating to these laws are important, too, if not more important, as the framework laws require specifications by way of decrees. For example, decrees set out the required standards of water quality and the regulations relating to applications for licences for the use of natural resources.

Federal v. regional legislation

Investors will need to remain up-to-date on the changing environmental legislation – as well as take into account that the subjects (i.e., the different regions) of the Russian Federation each have their own respective constitutions and legislation.

Since the nascence of the Russian Federation in 1992, there has been an ongoing battle between those wishing to decentralise and those wishing to centralise the Russian government. This battle was said to have been caused mainly by the question as to which level of the Russian state should benefit from natural resources: the federation itself, the region(s), or the self-governed municipalities?[29]

While President Yeltsin was in power, many differences developed between the federal and the regional legislation.[30]

However, under President Putin's government the differences became diminished due to the campaign for a single hierarchy of laws (*vertikal*). This campaign for 'legal equalization' is based on the presumption that federal laws are 'superior' to regional laws.[31] Thus, the supremacy of federal legislation currently prevails under Russian (environmental) law.[32] This means that applicable environmental principles and guidelines are stipulated in the federal codes or laws. Regional laws can supplement these federal codes or laws, but only within the federal framework legislation. In fact, it appears that there is very little room left for the regional legislation.

---

[28] Dated 30 December 2001, as amended several times every year.

[29] According to the Russian Constitution, natural resources fall under the joint jurisdiction of the federation and the regions (Art. 72 of the Russian Constitution). However, the Constitution does not specify the meaning of 'joint jurisdiction'.

[30] Since federal laws on the use of natural resources could not be enacted due to past conflicts between the Duma and the President, the Regions regulated this area of law by way of drafting regional laws. As a result, some Regions allowed private ownership of land, while others prohibited it. When the federal laws started to be enacted, the regional laws, which had been drafted earlier, contradicted them. Furthermore, regional constitutions contradicted the federal Russian Constitution, often proclaiming that natural resources are in the ownership of the region, see Nystén-Haarala, *supra* n. 8, p. 181.

[31] This presumption is understood to undermine the regional governments' incentive to redraft environmental legislation, see V. Kriukov, V. Seliverstov A. and Tokarev, 'Federalism and Regional Policy in Russia: Problems of Socio-Economic Development of Resource Territories and Subsoil Use', in P.H. Solomon Jr., ed., *The Dynamics of 'Real Federalism'. Law, Economic Development, and Indigenous Communities in Russia and Canada* (Center for Russian and East European Studies, University of Toronto 2004) p. 119.

[32] V. Leksin, 'The New Russian Federalism', in Solomon, *supra* n. 31, at p. 56. However, it should be borne in mind that even if the current Russian legal system resemble the simple and clear hierarchy of laws of the Soviet Union, in today's Russia there still exist considerable differences in the legislation of the different Russian regions. Text books tend to emphasize the supremacy of the federal legislation over the regional legislation. See e.g., M.M. Brinchuk, *Ekologicheskoe pravo* (Uchevnick, Moskva, Vyshee obrasovanie 2005) and Ikonickaya, *supra* n. 12.

With respect to federal environmental law, this is still changing and much of this change is still caused by the ongoing battle for power and ownership between the federation and the regions. For example, under the present legislation, the federation owns the forest. The income is, however, divided between the federation and the subject. The new Forest Code generally increases the power of the federation, however retains a possibility to delegate powers to the regional level. It remains to be seen, how this change in the law will be applied in practice. In the author's opinion, this change in the law suggests that after the federation won the battle with the regions and managed to centralize power as much as possible, the Russian government realized that centralization is not the answer to all the problems and is now interested in finding the most effective ways of governing natural resources, including by way of delegation of powers.

These changes seem to be having the effect of leading the Russian legal system back to the unitary model of arbitrary redistribution of budgetary funds that was characteristic of the economic system of the Soviet Union.[33] For example, the Law on Subsoil Resources (relating to oil, gas, mineral, groundwater resources) originally split the income derived from the use of subsoil resources between the federation, the regions and the municipal level. Thus, depending on the type of the raw material, the municipalities and the Regions received 60% to 80% of such income. In 2002, the law was amended in line with the budgetary reform and the direct payment requirements were replaced with a severance tax. This resulted in the municipal budgets being excluded from the revenue distribution scheme and the income share of the Regions diminished from 60% to 20% (in oil producing regions) and from 75% to 60% (in Regions producing other raw materials) – the reasoning for this change was that there was a need to ensure the equalization of budget funds per capita.[34] As a result, Russia's five main oil and gas producing regions lost a major share of their budget revenues.[35]

The battle for private ownership

Aside from the ongoing battle for ownership (in particular, over natural resources) within the different state levels, there is a further ongoing dispute as regards the merits of state ownership versus private ownership. Currently, former state enterprises are now mostly in private ownership, while natural resources are still owned by the state.

Economic success after the 1998 financial crisis caused a new wave of takeovers and disputes over ownership. The *Yukos* case is not the only, although the most famous and notable, dispute over ownership of valuable enterprises. For example, during the years 2004 to 2005, takeover attempts in the paper and pulp industry caused insecurity as regards property rights of owners, as well as investment aversive behaviour in such

---

[33] Solomon, *supra* n. 31, at p. 112.

[34] Ibid., at pp. 113-114.

[35] It is noted that the fees for the use of natural resources (sales tax), although vitally important for the municipalities and regions producing oil and gas, are comparatively low for companies.

enterprises fearing court proceedings, which may have resulted in the examination of their earlier privatization.[36]

Recently, the battle for private property rights over land and natural resources has grown stronger. For example, the most vividly discussed issue of the draft of the present Forest Code revolved around the issue of private ownership. However, the new code finally did not introduce private ownership of forests, but the main drafter, the Ministry of Economic Development and the Duma consequently chose to try and effectivize Russian forestry with other means.

The Russian Constitution, which came into force under the Yeltsin regime, and reflects the views of the then economic reformers, allows for the private, state and municipal ownership of natural resources (see Art. 9.2). However, the federal legislation provides that natural resources belong to the state and does not, so far, allow private ownership.[37]

The call to allow for private ownership of natural resources has become more pronounced with the increasing wealth of Russian industrialists. In earlier days, there was opposition to privatization based on fears that foreign investors would buy the nation's natural resources. Whereas today, those opposing private ownership do so also on the basis that they are afraid that the 'oligarchs' will further invest in natural resources.[38] During the drafting process of the present Forest Code the fear of privatisation was most pronounced amongst those who rent forest resources, but who do not have enough capital to purchase forest land.[39] However, those who were in favour for private ownership of forests argued, for example, that those who now rent forests do not have enough resources to invest in effective forest management or even utilize the rented areas effectively.[40] Thus, as the property rights issues regarding natural resources

---

[36] These fears were expressed in the author's interviews with paper and pulp mill managers in the Arkhangelsk region in February 2005, *supra* n. 1.

[37] According to the Forest Code, forests are in the ownership of the Russian Federation. The previous Water Code (of 1995) divided the so-called water objects between the ownership of the federation, the subject of the federation and municipalities. The present Water Code (of 2006) stipulates almost all water objects under the federal jurisdiction. According to the Law on Subsoil Resources, subsoil resources belong to the state and, nowadays, the licences for their use are permitted on the federal level. The opinion of the regional level, however, should also be taken into consideration and the relevant Regions participate in negotiations on the terms of licences.

[38] Many of those opposing private ownership of forests find their reasoning in the long tradition of state ownership of forests in Russia. (This reasoning is also manifested in the Decision of the Russian Constitutional Court dated 9 January 1998.) Opponents of private ownership do not wish to rely on industrialist taking care of forest resources, but argue that the latter strive for short-term profits; see for example as argued at <www.forest.ru>.

[39] When the author interviewed Finnish buyers and Russian sellers of round wood in those regions in May 2004, September 2004 and February 2005 *supra* n. 1, the fear that the oligarchs would take over the forest sector was quite common amongst small entrepreneurs in the Karelian Republic, the Arkhangelsk Region and the Leningrad Region. The interviews concerned contracting between trading partners. The results are published in S. Nystén-Haarala, 'Contract Law and Everyday Contracting', 49 *Scandinavian Studies in Law, the Proactive Approach* (2006) pp. 263-284. Property rights issues were touched upon by asking interviewees' opinions on how the Forest Code should be drafted and what effects the changes of the Code might have on their businesses.

[40] See the Russian NGO's Forest Club website <http://www.forest.ru/legislation>. It is noted that the ideological discussion regarding ownership has covered the more important issue of different manners in

are not yet clear in Russia, investments in the use of natural resources are considered risky.

Although these issues relating to ownership struggles appear rather remote in context of the mechanisms to promote 'green' business in Russia, they essentially represent the underlying tensions and reasons, which prevent more detailed focus on matters, such as developing key incentives under environmental law (or taxation law) to promote green business practices and green business *per se*.

Environmental permits, quality standards and inspections

Companies and investors which wish to get more involved in green business practices or operate green businesses in Russia may also wish to bear in mind that a typical feature of Russian environmental legislation is that federal laws are framework laws, which have to be supplemented by decrees and government regulations. These decrees and government regulations prescribe the rules applicable to environmental permits, quality standards and inspections.

It is often claimed that pollution could be diminished by way of introducing stricter regulations within Russian law. However, the Russian environmental law system is already based on strict regulations and inspections – the requirements relating to environmental licenses, as well as, for example, relating to the quality standards for air and water, are already quite strict. Nevertheless, such strict (and high) standards are often not realistically attainable and certain features of an 'antimodern' society appear to emerge as a result.[41] This considered, it is vitally important for investors in Russia to maintain good relations with the authorities and to enable them to understand that green business is as productive and lucrative a business as any other, perhaps more traditional, business. Suspicious attitudes can be overcome with relevant information being provided on the objectives of a 'green' business. In the author's opinion, especially environmental authorities often do understand the needs for green business, however, the challenge seems to be the lack of means to support such business incentives. In some cases, the author has encountered that feelings of shame or inferiority may emerge due to the drawbacks and shortcomings of the 'pre-conditions', which may result in negative attitudes towards enthusiastic 'green' investors. Therefore, it is vital that investors should avoid attempting to teach or adopt any practices which could be interpreted as being patronizing.

---

which to increase the efficiency of the use of forests and forestry, which in fact was the main concern of the Russian government in renewing the Forest Code.

[41] It is noted that strict laws and arbitrary enforcement of such laws are a typical feature of an 'antimodern' society. In an 'antimodern' society, unofficial relations are often used to circumvent and soften the strict laws, see R. Rose, 'Getting things done in an antimodern society: social capital networks in Russia', in P. Dasgupta and I. Serageldin, eds., *Social Capital- A Multifaceted Perspective* (Washington DC, World Bank 2000) pp. 147-171. Also available at <http://www.exclusion.net/images/pdf/778fimod_rose.pdf>. According to Richard Rose, an antimodern society is one in which formal organizations are numerous and important, but often fail to operate impersonally, predictably, and in accordance with the rule of law.

Licences and permits to use natural resources

The purpose of licences or permits to use natural resources is to ensure their rational use. The Law on the Protection of the Environment requires licensing for the use of the environment. However, the law does not specify what types of uses require a permit. Some activities, which require licences or permits, are regulated by the Law on Licensing, others by other framework legislation. For instance, the Water Code does not require a licence for the general use of water, but businesses are required to obtain a licence for the use of water with respect to their business activity. According to the Law on the Protection of the Environment, the requirements, which are set out in each given licence are regulated by law. These regulations are rather general at the 'level' of laws, but the licence requirements are regulated more specifically by way of decrees. These laws and decrees provide the framework for negotiations on the terms of the licence between the respective enterprise and the federal, regional and municipal authorities.

According to the Law on Licensing, the federal ministry or respective federal body, together with its regional organs, grants a licence for the use of a given resource. The procedure starts with an application. The terms of the licence for use of a given natural resource are decided at the federal ministerial level by the Ministry of Natural Resources and the Protection of the Environment ('MNR'). The relevant region of the federation, however, also participates in the negotiations relating to the terms of the use of the resource and payment requirements, which are incorporated in the licence. It is noted that the role of the Regions has recently diminished and this development has been reasoned due to them not enforcing licence holders' violations of their licensing terms and permits, during the time when the Regions had the main role in enforcing licence terms.[42]

Usually, the licence application must be decided upon within 30 days. However, the period for considering the application can be prolonged up to 90 days. Often the decision to grant a licence requires an onsite inspection. A refusal by the authority to grant a licence (or permit) needs to be accompanied by grounds for such refusal. The authority's refusal to grant a licence can be appealed before the 'arbitrazh court'.[43] In the case where procedural rules have been breached, municipalities, citizens and organizations (including businesses) can appeal to the arbitrazh court and apply for a licence to be granted to a given enterprise.

All licences are registered at the relevant ministry or other relevant federal licence granting authority. The licence is valid only after its registration. Licences for the use of subsoil resources are provided for 20 to 25 years, whereas licences for the use of

[42] According to an inquiry conducted by the MNR, of the over 20,000 licences that were issued in 2001, 1,020 licence holders violated their licensing terms and conditions. However, of these, only 40 licences were revoked (*Prirodoresursnye vedomosti* No. 2 2002, p. 109). The federal authorities have been said to have used such information as a cause for reducing the power of regional authorities and reasoning during the administrative reform for the returning of power towards the federal level (*vertikal*), see Kriukov et al., *supra* n. 31, at p. 106.

[43] The arbitrazh courts are the State Commercial Courts in Russia and they should not be mixed up with private arbitration proceedings.

water objects are granted for 3 to 20 years. Licences for industrial use of water are decided at the MNR. Licences concerning groundwater are decided according to the Law on Subsoil Resources at the Subsoil Agency, which also decides on licences for exploring and using oil, gas or minerals. The Federal Agency for Forestry and its regional offices take care of auctions for renting forests for forestry and grants the logging permits ('logging tickets').[44]

Taking care of drinking water, waste water and water for general purposes belongs to the duties of municipal authorities. Local authorities also specify sites, which are to be used for purposes of recreation, tourism and sports, as well as for sports fishery (with the consent of the territorial organs of the relevant federal organs). The municipality also has important functions in water policy as follows:

- Administration of the small areas owned by the municipality;
- Preparation of the local programmes, which must be co-ordinated with the regional and the federal programmes;
- Setting rational terms for the rational use of water and its protection from pollution;
- Providing information to local citizens;
- Keeping records relating to water management, monitoring and evaluation.

The Land Code permits the purchase of land for purposes of housing, industrial use, agricultural use and recreation. Foreigners, however, have no right to purchase land. In practice, the leasing of land is the usual form for using land also for Russian citizens. If the land is used for industrial or agricultural purposes, the relevant entity will be required to pay rent over 49 years or a shorter period. The Land Code also allows privatized enterprises to decide whether they wish to purchase or rent the land lot upon which the enterprise is situated.[45]

As regards industrial activity, permits for discharging specific polluting substances into the air are granted for 3 years or less. In order to obtain such a permit, Russian law prescribes that an inspection be made of the substance(s) being emitted into the air, and that an environmental audit be performed with respect to the emitting company.[46]

Quality standards and emissions standards

In Russia, quality standards, environmental standards and the rules relating to environmentally damaging activity are, typically, strict. Both the ISO standards, as well as

[44] As mentioned earlier the Forest Code also allows delegation of powers to the Regions. Under the new code the regional authorities give permissions for cutting of forest for private use of citizens, while industrial use is under the decision power of federal authorities.

[45] This decision is usually considered difficult as it is near impossible to predict whether the land tax applicable to a given property will be higher in the future than the required land rental payments.

[46] It is noted that some enterprises regard such permits to discharge given substances within the agreed limits as an official 'permit to pollute'. By way of these permits, the polluter pays principle is integrated within Russian law, however, it appears that revenues from such payments are not all re-invested into improving the quality of the environment. Investments in the environment constitute 0.4% of the revenues of the federal budget (see Brinchuk, *supra* n. 32, p. 252).

Comecon- and GOST-standards are applied in Russia. The latter standards are specifically Russian standards, which at times can be stricter than EU quality standards. These standards establish technical regulations and they are solely regulated by the Federal Law on Technical Regulations of 27 December 2002.

The rules on emissions limits are set out in governmental decrees.[47] The high standards required for water and air quality are strengthened by way of strict control mechanisms and high penalty fees for breaching the applicable standards or terms of licences. Inspections and controls are undertaken at the municipal, regional and federal levels. All these governmental levels can order that fines be applied for deviations from the relevant standards. It is noted that in practice, the requirements of the different authorities can be contradictory.

Administration of environmental matters – inspections

Steps have been taken to try to coordinate and simplify the administration of environmental matters. Joining several ministries and state committees under the MNR was one attempt to rationalize the complicated administration of environmental matters.[48]

Within the MNR there are different departments dedicated to different natural resources, and there is a separate department for the protection of nature. The regional organs of the ministry are divided in an identical manner. For forestry and subsoil resources there are also special federal agencies and their regional offices. The inspections, which are taken care of by special inspection organs (the *Rosnadzors*)[49] can be regular (planned) or irregular. Regular inspections (every second year) are undertaken also in cooperation with the Sanitary-Epidemological Control Authority (a federal organ) together with the State Committee of Fisheries or the State Land Control Authority. Regional and municipal authorities participate in monitoring the due implementation of the terms of licences.

An 'irregular' inspection is started at the initiative of either the MNR or any other relevant federal organ, the State Prosecutor's Office (*Federal'nyj Prokuror*), by way of a complaint of a citizen, a NGO regional body or municipal bodies.

The inspection of enterprises and other entities was previously regulated only by way of internal orders. Currently, the Federal Law on Environmental Expertise 1995[50] sets out the general principles for inspection, including the requirement of legality, the requirement that inspectors are independent, and so forth. The more detailed inspec-

[47] Government Decree on the Procedure on Setting Ecological Norms for Emissions of Polluting Ingredients to the Environment, Limits of Using Natural Resources and Defining Emissions of 3 August 1992, as amended in June 2000.

[48] By way of a presidential decree in 2001. See also chapter 5 (*Razbash*) for detail as to the ongoing administrative reforms.

[49] The Federal Service on Ecological, Technological and Nuclear control (*Rostehnadzor*), the Federal Service on Control for Using the Nature (*Rosprirodnadzor*), the Federal Service on Veterinary and Phytosanitary Control (*Rossel'hoznadzor*) and the Federal Service on Control on Human Health and Social Well-being (*Rospotrebnadzor*).

[50] Dated 23 November 1995.

tions are regulated by several ministerial instructions, whereas the procedure itself is regulated by way of a governmental decree.[51] After an inspection, a 'protocol' is added to the inspection report, which describes the offence, the damage caused to the environment and the according reparations, which are required. The inspection authority then also determines the applicable fine(s), which the ministry will already have confirmed within the terms of the licence.[52] The findings of the inspection report and the applied fine can be appealed at an arbitrazh court.

Environmental programmes

The federal target programmes[53] are a very important means for developing environmental law and regulations, as well as implementing them. These programmes are usually coordinated between the federal, regional and local level and they can be compared with the EU action programmes or national action programmes of the EU member countries.

Green businesses may thus wish to coordinate their activities within these environmental programmes in order to obtain positive attention from the authorities, both on the federal and municipal levels. There may, for example, be federal environmental programmes, which could include possibilities to obtain funding for investments in environmentally friendly technology. It should, however, be noted that often, especially local and regional, authorities do not provide for adequate (if any) funding for these programmes, and most of the money is intended to be collected with the help of enterprises.[54]

Tax legislation and incentives

Russian tax legislation is rather unclear and rapidly changing[55] and those companies interested in doing green business will note that there are currently no specific tax incentives in place to promote such business (e.g., for the putting in place or use of

[51] Dated 11 June 1996.

[52] The Code for Administrative Offences of 2001 sets the framework as regards the applicable fines.

[53] The scope of federal target programmes can either cover the whole federation or merely a certain area. The leading environmental programme is The Federal Target Programme for 'Ecology and Natural Resources 2002-2010'. The programme is financed by the budgets of different levels. Some parts have been reduced to end already in 2004 or 2005. A new programme or act should replace those parts. Green business can find support also from other target programmes, such as The Federal Programme for Development of Resources of Agriculture or the Federal Nuclear Safety Programme (2007-2010). Federal programmes are usually confirmed by a government decree.

[54] This is how interviewed business managers (2004-2005) experienced the situation, *supra* n. 1.

[55] Tax evasion was widespread in Russia a few years ago, whereas nowadays companies complain that even ordinary tax planning is difficult to achieve. These complaints may in the opinion of the author be partly exaggerated and affected by the fear caused by the *Yukos* case. Business managers, which the author interviewed in 2004 and 2005 (see *supra* n. 1), were puzzled with some opinions of politicians who regarded tax planning as a means to circumvent the prevailing tax legislation.

environmentally friendly technology).[56] It is noted that in free economic zones the relevant regional authorities can offer incentives, which are not available under general Russian tax legislation. This chapter, however, does not discuss the special case of free economic zones.

In context of tax revenues, investors and businesses ought to bear in mind that the municipalities, which bear most of the costs of health care, education and municipal infrastructure, are under pressure to build up municipal services. However, municipalities appear to receive the least amount of tax revenues and the revenues received from leasing land are almost their only direct source of income. Other incomes depend on the negotiations with the federal and the regional governments.[57] Thus, large cities, such as Moscow and Saint Petersburg, are in a much better financial position than cities located in more remote areas. The municipalities of oil and gas producing areas used to be rich before revenues from the use of natural resources were redirected directly to the federal budget. As a result, almost all Russian municipalities are in constant cash crises.[58] Often, they have difficulties to pay even the salaries of their employees, such as teachers, nurses and medical doctors. Therefore, it is often the case that large enterprises are asked to assist the municipalities.[59] In Soviet times, state enterprises supported the entire local community infrastructure, ranging from the sewage water system to kindergartens. Nowadays, the municipal infrastructure is run by the municipality itself, but often the municipalities ask larger investors or companies for assistance with developing the local infrastructure. As a result, some companies loan money to the municipality and some arrange for the services themselves at their own cost.

As the municipalities are often short of funds, they may try to get into close cooperation with new foreign investors. Since the land rental payments constitute the main source of income for most municipalities, they may try to set the rent as high as possible. Helping municipalities with their financial challenges is often considered as a duty of any larger enterprise operating in a particular municipality – furthermore and in any event, it is often considered as a means to build a good public image of the company.

[56] The absence of such tax incentives may have historical reasons as many tax reductions were introduced during the end of the Soviet period and at beginning of the 'new Russia'.

[57] Idem., n. 34. The budgetary system is still quite arbitrary. The Law on Financial Principles of Municipal Self-Governance (dated 25 September 1997, as amended 28 December 2004) sets out the minimum share of tax revenues allocated to municipalities relating to each source of taxation. These shares only partly cover the municipalities' expenditures, whilst the remainder of the revenue shares need to be negotiated with the federal and regional authorities. There are ongoing ambitious reforms underway relating to municipal self-governance, the main goal of which is to strengthen self-governance. The reforms were planned to enter into force at the beginning of 2006, but they were postponed until the beginning of 2009 mainly due to the lack of financial resources, see *Rossiskaya Gazeta*, 18 October 2005. The entering into force of the Law on General Principles on Organizing Local Self-governance in Russia (dated 6 October 2006) in the beginning of 2009 abolished the Law on Financial Principles of Municipal Self-Governance.

[58] Idem., n. 34 and the discussion at pp. 6-7.

[59] The municipal cash crisis is mentioned in both interviews of business managers and municipal authorities conducted by the author. The information on the municipal budgets is based on budgets shown by interviewed municipal authorities in 2004 (*supra* n. 1).

Credits for green business

There are only few low-cost credit facilities available for environmental projects (see for example those loans provided by the NPAF, see Chapter 15 (by Maksimenko and Gorkina)). Russian enterprises may, therefore, try to obtain international funds for environmental projects, such as for the development of purification plants or green business ideas. International funds, however, finance mostly very large high-cost projects, which involve both national and international private and public partners. Previously, enterprises could obtain credits for environmental purposes from the Federal Ecological Fund (*Ekologickeskij Fond Rossiiskoi Federatsii*) or territorial ecological funds, which were established in 1992 by the Russian government in connection with the collection of environmental fines. However, during the budgetary reform of 2001, these numerous state funds, which existed outside the budget, were abolished. Thus, today, environmental fines flow directly to the State budget.

Small enterprises

Although there are no tax incentives to promote environmentally sound operations or green business itself, Russia has developed a subsidy system for small enterprises as part of other policies to promote the development of a balanced market economy. Small enterprises are defined as enterprises with an annual turnover of less than 15 billion roubles. Small business accounts only for approximately 4% of the Russian GDP, and employs only 11,5% of the labour force (2001 figures), which constitutes much less than half the proportion of people employed in other developed countries.[60] Although small businesses are entitled to benefit from tax incentives, these incentives are rarely utilised.[61]

Strategies of enterprises for promoting environmentally friendly investments

As mentioned above, a key strategy for enterprises and investors seeking to do green business, is to maintain open and friendly relations with the Russian authorities. Doing

---

[60] <http://www.waytorussia.net/features/stats>. The turnover of UK small businesses reaches as much as 38% (2003) and UK small business employs 46.2% of the working force. The official statistics might lower the real turnover details, as many Russian businesses prefer to 'hide' the real figures to avoid taxes.

[61] This is so as enterprises regard the tax incentives as non-beneficial in practice due to bureaucracy issues and the strictly specified terms of the incentive connected with the application process, as explained to the author in 2005 by managers of small and medium sized companies in Murmansk and Arkhangelsk areas. The interviewed managers working in the fishing or forest industry assumed that the incentives might be given to the service sector (*supra* n. 1). In this context, it is noted that the Federal Fund for Support of Small Business, which was created in 1995 was closed down this year (2005). The reason for closing it was that the fund proved to be an ineffective bureaucracy, which was unable to reach out to small businesses. Indeed, a survey by the Ministry of Economic Development and Trade in 2003, found that only 5% of entrepreneurs had come across the fund (see P. Rutland, *Business – State Relations in Russia* (draft paper developed at Center for Security Studies, Swiss Federal Institute of Technology, Zurich 19 March 2005).

green business does not make it easier when it comes to inspections and monitoring, as the Russian legal system does not provide for any specific incentives or mechanisms to promote green business.

Currently, and within the environmental, and general, legal framework as described above, the following strategies can be employed by companies wishing to promote green business, or environmentally friendly investments, in Russia:

- Environmental (ecological) Plans
- Infrastructure assistance
- Certification
- Kyoto (climate change) related activity[62]

Environmental plans

A good means of promoting environmentally friendly investments are the adoption of the yearly environmental plans or 'plan for ecology' (*ekologicheskij plan*), which are required by law. This plan must include a company's envisaged technological changes and purification plans. Companies devise these plans independently and then request the state authorities to enforce them. In the case that the plan cannot be implemented, for example due to some disturbance on the markets or due to any other economic fluctuations, the company may wish to ask for the extension of any time periods within which it should achieve its objectives. This extension of time is, however, not easily granted by the authorities. Thus, companies are advised to be very careful and realistic in developing their plans. It is considered better not to set too high targets, but instead, companies may wish to set relatively modest (and realistic) targets, which can be realistically achieved or even exceeded.

Developing such (realistic) environmental plans can result in portraying a positive 'green' image to the federal and municipal authorities. It should, however, be borne in mind that the authorities do not assist companies in reaching their targets, nor are they likely to understand if difficulties appear in the implementation phases.

Infrastructure assistance

As mentioned above, companies and investors are often asked to assist the state authorities (e.g., municipalities) with their financial challenges or by way of infrastructural improvement projects. This said, there appears to be a fine line as to what can be considered infrastructural assistance and what could be deemed as corrupt practices. It is well documented that corruption is still a problem in Russia.[63]

---

[62] This item is not discussed in this chapter. See chapter 13 (*Yulkin*) for a discussion on the Kyoto Protocol and climate change issues in Russia.

[63] Transparency International produced a report on 163 countries, the information of which was collected during the year 2006, and Russia 'scored' at the 121th place, with 2.5 points (point scores are given between 0 to 10). By way of comparison, the least corrupted countries, Finland, Iceland and New Zealand were awarded 9.6 points. See <http://www.transparency.org>.

The Russian government is trying to fight bribery and corruption, for example, by routinely changing the inspectors and other state officials and by moving them to other districts or posts to prevent the development of corrupt relations with companies. The result, however, can be that the 'costs' relating to bribery become higher as the circle of those demanding for the payment of bribes becomes wider.

Businesses complain about the increasing bureaucracy and strict means of governmental control.[64] However, this increased control is to a great extent due to the government aiming to introduce more order and legality into Russian business.[65]

The Medvedev administration (as did the Putin administration) encourages companies to participate in the building of Russian society. Thus, essentially, socially responsible business investments and environmental investments are welcome and encouraged in Russia.

However, within such environmental investment and socially responsible business activities, establishing good relations with municipal and state authorities can also cause difficulties. The municipalities are poor and the main solvent actors in society are generally companies. Many corporate leaders, thus, feel that the burden of social responsibility on companies is too heavy.[66] On the one hand, these circumstances allow for a strong negotiating position with the municipal and regional authorities, which may become interested in supporting start-up green or socially responsible businesses on the basis that these will result in generating more local employment and add to the fabric of society. On the other hand, the 'wish list' of municipalities may become too long. Investors may feel that they are under pressure and tied to constant high expenses. This situation is viewed by some as promoting the socially responsible behaviour of investors, however, it has also been criticised due to the level of influence by the municipalities and on the basis that it is sometimes difficult to differentiate from corrupt practices. Therefore, some may regard it as participating in corrupt practices when a company pursues socially responsible business activities or implements environmentally favourable projects. Thus, even if it is recommendable that companies invest in local infrastructure for purposes of trying to obtain goodwill from the authorities, such projects should be run carefully. In order to avoid becoming involved in corruption or being subject to misuse, the reliability of the relevant authorities should be examined before commencing project work and by using reliable information networks.

---

[64] Increased bureaucracy was one of the main problems, which managers mentioned in interviews with the author in the Arkhangelsk, Murmansk and Leningrad regions, as well as in the Karelian republic in 2004-2005 (*supra* n. 1). Furthermore, especially the working practices of the tax authorities are often a source of complaint. See also Kits et al., *supra* n. 10, in which the level of professional knowledge and skills of tax authorities is criticised.

[65] However, it is often complained that honest enterprises may have to suffer due to the weak quality of knowledge and commitment of the state authorities, see Kits et al., *supra* n. 10.

[66] This was one of the findings of the author's interviews of Russian managers in 2004- 2005 in the Arkhangelsk region and Karelian Republic (*supra* n. 1). They all regarded heavy social responsibility as a burden for companies, although they also considered that helping the community and poor people is their duty in present circumstances.

'Green' certification

Although it is important for companies and investors intending to pursue green business (or green business practices) that they maintain good relations with the authorities, for example by way of infrastructural assistance or by way of implementing realistic environmental plans, there are other means by which to engage in green business practices.

Voluntary 'green' certification is a new mechanism promoted especially by NGOs to encourage green business operations. Voluntary national state certification of environmentally friendly products, production processes and services can be obtained according to law.[67]

Obtaining 'green' certificates for products can help boost sales in such markets where consumers are aware of and care for environmental issues. Marketing products, which are produced in Russia with environmentally favourable methods might also be a success in Western markets. Although in-depth market research would need to be undertaken, it appears that where Russian businesses target Western markets, the use of non-obligatory international certifications, rather than domestic Russian certification, may be more recommendable, simply on the basis that the latter are more recognizable by the Western consumer. In Russia, the market for environmentally friendly products is still rather uncertain, because 'green' certification is a new phenomenon. It is noted that some Russian companies work with international environmental NGOs in the hope of gaining a better image in the eyes of Western consumers.[68]

An example of green certification is forest certification, which has advanced to the pilot project stage.[69] Typically, forest certification is undertaken by Russian companies, which are involved in the whole chain from leasing and cutting forests to producing pulp and paper. In principle, there is both voluntary and mandatory certification in Russia, since the new Forest Code of 2006 (Article 71) introduces the concept of mandatory forest certification.[70] Currently international co-operation between environmental NGOs and companies resulted in flourishing voluntary certification and the choice of the Forest Stewardship Council ('FCS') certification as the certification system to be applied in the Russian pilot projects.

NGOs seem to be satisfied with the opportunity to cooperate with companies and directly influence their practices. In the FCS certification process, NGOs are involved in monitoring, which has resulted, *inter alia*, in public hearings, improvement of living

---

[67] See as regulated by the Law on Technical Regulations of 27 December 2002, as amended December 2007.

[68] See for example at <http://wood.ru>, 4 May 2005 *Kompaniya Evropartner mozhet stat' auditorom v. sistemi FSC.*

[69] Information about forest certification in Russian can be found at <http//:www.forest.ru/eng/sustainable_forestry/> (Russian NGOs Forest Club) as well as at <http://www.wood.ru> (Russian Forest Industries).

[70] The Federal Law on Certification of Products and Services dated 27 December 1995 mentions that there can be both mandatory and voluntary certifications Certification is also mentioned in Article 31 of the Law on the Protection of the Environment. Both acts regulate that the conditions and r[illegible]ements of obligatory certificates are decided by the Government of the Russian Federation.

conditions of local people, environmental education at local schools. With the help of NGOs, companies might even more easily find effective ways to contribute to the improvement of social and environmental issues in their working areas than, for example, with the help of authorities. Companies, however, will need to balance between cooperation with NGOs and authorities, as the Russian Government appears to want to maintain a leading role also in developing and directing business.

However, it seems that even if the Federal Government sets and controls the legal environment for business in Russia, local cooperation and flexible local solutions are always the key for establishing good business-community relations and good reputation for companies.[71] In advertising green business informal means of approaching NGOs might be a good chance to get publicity. Local newspapers are also often easy to reach.

Concluding remarks

Investments into green business and environmental projects in general are not very abundant in Russia. Nevertheless, although the government has not yet put in place fully fledged programmes or laws to support such activity, there are several incentives available to businesses which allows them to get involved in green business practices or environmental projects.

Doing green business in Russia is still in its pioneer stages and exists mostly in the form of companies voluntarily investing into such projects and activities. In this respect, public relations work, effective marketing, as well as cooperation with environmental NGOs and the Russian authorities will be of key importance to companies wishing to undertake green business. As a general rule, doing business in Russia requires not only quality products (and services), but also skills needed in context of operating in a transition economy, such as patience with abundant bureaucracy, constant alertness to quickly changing laws and regulations, as well as good networking and social abilities. Despite the challenges, the Russian market is open for green business and those companies, which manage to establish their operations early and are ready to work patiently and in a transparent way, may well expect to gain long-term returns.

---

[71] See Nystén-Haarala, *supra* n. 1. The author has started a new research project with Russian and Finnish researchers and funded by the Academy of Finland 'Trust in Finnish-Russian forest industry relationships (2008-2010)' studying how trust is and could be built at the local level in logging areas of forest companies. Voluntary certification is one of the ways of building trust, as is being investigated in the ongoing research project.

Chapter 8
ENVIRONMENTAL POLICY AND THE INTERNATIONAL COMPETITIVENESS OF THE RUSSIAN ECONOMY[1]

Evgeny A. Shvarts[2] and Ivetta Gerasimchuk[3]

This chapter provides an overview of how Russian environmental policy could act as an instrument to enhance Russia's international economic competitiveness. In particular, this chapter identifies and analyses some core environmentally-related problem areas which currently constrain Russia's international competitiveness. This chapter also provides and discusses a series of 'green' solutions to ameliorate the international competitiveness of the Russian economy.

INTRODUCTION – RUSSIA'S COMPETITIVE ADVANTAGES

Russia's main global competitive advantages are its human capital and environmental capital, including its natural resources. Both are interlinked and depend on one another – as does the health and wellbeing of the Russian people to a large extent depend on the state of the environment.

Recently, the type of capital which has been securing Russia's global competitive advantage (including the export of oil and other natural resources) have been negatively affected. Furthermore, although Russia's GDP increased by an average of 6.5% per annum (between 1999 to 2008), Russia's population and Russia's net adjusted savings[4] have decreased. According to the World Bank, the net adjusted savings rates in Russia in 2006 were negative and dropped to – 13.8%, mainly due to the depletion of natural resources (see Table 1 below).[5]

[1] The authors take this opportunity to thank WWF UK and UK Department for International Development ('DFID') for their financial support of WWF Russia.

[2] Director of Conservation Policy, WWF Russia; Email: *eshvarts@wwf.ru*.

[3] Scholar-in-Residence/Lecturer in Environmental Economics and Management at the Cyprus International Institute of Management ('CIIM') and Senior Advisor, Trade and Investment Programme, WWF Russia; Email: *ivetta@ciim.ac.cy*.

[4] Calculated after taking into account the depletion of natural resources and damage caused by pollution.

[5] Although the same indicator was positive in China (36.1%), India (20.6%), the Euro Zone (12%) and in the USA (4.1%).

W.Th. Douma and F.M. Mucklow, eds., Environmental Finance and Socially Responsible Business in Russia
© 2010, T·M·C·ASSER PRESS, The Hague, The Netherlands and the Authors

Table 1. Adjusted Net Savings for Selected Countries in 2006.[6]

Country	Adjusted Savings (%)	Country	Adjusted Savings (%)
Japan	15.8	Russia	–13.8
Germany	12.1	Czech Republic	14.7
France	11.4	Poland	7.8
Great Britain	6.9	Ukraine	4.1
Canada	5.4	China	36.1
USA	4.1	India	20.6
Norway	9.2	Brazil	3.5
Euro Zone	12.0	Worldwide	8.3

Thus, there is significant scope to seek and identify new ways in which Russia's international competitive advantage can be boosted. In this respect, the 'greening' of the Russian economy offers new and interesting ways in which Russia could re-launch and further strengthen its strategic advantage on the global markets. Not only do environmental issues play an increasingly large role in the world economy, but they are becoming even more important with growing problems associated with increased pollution, accelerating climate change and depleting levels of natural resources (including water resources).

Due to its geopolitical status, Russia can be considered an 'environmental super power', as it possesses approximately 9% of all 'biocapacity' on planet Earth (1.2 billion out of 13.6 billion 'gha').[7] Indeed, unlike China, India, EU countries and the USA, which are 'ecological debtors', Russia is an 'ecological creditor'. Thus, taking into consideration environmental scarcity and especially in these financially challenging times, Russia's natural and environmental capital should be used to its benefit.

Russia could follow the way in which Brazil takes advantage of its biocapacity. Brazil is the only country which outruns Russia with respect to its biocapacity (in absolute figures) and is a country which actively uses its advantage of being an 'environmental super power' in the world political and economic processes.

However, Russia's tendency to deplete its natural resources and environmental capital undermines its ability to capitalise on its biocapacity – and so undermines its ability to secure its role as a key player in global energy security. The elements which restrict both Russia's ability to capitalise on its biocapacity and export potential are as follows:

- high energy and resource intensity of the Russian economy
- ineffective usage of natural and environmental capital
- insufficient investment into intensive methods of domestic mineral production

[6] The Little Green Data Book 2008 (World Development Indicators 2007). World Bank, Washington DC, 2008. <http://siteresources.worldbank.org/ENVIRONMENT/Resources/LGDB2008.pdf>, see pp. 2, 15, 44, 51, 58, 68, 83, 88,103, 112, 161, 170, 175, 213, 215, 216.

[7] A 'gha' ('global hectare') is the average global indicator of the biological capacity to regenerate resources and to absorb wastes. Calculated on the basis of: Living Planet Report 2008. WWF, Gland, (2008) , pp. 32-39, and see <http://assets.panda.org/downloads/living_planet_report_2008.pdf>.

Thus, in order to boost the international competitiveness of the Russian economy and in order to achieve the goals set out by the '*2020 Concept of Long-Term Social and Economic Development of the Russian Federation*',[8] Russia's environmental capital must be preserved and replenished, instead of being depleted at current and constantly growing rates of usage.

The Presidential Decree '*On Certain Measures to Increase Energy and Environmental Efficiency of the Russian Economy*', dated 4 June 2008 (the 'Energy Efficiency Decree')[9] and the related 'Main State Policy Directions Aimed at Increasing the Energy Efficiency of the Power Energy Sector through the Use of Renewable Energy Sources for the Period until 2020', approved by Russia's Prime Minister on 20 January 2009,[10] each undoubtedly represent a step towards enhancing the competitiveness of the Russian economy through launching 'greening processes'. However, a breakthrough in the sphere of capitalising on Russia's enormous biocapacity would require many more efforts. In this regard, the current financial crisis provides an excellent opportunity for Russia to reconsider and re-launch her environmental (and biocapacity) strategies and to re-evaluate the risks inherent in old-fashioned (unsustainable) economic development strategies.

The below sets out some specific socio-environmental problems and characteristics of the Russian economy. Thereafter, economically rational solutions thereto are discussed, which could transform Russia from being a global supplier of natural resources into being a 'natural bank' of the new global economy.

Environmental problems in the Russian economy

The environmental quality of approximately 15% of Russia's territory, which is occupied by 60% of the Russian population, is unsatisfactory. It is reported[11] that Russian residents on average lose up to 1 year of their life due to air pollution – and up to 4 years in the most contaminated cities. Furthermore, air pollution accounts for up to 8% of the total death rate in Russia.[12] Approximately 10 million people in Russia use drinking water which contains one or more harmful substances in quantities higher than maximum permissible values. The extent of economic damage resulting from the impact of air and water pollution on Russia's national health is estimated to be at least 4-6% of GDP.[13]

---

[8] See <http://www.government.ru/content/governmentactivity/rfgovernmentdecisions/archive/2008/11/17/2982752.htm>.

[9] See <http://document.kremlin.ru/doc.asp?ID=046255>.

[10] See <http://www.government.ru/content/governmentactivity/rfgovernmentdecisions/archive/2009/01/08/1528437.htm>.

[11] S.N. Bobylev, V.N. Sidorenko, Y.V. Safonov, S.L. Avaliani, E.B. Strukova and A.A. Golub, *Macroeconomic Assessment of Environmental Pollution Health Effect in Russia* (Moscow, World Bank Institute, Nature Protection Fund 2002).

[12] Ibid.

[13] Ibid.

When comparing to other BRIC[14] countries, reduced levels of environmental responsibility of Russian businesses results in Russian goods (especially goods with a high share of added value) having difficulties in penetrating international markets which are known for their high levels of consumer environmental awareness. For instance, export of motor fuel from Russia is limited as external markets set much stricter technical and ecological requirements to such fuel than does the Russian domestic market.[15]

The environmental characteristics of the Russian economy directly affect the value and availability of foreign loans for Russian businesses, which are of vital importance during a financial crisis. Both international finance institutions (e.g., the International Finance Corporation ('IFC') or the European Bank for Reconstruction and Development ('EBRD') and major private banks operating in accordance with the Equator Principles[16] demand that all projects financed by them be in strict compliance with international environmental standards. Acting in accordance with these requirements, the EBRD, for example, acknowledged in 2008 that its participation in financing the Boguchanskaya hydroelectric power station was a mistake.[17] Environmental risks are increasingly being taken into account not only by the finance institutions of the OECD countries, but also by banks in other nations, including in Brazil, South Africa and China.

The above mentioned problems and socio-environmental characteristics applicable to the Russian economy need addressing. If they are addressed effectively and swiftly, they promise to significantly ameliorate Russia's international competitiveness. The below discusses possible solutions and ways in which to enable the 'greening' of the Russian economy and so to create a more internationally competitive and modern Russia.

Modernization of the Russian Environmental Management System (EMS)

It is impossible to increase the international competitiveness of the Russian economy by way of 'greening', without thorough modernization of the entire environmental management system ('EMS') and of Russian environmental institutions. However, even '*The Environmental Doctrine of the Russian Federation*',[18] a strategic document which was adopted in 2002, is on average progressive, but does not set up the mechanisms required for such modernization.

---

14 Brazil, Russia, India and China.

15 According to the Russian Federal Customs Service, in 2007, Russia exported oil for US$ 114 billion and diesel fuel and petroleum – for only US$ 3.5 billion and US$ 21.5 billion, respectively. See <http://www.customs.ru>.

16 See Annex II.

17 See <http://www.bellona.ru/articles_ru/articles_2008/ebrd_annual_meeting_2008>.

18 See <http://www.mid.ru/ns-dmo.nsf/a1c87897b58a9d2743256a550029f995/432569f10031eb9343256c9a002f10ca?OpenDocument>.

Comparative analysis of environmental management practices of economically developed countries, large developing countries (such as Mexico, China and India), former Soviet Republics (such as Kazakhstan and Ukraine) and in Russia, evidences that Russia is lagging behind other nations with respect to some important EMS components. Ill-conceived restructuring of environmental agencies and unwarranted changes in Russian laws over the last ten years have contributed towards Russia lagging-behind in this respect. The main differences between Russian EMS and best examples of international EMS practices are:

- Russian governmental officers' limited knowledge of environmental problems/priorities;
- abundant regulations and standards coexisting with gaps and contradictions in environmental laws and bye-laws;
- very limited use of environmental assessment and environmental impact assessment ('EIA') procedures for projects in Russia, and their low efficiency in protecting environmental and social-economic interests of the population, as well as of small and medium-sized businesses;
- vague and non-rational distribution of responsibility among Russian federal, regional and municipal agencies exercising environmental control;
- undeveloped system of general public involvement in decision-making processes and absence of an open public system enabling the reconciliation and protection of various social groups' interests.

The political powers of Russia are aware of the need to improve the situation. In January 2008, a special meeting of the Russian Security Council was held, at which Vladimir Putin (the then President of Russia) declared that the goal was '*... to create the conditions required to ensure that the future growth of the Russian economy is based on the highest environmental standards*'.[19]

Further public demand for the improvement of the state of EMS in Russia was reflected in the Energy Efficiency Decree.

In order to comply with the decisions made by the Russian Security Council and with the Energy Efficiency Decree, the Ministry of Natural Resources and Ecology of the Russian Federation ('MNR') suggests the implementation of the following measures to impromve EMS in Russia:[20]

- transition from subjective environmental regulation to unified principles, which develop regulatory standards based on best available technologies, firstly, through the abandonment of the current system of standards (limits) of emissions and discharges for existing enterprises, set on a temporary basis;
- introduction of an integrated environmental permit issuance system, including a simplified procedure for small and medium-sized business enterprises;

---

[19] See <http://president.kremlin.ru/text/appears/2008/01/158674.shtml>.

[20] See Minister Yuriy Trutnev's speech in the State Duma during Governmental hour, 3 December 2008, <http://www.mnr.gov.ru/part/?act=more&id=3093&pid=729>.

- transition from a 'permissive' natural resource use system to a system whereby declarations are issued with respect to entities whose business activities adhere to environmental standards and have no significant impact on the environment;
- drastic increase of the role of environmental audit;
- increasing the level of penalties (or charges) being imposed on entities which have a negative impact on the environment;
- replacing the current system, whereby penalties are imposed for excessive pollution, with a new system, whereby polluters must utilise the court system to recover their costs for reparations or to claim for compensation with respect to damage caused through the court system;
- creation of a system of economic incentives (tax and non-tax benefits, environmental insurance).

Changes in the Russian EMS, as proposed by MNR, appear to be reasonable and are aimed at bringing the Russian system into compliance with international practice. However, the MNR has a certain conflict of interest, as it acts not only as the Russian environment protection agency, but also as a licensing authority. In addition, the main environmental problem in Russia is not so much that Russian environmental legislation is weak, but that it is poorly enforced and poorly administered at the regional and local level. According to a source at MNR itself, one cannot find '*a single company or a single well in Russia operating in strict compliance with environmental laws*'.[21]

Therefore, alongside state environmental regulatory reform in Russia, the issue of utmost importance is 'greening' the other factors, which determine the practices of users of natural resources, namely:

- developing civil society institutions;
- increasing the environmental awareness of consumers (both individual and institutional);
- implementing voluntary market-oriented mechanisms aimed at businesses' environmental responsibility.

Voluntary market-oriented mechanisms of environmental responsibility

Market-oriented voluntary mechanisms of environmental regulation, such as certification and non-financial reporting verified by an independent 'third' party, are much more resistant to risks of corruption and/or administrative and political pressure. Those mechanisms in many cases could be effective alternatives to administrative regulation (so-called 'administrative barriers') in environmental and nature conservation.

---

[21] *Kommersant Daily*, 10 February 2004, p. 20.

Forest certification

Russia has accumulated quite significant positive experience in the sphere of forest certification measures. The use of the voluntary forest certification system developed by the Forest Stewardship Council ('FSC'), has substantially increased the competitiveness and investment attractiveness of Russian wood and timber companies (for instance, OAO Ilim Group, OAO Terneyles). For the past 10 years, Russia has ranked second worldwide in terms of aggregate area of forests certified under the highest environmental and social standards: 19,853,025 hectares[22] or approximately 15% of all forests leased out for timber industry purposes (Canada is ranked first with more than 26.5 million hectares of FSC-certified forests). However, voluntary forest certification in Russia has been developed mainly due to the active leadership of environmental non-governmental organizations within Russia, and until 2006, such international voluntary certification, in fact, had no support from the state management bodies regulating Russian industry.

International management standards and non-financial reporting

Having limited experience in competing on the global markets, Russian companies have been slow to implement international corporate management standards and environmental action plans, such as the ISO 14000 series standards. Thus, according to information available at the end of 2007 (see Table 2), Russia ranks 50th worldwide with only 267 issued certificates, far below China (ranked first with 30,489 certificates) and Japan (ranked second with 27,955 certificates). Russia is substantially lagging behind many comparable countries with a developing economy.[23] Please note that the highest growth rates of the number of EMS certificates are recorded in countries whose economies are developing at the fastest pace due to exporting products with a high added value share – i.e., in China and India (see Table 2 below).

Russian companies also fall behind with respect to implementing other voluntary market-oriented non-corrupted mechanisms of environmental responsibility, such as producent non-financial reports certified by an independent 'third' party. No Russian banks or investment companies acceded to the Equator Principles and the UN Principles for Responsible Investment, which are unquestionably the most widespread mechanisms to ensure financial institutions act in an environmentally sound and socially responsible manner. Furthermore, to date, no Russian company has provided its carbon emissions data in response to the global CDP (Carbon Disclosure Project) information request.

Measures to promote voluntary market mechanisms

The successful development of voluntary market-oriented mechanisms to enable and promote environmental responsibility in Russia requires that the national legislative

[22] See <www.fsc.ru>, accessed in February 2009.

[23] For example: India ranks 12th with 2,640 certificates; Brazil ranks 15th with 1,872 certificates; South Africa ranks 36th with 530 certificates, Turkey ranks 18th with 1,402 certificates.

Table 2. ISO 14001 Environmental Management Standard Certification.[24]

Country's ranking in 2007	Country	Number of certificates at the end of year						
		2001	2002	2003	2004	2005	2006	2007
1	China	1,085	2,803	5,064	8,862	12,683	18,842	30,489
2	Japan	8,123	10,620	13,416	19,584	23,466	22,593	27,955
3	Spain	2,064	3,228	4,860	6,473	8,620	11,125	13,852
4	Italy	1,295	2,153	3,066	4,785	7,080	9,825	12,057
5	Great Britain	2,722	2,917	5,460	6,253	6,055	6,070	7,323
6	Republic of Korea	880	1,065	1,495	2,609	4,955	5,893	6,392
7	USA	1,645	2,620	3,553	4,759	5,061	5,585	5,462
8	Germany	3,380	3,700	4,144	4,320	4,440	5,415	4,877
9	Sweden	2,070	2,730	2,330	3,478	3,682	3,759	3,800
10	France	1,092	1,467	2,344	2,506	3,289	3,047	3,476
12	India	400	605	879	1,250	1,698	2,016	2,640
15	Brazil	350	900	1,008	1,800	2,061	2,447	1,872
18	Turkey	91	135	240	338	918	1,423	1,402
36	South Africa	169	264	378	393	540	485	530
50	*Russia*	*12*	*23*	*48*	*118*	*185*	*223*	*267*
	Worldwide	36,464	49,440	64,996	89,937	111,162	128,211	154,572

and regulatory framework provides for a relatively high level of requirements, which would prevent companies from gaining an unfair competitive advantage by intentionally adopting 'environmental dumping' strategies (i.e., performing below the voluntary environmental responsibility standards). Furthermore, entities that do not comply with environmental requirements should not gain an unfair competitive advantage, as has been recorded in some sectors, as follows:

- *Forestry sector*: Some major Russian timber producers implement timber tracking systems to prevent illegal timber trade and actively participate in the voluntary forest certification in order to secure their global competitiveness. However, these major timber producers have been losing out to companies which deliberately process timber of knowingly illegal origin, as such 'illegal timber' is much cheaper (as it is produced without the producer paying relevant payments at auctions for cutting permissions, taxes, and implementation of social security measures, etc).
- *Oil sector*: Due to the postponement of the introduction of the new fuel standards in Russia, Lukoil, which switched to Euro 3 fuel standard as far back as 2006, has been at a disadvantage to companies (e.g., state-owned Rosneft) which have not invested into environmentally-friendly oil refining processes.[25]

---

[24] The ISO Survey of ISO 9001:2000 and ISO 14001 Certificates 2004 / International Standardization Organization. Geneva, 2007, pp. 24-27. <http://www.iso.org/iso/iso_catalogue/management_standards/certification/the_iso_survey.htm.>

[25] A. Knizhnikov, N. Pusenkova and E. Solntseva. *Social and Environmental Aspects of Russian Oil Refining* (Moscow, World Wildlife Fund (WWF) 2008). See <http://wwf.ru/resources/publ/book/302/>.

ENERGY AND ENVIRONMENTAL EFFICIENCY

A further effective measure through which Russia could ensure an initial surge and the 'economy of scale' effects derived from upgrading and 'greening' domestic industrial entities and business processes, is to implement requirements relating to energy and environmental efficiency of goods and services purchased by the State at the national, regional and municipal levels.

Foreign experience might prove very useful in this respect. For example, the European Union has integrated environmental strategies into EU public procurement policies at the level of the European Directive 2004/17/EU (dated 31 March 2004), and into EU green procurement policies at the level of most member states (Austria, Belgium, Great Britain, Germany, Greece, Denmark, The Netherlands, and France). Similar practices exist in many other countries, including Canada, Japan, New Zealand, Mexico and the USA.[26]

In order to promote and to ensure that Russian goods and services become energy and environmentally efficient, practical measures should include:

- incorporating energy efficiency requirements in technical regulatory acts (e.g., requirements specific to the energy/fuel consumption by machines and equipment, heat loss from buildings, water consumption by water treatment plants) and further developing regulatory instruments relating to environmental responsibility and security (e.g., legality of products of natural origin and sustainable forest use confirmed by voluntary forest certification);
- supplementing the public procurement laws and regulatory acts, resulting in the incorporation of provisions which give priority to goods complying with the highest standards of energy efficiency and legality, as well as to goods originating from environmentally sustainable sources as confirmed by international and/or internationally recognized certificates (e.g., FSC – for forest industry products);
- implementing the labelling of goods which grades levels of energy efficiency, including mandatory labelling of energy-consuming equipment and machines by levels (classes) of energy efficiency;
- putting in place mandatory procedures, whereby government-owned corporations and fully-owned state enterprises must publish non-financial reports on sustainable development and such reports to be certified by an independent 'third' party;
- inducing companies, which are partially or fully owned by the Russian government, to launch voluntary environmental responsibility mechanisms via the governmental representatives on the board of directors.

[26] Note that in the USA, special additions were adopted in 2008 relating to wood of illegal origin coming from Russia – see the revised Lacey Act 1990 (as amended in 2008) which is a US conservation law.

INTERNATIONAL ENVIRONMENTAL LAW

As in many other countries, international environmental law mechanisms are often regarded in Russia as attempts to limit national sovereignty and economic development. This explains, for instance, why it took Russia so long to ratify the Kyoto Protocol to the United Nations Framework Convention on Climate Change (Russia signed the Kyoto Protocol as late as 2004). For the same reasons, Russia continues to avoid the ratification of some environmental conventions, such as the 1991 Convention on Environmental Impact Assessment in a Transboundary Context (the 'Espoo Convention');[27] the Protocol on Strategic Environmental Assessment to the Espoo Convention (the Kiev (SEA) Protocol, 2003);[28] the UNECE Convention on Access to Information, Public Participation in Decision-making and Access to Justice in Environmental Matters (the Aarhus Convention, 1998);[29] as well as the Convention on Migratory Species (the Bonn Convention, 1979).[30]

However, if implemented competently, international environmental conventions can serve as an instrument to protect Russia's national economic interests. For example, the ratification by Russia of various international environmental conventions would facilitate the resolution of a number of pressing environmental problems in Russia, such as caused by:

- the transboundary transfer of air pollution from Poland and Ukraine to the territory of Russia;
- pollution resulting from the production and transportation of oil from the Ukrainian Black Sea shelf;
- environmental risks related to the development of oil and gas production facilities in the Norwegian sector of the Barents Sea;
- pollution of the Amur basin by Chinese industrial enterprises' discharge.

The importance of Russia's participation in international environmental protection was illustrated during the large-scale pollution of the Amur's tributary, the Sungari River, when on 13 November 2005, a series of explosions at the petrochemical plant in Jilin Province of China resulted in the contamination of the Amur river due to the accidental release of approximately 100 tonnes of toxic pollutants. It was the first serious case of transboundary pollution of a significant part of Russia's territory, that demonstrated to the Russian government the importance of international environmental agreements to protect national social and environmental interests.

Moreover, due to its constantly increasing integration into the world economy, even without ratifying relevant international environmental treaties, Russia is required to comply with many provisions contained therein. For example, in the course of imple-

---

27 See <http://www.unece.org/env/eia/welcome.html>.

28 See <http://www.unece.org/env/eia/sea_protocol.htm>.

29 See <http://www.unece.org/env/pp/>.

30 See <http://www.cms.int/documents/convtxt/cms_convtxt.htm>.

menting the Nord Stream project,[31] due to pressure from some opponents in the EU (especially governments of Poland, Lituania, Finland and Sweden) to this project, the Russian government was obliged to comply with the requirements of the Espoo Convention. Although Russia has not ratified the Espoo Convention itself (it was signed by the USSR in 1991), relevant Russian ministries were determined to meet the provisions of this convention in order to ensure the implementation of the Nord Stream project. Furthermore, Russian state-owned companies will also be required to comply with the provisions of the Espoo Convention in the course of construction of the Burgas-Alexandroupoli oil pipeline for the same reasons of aiming to maintain Russia's competitiveness as an oil and gas supplier to Europe.

Russia's accession to existing international legal instruments for the protection of the environment and its active participation in drafting new international agreements, specifically, 'post-Kyoto' instruments to enhance energy efficiency and to mitigate climate change, will allow Russia to protect its national interests consistently and, as an 'environmental super power', to make the best use of its competitive advantages in the world economy.

Conclusion

The global financial crisis that erupted in the second half of 2008 has brought to the fore that uneven distribution of risks lead to economic instability and that more thorough risk assessments, including of environmental risk assessments, need to be undertaken. Thus, the diversification and reduction of risks (including environmental risks) and an innovative approach to the use of all types of capital, including natural (environmental) capital and natural resources, will assist the prevention of future (financial or other) crisis events globally and in the Russian economy.

The 'green' economic development (or sustainable development) of Russia may become the most important instrument to modernize the Russian economy. It will enable Russia's transition towards an innovative, socially and environmentally responsible nation and assist with the achievement of the goals as set out in the *2020 Concept of Long-Term Social and Economic Development of the Russian Federation*. Furthermore, as the 'environmental footprint' of goods and services become increasingly important in the global market (especially in developed countries) – the relevant producing business' environmental responsibility, as well as the energy efficiency (or 'carbon footprint') of its goods and services become very important factors when selling to end consumers. Thus, the 'green' economic transformation of Russia will facilitate the diversification of Russia's exports (including those which contain a high share of processed natural resources and value-added goods) and enhance Russian companies' competitiveness on the global markets.

---

[31] This projects relates to the construction of a gas pipeline along the bottom of the Baltic Sea. See <http://www.nord-stream.com/en/>.

Chapter 9
ENVIRONMENTAL EDUCATION AND CAPACITY BUILDING IN RUSSIA

Tatiana V. Guseva[1]

Modern environmental education in Russia forms a mosaic of diverse approaches implemented by various institutions. In general, the educational approaches are either rooted in methods developed in the former Soviet Union in the 1970s to the early 1990s, or they are linked to international education initiatives which have been recognised by Russian education specialists since the 1980s. It is very difficult (and not really needed) to differentiate between these two categories of environmental education as they have come to overlap over the last ten years. Nevertheless, specific procedural features of environmental education and its results can be explained based on the suggested classification of (a) Soviet environmental education and (b) such education which is linked to international initiatives.

This chapter first reviews the development of these two different streams of environmental education in Russia (and the former Soviet Union) and then discusses the different types of capacity building activities which have prevailed in Russia. The aim of this chapter is to provide businesses, which have an interest in environmental finance and socially responsible investments, with an insight into the general state of affairs of environmental education in Russia and to provide such businesses with an overview of possible avenues of action.

Traditional (USSR rooted) environmental education

In the 1970s, environmental education was provided mostly at science oriented universities (these universities are often referred to as the 'classical' universities in Russia, as opposed to the technical universities recently transformed from former polytechnic institutes) within academic specialities such as biology, geography, geology, hydrometeorology, and hydrology. The technical higher education establishments offered short courses in environmental protection and training of specialists in wastewater treatment, land improvement, physicochemical analysis of air, water, and soil.

At the beginning of the 1980s, first departments of 'industrial ecology' were established. It is important to note that the term 'industrial ecology' was first introduced in

---

[1] Professor, D. Mendeleyev University of Chemical Technology of Russia, Moscow, Russia; Email: *tguseva@muctr.ru*.

W.Th. Douma and F.M. Mucklow, eds., Environmental Finance and Socially Responsible Business in Russia
© 2010, T·M·C·Asser Press, The Hague, The Netherlands and the Authors

the USSR and then became an internationally accepted discipline. In 1994, industrial ecology was defined by a distinguished international research team as 'a new example concept in which the interactions between human activities and the environment are systematically analysed'.[2] Along with industrial ecology departments, schools of environmental engineering (which mostly focused on 'end-of-pipe'[3] approaches) continued to strengthen in the 1980s and at the beginning of the 1990s. In the mid 1990s, risk management and process safety training courses also emerged, especially in context of occupational health and safety, as well as in context of civil defence courses.

In Russian society, the demand for environmental education was very high between 1985 to 1995, when competition among new university applicants reached 8-10 people per one place. In other words, as soon as it had been officially recognised that environmental problems not just existed, but needed to be solved, the young generation began to show an interest in this new field of science and practical activities. To a certain extent it was a sign of the growing responsibility felt by the younger generation for the fate of their country. During somewhat ten years, environmental courses formed a compulsory component of the national secondary school curriculum. By 1991, there were already more than 50 departments training specialists in environmental protection for industry, agriculture, municipal services, as well as for the authorities. Furthermore, improvements in transparency and availability of environmental information[4] created opportunities for better training in various fields, such as environmental monitoring and enforcement, and economic instruments of environmental protection. Yet, for decades the environmental education provided by technical higher education establishments remained technocratic and emphasised the 'end-of-pipe' methods of environmental protection. Science oriented universities focused their attention on ecosystem studies, water chemistry, toxicology, environmental monitoring and economic issues – they did not introduce courses related to environmental aspects of production processes until the late 1990s.[5] This resulted in the formation of rather strong schools of environmental scientists (theoreticians) and engineers. However, technological developments were (and to a large extent still are) seen separately from enabling environmental improvements. Moreover, it was thought that environmental (end-of-pipe first of all) solutions were too expensive and unaffordable for enterprises and that requirements of economic growth of the country were much more important than the

---

[2] R.H. Sokolow, C. Andrews, F. Berkhout and V. Thomas, eds., *Industrial ecology and global change* (Cambridge, University Press 1994).

[3] The traditional means of combating pollution has been by means of end-of-pipe systems, i.e., treatment of wastes and polluting streams. This end-of-pipe approach, while still essential for many industries and for many technologies, should only be used as a last resort and Cleaner Production opportunities should be investigated first. End-of-pipe systems may involve treating water, air, noise and solid wastes <http://www.icett.or.jp/>.

[4] The first open (publicly available) report on the state of the environment was published in the USSR in 1988.

[5] Recently, industrial ecology became one of new core disciplines taught to younger environmentalists at the Moscow State University named after M.V. Lomonosov. This new course on industrial ecology has been developed by teachers of the first Department for Industrial Ecology established at D. Mendeleyev University of Chemical Technology in 1983.

newly introduced environmental standards. Thus, educational disintegration led to the disintegration of practical activities and the environmental situation continued to deteriorate in many regions.

Methodological Councils

In order to systematise the curriculum and to improve the quality of environmental education, at the beginning of the 1990s, the Russian Ministry for Education (nowadays referred to as the RF Ministry for Education and Science) established special independent Methodological Councils ('MCs'). These bodies function as unions of leading higher education establishments. The MCs jointly develop the requirements of training programmes and special courses, and consider the applications of newer institutions and departments which show a willingness to start training environmentalists. There are three main independent (and somewhat competing) MC branches, as follows:

- The first MC is referred to as the 'Ecology' MC (*Экология*). It focuses on education relating to 'Environment', 'Nature Use', 'Ecology' and 'Geoecology', and co-ordinates training programmes of related departments in 72 'classical' universities. Until recently, the 'Ecology' MC has been working to establish a new section on the 'Education for Sustainable Development'.
- The second MC is focused on more technical and technological issues and is referred to as the 'Environmental Protection and Rational Use of Natural Resources' MC (*Охрана окружающей среды и рациональное использование природных ресурсов*). Its function is to co-ordinate the training programmes of over 50 related departments in the polytechnic institutions.
- The third MC deals with risk management training and is called (a little peculiarly) the 'Safety of Life' MC (*Безопасность жтзнедеятельности*). It co-ordinates the training programmes of nearly 60 departments (chairs) focused on training students in occupational health and safety, process safety and similar subjects.

In addition to the MCs, there are further higher education establishments, which train students in medicine, law, and economics that offer courses in environmental issues. Depending on the particular school's quality of education, graduates can emerge as very strong environmental specialists.

Educational standards

The first series of educational standards was developed in Russia in the 1990s. These educational standards relate to the descriptions of knowledge and skills of graduates as well as contents and duration of courses of 5 or 5.5 year programmes, which are similar to Master in Sciences (MSc) degrees.

If higher education establishments wish to be accredited, they must comply with these standards and provide the necessary student facilities (e.g., lecture halls, libraries, laboratories). Special commissions assess preparedness of such establishments to training students. Based on the recommendations of these commissions, the educational authorities issue licences, which specify what particular professions can be obtained by students studying at the universities.

The introduction of these educational standards helped to regulate and control the growth at the end of the 1990s of numerous low quality educational establishments (which in some cases existed mostly on 'paper' nearly selling diplomas, rather than providing actual teaching). Furthermore, they formed a good basis for the further development of environmental education in accordance with the Bologna Process[6] ideas. The ultimate goal of the Bologna Declaration is to create a common European Higher Education Area by 2010 with a view to improving the competitiveness and attraction of European higher education in relation to other continents. Russia signed the Bologna Declaration and in accordance with its requirements started reforming university degrees to make them compatible with European Higher Education degrees.

Environmental science in Russia

Since the 1980's, the number of MSc, PhD and Doctor of Science[7] dissertations in the field of environmental protection and the rational use of natural resources has been growing dramatically. Recommendations of many dissertations were implemented in industry and agriculture; whilst other findings helped develop environmental legislation and enforcement mechanisms.

However, during 1999-2000, the Russian Ministry for Education took the decision to eliminate the 'Environmental Protection and Rational Use of Natural Resources' branch from its list of research areas in which students can obtain a PhD and Doctor of Science degree. At the same time, environmental courses were excluded from the list of compulsory disciplines of the national secondary school curriculum. Moreover, at that point, the Ministry for Natural Resources (nowadays referred to as the RF Ministry for Natural Resources and Environment) was no longer required to contribute towards the development of environmental education.

This means that, officially speaking, environmental science does not exist as an academic discipline in Russia today. However, piecemeal research in this area survives (and even shows some progress) within the academic branches formally termed 'Ecology', 'Geoecology' and 'Economy of Nature Use'. This drastic change led to the weakening of environmental education in schools and interdisciplinary research, without which it is difficult to imagine how strong professional (higher) environmental education can be maintained or developed. The economic crisis of higher education establishments and the unsuccessful (ongoing) reform in education,[8] have speeded up the

---

[6] See for example, <http://www.minedu.fi/minedu/education/bolognaprocess.html> or <http://www.eua.be/eua/en/policy_bologna.jspx>.

[7] The Doctor of Science degree is an academic degree ranking higher than a PhD and requires the development of a new branch in science or a solution of an important scientific or practical problem; it is one of the pre-conditions for full professorship in Russia and some NIS countries.

[8] See for example *Transition*, The Newsletter About Reforming Economies, The World Bank, January/February/March 2003, <http://www.worldbank.org/transitionnewsletter/>, A. Chowdhury and I. Verbina,

deterioration of the traditional environmental education system, which prevailed in the USSR. This has also contributed to slowing down development of newer approaches based on Cleaner Production, Environmental Management System and Resource Efficiency concepts.

This means that there has been a clear reversal of the development of the traditional (USSR rooted) environmental education in Russia. Many institutions still suffer from the lack of funding and poor management, which means they cannot afford or obtain the required expensive research equipment (e.g., analytical laboratories, pilot installations). However, thanks to the efforts of dedicated teachers, the result of higher environmental education (and, thus, knowledge of graduates) remains at a good level. Those higher school establishments, which obtained the status of federal, innovative or research universities and substantial financial support from the state, are in a much better position and develop educational programmes rather successfully. Many of them (for example, both classical and technological universities of Tomsk) have opened excellently equipped laboratories and started inviting European professors to deliver new courses (including environmental ones).

In 2008, the Russian government started paying stronger attention to the issues of energy and environmental efficiency. Presidential Decrees were passed and legislation drafted both at the federal and regional levels. These documents require (and thereby open opportunities) for strengthening environmental education and establish clear links between technical regulation, energy and environmental efficiency and economic progress. Under the current global crisis conditions it is difficult to predict how rapid and efficient the new developments in environmental education will become, but the prerequisites for future improvement have been set.

'Imported' approaches of environmental education

The above word 'imported' may sound unpleasant to Russian environmental education specialists. One could, instead, describe the environmental education approaches as having been 'transferred from', 'inspired by' or even recently developed from international initiatives. However, regardless of how such approaches may be described, it is obvious that these education methods are strongly linked to internationally developed concepts, such as the concept of sustainable development, environmental management, or Corporate Social Responsibility.[9] This can easily be demonstrated by looking at the Russian language terminology in related fields: nearly all relevant environmental educational terms are (either successfully or poorly) translated from the English language.

'Reforming Russia's Education System', at <http://www.cdi.org/russia/johnson/7171-15.cfm>; see also B. Eklof, L.E. Holmes and V. Kaplan, eds., *Educational Reform in Post-Soviet Russia: Legacies and Prospects* (The Cummings Center Series, 2005).

[9] Please see chapters 19 (*Kostin*) and 20 (*Dayman*), for more detail about Corporate Social Responsibility.

New environmental programmes

The rapid adoption in Russia of 'modern instruments of environmental regulation'[10] resulted in the development of new higher education programmes. These include, for example, the development in educational institutions of the availability of Master degrees in Environmental Auditing or the possibility to specialise in Environmental Management (within Master degrees in Management). Students are able to undergo PhD research in the 'Economics of Nature Use' if such research develops new approaches in Environmental Impact Assessment ('EIA').

Nevertheless, in all cases, although such educational programmes are environmental in nature, dissertations are assessed by experts who, in Russia, are traditionally considered to be economists, and not by environmental specialists as might be expected. This is another sign of the disintegration of environmental education and research discussed earlier.

Furthermore, unfortunately, many economics-oriented higher school establishments offering degrees in Environmental Management train their students in the various related areas, but not in Environmental Management Systems ('EMS') or in the area of EIA.[11] At technical establishments, Environmental Management courses focus on mathematical methods used to assess the dispersion of pollutants into the environment, and do not address environmental management procedures, *per se*. This means that, although programmes at universities incorporate environmental terminology[12] and environmental elements, key focus areas, such as EMS and EIA, remain largely outside the scope of environmental education in Russia.

Education in sustainable development

In Russia, education relating to the concept of sustainable development forms another field, which has undeniably been influenced by international trends. There is serious competition between the 'classical' and technical universities in this area of education, but both institutions lack the necessary holistic approach. Quite often sustainable development programmes are formed by simply adding certain social subjects to the lists of environmental disciplines (this is typical of geographic, chemical and biological

---

[10] Modern (or rather market) instruments of environmental regulation include first of all Environmental Impact Assessment, Environmental Auditing and Environmental Management Systems. One can understand modern as a word used in Russian to emphasise that these instruments are not rooted in the old command-and control system.

[11] In Russia, the term 'EIA' is often used to describe environmental review, which is the last (quality assessment) stage of EIA, which in Russia is conducted by a special commissions formed by the environmental authorities.

[12] Although steps have been taken towards the integration of such new environmental education programmes, it is worth mentioning that both older and new environmental terms co-exist in education and research. In many cases these terms contradict one another. For example, 'environmental management', as a term, is accepted at the bachelor and Master levels, whilst only the term 'environmental administration' (which is closer to command-and-control administration procedures) is only used in context of the requirements, which must be met by students undertaking their PhD or Doctors of Science dissertations.

faculties). Sometimes, environmental courses are included in socio-economic ones. However, this also causes difficulties due to the weak environmental and natural sciences training of teachers at the secondary schools level and due to the lack of needed experience of university teachers employed at the economics faculties. It is difficult to predict how this situation will develop.

Thus, as the concept of sustainable development itself is neither understood well enough in Russia, nor strongly supported by the government, it is understood that it is likely that this branch of education can progress only with significant support of international organisations and donors, including foreign investors or businesses interested in environmental investments or SRI.[13]

Next steps

The consequences of the division between older (technocratic) and newer (interdisciplinary) education approaches can be seen in the system of environmental decision-making which influences all levels of society – from individual settlements and small enterprises to the regions and larger companies, as follows:

- technical and technological solutions of environmental problems are considered in isolation of managerial solutions; and
- reactive decision-making continues to prevail, whilst proactive decisions are often undervalued.

Leaders of many international environmental projects implemented in Russia within the framework of the EU, EBRD, World Bank, or USAID funded projects agree that, normally, technical Russian experts are very effective and contribute much towards the overall progress of a project. However, at the same time, the management related components (such as Management Systems Strengthening, Good Governance, and Corporate Development) always require international experts, and sometimes it is extremely difficult to reach a reasonable compromise and agreement between the positions of Russian and international specialists.

These realities considered, there is a clear need for further environmental capacity building in Russia, which is investigated in the next section below. Clearly, this need for capacity building (or training, as such) also represents an area where foreign investor businesses, which have specific expertise in training of experts in management systems strengthening, good governance or corporate development, may see great opportunities for investment and activity.

[13] It is noted that since 2000-2002, most international donors have not supported environmental projects, including those aimed at the improvement of environmental education.

Environmental capacity building

When addressing the needs of environmental capacity building in Russia, the first question to consider is what are the target groups which require capacity building[14] in environmental issues? And secondly, which capacity building approaches can be regarded as well-established and which require additional support?

For purposes of this discussion it is logical to divide all stakeholders into four major target groups, namely:

- *Producers* (both managers and engineers);
- *Regulators* (acting at national, regional and local levels);
- *Consultants* (including consulting companies themselves, as well as traditional research and educational establishments); and
- *Non-Governmental Organisations* ('NGOs') (which reflect the interests of the general public and which function as the most active part of such).

This division may appear unnecessary, but nevertheless useful in order to analyse the effects of past capacity building efforts and in order to assess future needs.

Producers

The producers are represented by graduates from technical and economic higher school establishments, as well as from the so-called professional schools or technical colleges.[15] In the past, all mid-level managers, technicians and engineers (within this producer group) were obliged to regularly undertake professional development courses. The effectiveness of such courses varied substantially: from formal attendance of courses resulting in hardly any educational benefits, to serious capacity building courses resulting in significant promotion and performance improvement of the individuals attending the relevant course.

In Soviet days, environmental courses rarely played a serious role in producers' professional development. Nevertheless, in many regions, the attendance of environmental education courses (including legislation, permitting procedures, 'polluter pays' principle, etc.) was required by the environmental authorities as a condition for their approval of environmental permits to given industrial and agricultural enterprises.

Today, people who fall within the producers group mainly attend courses in environmental analytical chemistry. This is largely market-driven as entrepreneurs often seek the accreditation of analytical laboratories with respect to proving their corporations' compliance with their environmental permits' conditions.

---

[14] Capacity building refers to activities that improve an organisation's ability to achieve its mission. It may relate to almost any aspect of its work: improved governance, leadership, mission and strategy, administration programme development and implementation, etc. For individuals, capacity building may relate to leadership development, advocacy skills, training/speaking abilities, technical skills, organizing skills, and other areas of personal and professional development.

[15] In Russia, there are 1.5 or 2 year programmes available to obtain a technical profession needed to work as a technician or a shift manager.

EMS-related training is conducted by consulting companies by way of helping entrepreneurs to develop and implement EMSs. Though all courses are based on ISO 14000 standards, quality of training varies a lot depending on the experience of consultants and attention paid by the top management to EMS implementation.

In general, capacity building approaches regarding producers need to be systematised and strengthened. Partially it might be achieved with the support of environmental enforcement authorities aiming to improve performance of their regulatees. But this is hard to imagine since in most Regions these authorities have been weakened by the recent administrative reform and have no capacity themselves to support the environmental training of entrepreneurs. A more realistic solution could be the wider promotion of integrated management approaches based on the assumption that environmental management forms an essential part of the overall management system. In this case, top managers themselves would drive the professional development of their employees in environmentally sound practices.

Regulators

The regulators group is represented by graduates of various higher school establishments who either obtained degrees in environmental disciplines or gained relevant professional experience.

Regulators are generally overloaded by routine work and, therefore, rarely undertake any professional development courses.[16] In many cases those who have received training have left the environmental authorities in search of better-paid and more sustainable positions. Thus, the number of experienced and well-educated regulatory specialists has decreased.

Many previously well-established international capacity building and institutional development oriented projects and programmes, which were quite widespread in the 1990s, have been abandoned. Furthermore, the leaders of the remaining international projects face serious problems when trying to find appropriate and interested beneficiaries representing environmental authorities. It is understood that these changes are interlinked with the continuing restructuring of the environmental protection and natural resource management authorities. The Strategic Programme Fund ('SPF') of the British Foreign and Commonwealth Office is a fortunate exception. The SPF (formerly GOF – or the Global Opportunities Fund) runs climate security, energy security and economic reforms programmes in Russia (along with 21 other countries). In particular, SPF funded projects involve regulators, researchers, educators and the business community in studies, training and practical pilot projects aimed at improving energy efficiency. In addition, SPF funded projects help identify gaps in energy efficiency, as well as find and implement practical solutions that comply with key international and national (Russian) priorities.

It is noted that higher school establishments also require capacity building, but at the same time can offer certain experience in capacity building themselves. These in-

[16] Between 1998 and 2005 the number of environmental inspectors decreased by 4-5 times.

stitutions continue to run professional development programmes for specific sectors in industry, provide support to secondary school teachers and supply other groups with trained specialists. However, in context of the approach taken towards capacity building of target groups, one encounters the initial problems of environmental education in Russia. Technically, Russian educators can offer much. However, they lack a holistic approach, possess little experience in the field of management education and have no instruments needed to promote modern solutions to environmental problems. This exemplifies how environmental education and capacity building are very much interrelated. Both fields are currently unsustainable. The problems faced by the educational system cause difficulties in the development of various economic sectors.

There is hope, however, that the newly formed Ministry for Natural Resources and Environment will develop into a strong authority taking over the functions of the older bodies and improve the overall performance of environmental regulation and – via this – of environmental education.

Environmental consultants

The environmental consulting business is still very young in Russia. It feeds from research and project design institutions, universities, environmental authorities, and sometimes from advanced NGOs.[17]

To a certain extent the environmental consulting businesses in Russia suffer from problems typical of other target groups, but on the other hand environmental consultants form a dynamic and self-developing community. They gain much experience through international projects and are gradually becoming strong players, contributing towards capacity building of producers and, sometimes, regulators – for example in EMS training and capacity building. In this particular area of training, the Russian environmental consulting companies do much better than higher school establishments. Nevertheless, the latter continue to try to form an independent niche in EMS training and capacity building. It is noted that the performance of both individual experts and environmental consulting companies varies significantly.

Nevertheless, it is unlikely that recent attempts to form associations of environmental auditors, EMS consultants etc., will lead to any serious improvement in this field. Capacity building of this target group could be organised through a formation of a national EMS certification system.[18]

Environmental NGOs

Environmental NGOs in Russia are a very diverse group: there are older research oriented organisations, such as the Society of Naturalists established in the 19th century,

---

[17] Advanced NGOs are those non-profit organisations that are well-established, work highly professionally and offer advice to regulators, entrepreneurs and NGOs, but remain dependent on grants rather than contracts. Such organisations as the Moscow based WWF Russia, RREC, Ecoline, or the Society for Naturalists (in Saint Petersburg), CETI (in Yekaterinburg) are recognised as advanced environmental NGOs.

[18] See chapter 16 (*Dayman*).

and young local groups of teachers and teenagers or university students. In the 1990s, environmental NGOs received strong support from US and European donors which helped to build NGO capacity and to clearly formulate their interests. In return, NGOs became active in disseminating environmental information and raising awareness of the general public.

As is mentioned above, international funds supporting environmental initiatives (including education) have become scarce, and national donors (for example larger corporations, regional governments) are not prepared to step into the shoes of international donor agencies. Under these conditions, only stronger and dedicated NGOs can survive. Some of them (for instance, those which actively participated in the training programmes of the US Academy of Educational Development) transformed themselves into rather professional organisations and began offering training courses to smaller NGOs and other target groups.

Business and environmental education and capacity building

Until recently, businesses operating in Russia have been supporting environmental education and capacity building in Russia. The picture is still patchy and while there are Regions which gain much from businesses supporting environmental education (for example the Perm oblast where Lukoil petroleum company invested into the development of education), one cannot say that the business community has succeeded international donor organisations which provided environmental grants to NGOs and higher education establishments in the 1990s. Currently, quite promising steps in Russia are being made by BP which supports leading Russian universities and plans to run a capacity building programme on 'Energy and Environment'. Furthermore, British American Tobacco supports a programme on 'EcoLogical Technologies' for young talented environmentalists. These initiatives are strongly linked with the way these multinational companies see their corporate responsibilities and it is necessary to mention that both BP and BAT are open to suggestions of higher education establishments and NGOs concerning environmental performance of their companies and the further improvement of their relations with local communities.

Concluding remarks

There is a need for larger-scale capacity building projects in Russia, which involve the representatives of all sectors of society and which contain strong training components. Such projects could be regionally orientated and a successful example of such a project exists in the Tomsk region,[19] which displays the significant contribution which regulators, producers, NGOs and universities can provide in capacity building, as well as the

[19] 'Environmental Management in the Tomsk region' project funded by the UK Department for International Development.

sustainability of the results of such projects. At the same time, projects can be 'sectoral' in nature – for example, the water utilities[20] sector that received international support in 2002-2005.

Fortunately, most information relating to the current state of the Russian educational system, as well as relating to ongoing (and sometimes completed) capacity building projects and environmental information resources supported by regulators and NGOs, is available via numerous publications and through the internet.[21] This forms the necessary basis for thorough decision-making of investors planning for future steps, assessing the strong and weak points of Russian stakeholders, and looking for partners and experts. Furthermore, it provides an ample base of primary information for businesses which are interested in environmental education and capacity building in Russia.

[20] Municipal Water and Wastewater Project (Technical Assistance for Management Strengthening and Operational Improvement Programme) funded by several European and international donors. It is too early to say whether this project has had positive results, but its scope certainly focuses on management improvement and includes a strong training component.

[21] See <www.ed.gov.ru>; <www.mnr.gov.ru>; <www.rusrec.ru>; <www.ecoline.ru>; <www.14000.ru>; <www.14001.ru>; <www.ineca.ru>; <ceti@mail.e-burg.ru>; etc.

Chapter 10

GREEN PUBLIC PROCUREMENT: USING EUROPEAN EXPERIENCES IN RUSSIA

Ekaterina N. Khmeleva[1]

Currently in Russia, procurement for federal and municipal needs ('public procurement' or 'government contracts') represent a substantial part of the Russian GDP. For example, in 2008, public procurement on the federal and regional levels accounted for approximately 26% of annual expenditures of the state budget.[2] If one were to take into account municipal procurements, the share of federal state funds allocated for these purposes is much higher.

The development and implementation of environmentally responsible public and municipal procurement ('green public procurement') laws and policies, which are in line with generally accepted international practices, is an efficient way to stimulate the use of better (and cleaner) technologies, the application of innovative activities and 'green' investments, as well as promote the use of renewable resources.[3]

This considered, environmentally responsible or 'green' public procurement policies within Russia could significantly assist in ensuring the development of an efficient public procurement system. Indeed, President Medvedev called for the need to radically improve the public procurement system and that 'public procurement should be a means of encouraging technological modernisation and competitive effective producers'.[4]

Before venturing into a discussion of the public procurement laws in Russia, this chapter briefly reviews how other countries and, in particular, the members of the European Union ('EU'), have developed their respective green public procurement policies.

---

[1] Candidate of Legal Sciences, Environmental Law Program Officer, WWF-Russia; Email: *EKhmeleva@wwf.ru.* The author takes this opportunity to thank WWF UK and UK Department for International Development (DFID) for their financial support of WWF Russia.

[2] See <http://www.gks.ru/metod/torg.html> for the amount of state procurement and <http://www.gks.ru/bgd/free/b01_19/IssWWW.exe/Stg/d000/i000610r.htm> for annual expenditures of the state budget in 2008. It is noted that public procurement laws regulate how public authorities spend public money to buy goods and services or to order works, from the most basic paper clip to ship building contracts or waste water treatment plants (see <http://ec.europa.eu/trade/creating-opportunities/trade-topics/public-procurement/index_en.htm>).

[3] See <http://ec.europa.eu/environment/gpp/index_en.htm> and <http://www.thevalueofgreen.com> for general information on green public procurement.

[4] See Johnson's Russia List, 25 May 2009: '[Medvedev] Opening Remarks at Meeting with Government Cabinet Members and Chairmen of the Federal Assembly Upper and Lower Houses on Budget Policy in 2010-2012', see at <http://www.cdi.org/russia/johnson/2009-97-6.cfm>.

W.Th. Douma and F.M. Mucklow, eds., Environmental Finance and Socially Responsible Business in Russia
© 2010, T·M·C·ASSER PRESS*, The Hague, The Netherlands and the Authors*

Green public procurement policies outside Russia

Green public procurement policies have been developed by many countries of the EU – notably Austria, Belgium, the United Kingdom, Germany, Greece, Denmark, The Netherlands, France – and in Iceland. Similar regulations and policies as existing in the EU and in Iceland have developed in the USA and in Japan.

European Union

In the European Union, the volumes of public procurement (using public funds from all levels) constitute 16.3% of the EU's GDP and, thus, plays an important role in the EU economy.[5] In the EU, public procurement is considered as a key driver of public policies, regional integration and an important component of development (e.g., fight against corruption) – and it is understood that 'opaque and discriminatory procurement procedures and practices therefore significantly distort trade and reduce potential growth'.[6]

Accordingly, the EU members adopted various legislative provisions on public procurement,[7] which are aimed at modernising and facilitating the award of public contracts. It applies tools such as the TED (Tenders Electronic Daily) database, the single classification system (evidenced by the common vocabulary for the public contracts) and the System of Information on Public Procurement ('SIMAP'). The EU has also signed the multilateral Agreement on Government Procurement (the 'AGP') and negotiated an international award procedure within the World Trade Organization ('WTO').[8]

The following two directives provide the rules in the sphere of green public procurement, which harmonise EU procedures and promote fair competition amongst corporations through transparent selection procedures:

- *Directive 2004/18/EC* of the European Parliament and of the Council of 31 March 2004 on the coordination of procedures for the award of public works contracts, public supply contracts and public service contracts (as amended) (the so-called 'traditional Directive');[9] and
- *Directive 2004/17/EC* of the European Parliament and of the Council of 31 March 2004 coordinating the procurement procedures of entities operating in the water,

---

[5] See <http://europa.eu/publicprocurement/index_en.htm>.

[6] See <http://ec.europa.eu/trade/creating-opportunities/trade-topics/public-procurement/index_en.htm>.

[7] See for an overview of the various relevant EU provisions on public procurement: <http://europa.eu/legislation_summaries/internal_market/businesses/public_procurement/index_en.htm>. See also <http://ec.europa.eu/environment/gpp/pdf/national_gpp_strategies_en.pdf> for an overview of national action plans on green public procurement of the EU Member States.

[8] It is noted that public procurement is exempt from the basic rules of the WTO. In addition to the EU community, 12 countries have signed the AGP, which is currently being reviewed by its signatories. See <http://www.wto.org/english/docs_e/legal_e/gpr-94_e.pdf> for the full text of the AGP.

[9] See Official Journal L 134 of 30.04.2004, pp. 114-240.

energy, transport and postal services sectors (together with the traditional Directive, the 'Directives').[10]

These two Directives cover all sectors in which public contracts are entered into. Subject to some exclusions, reservations and public contract value thresholds, they establish the rules, criteria and procedures pursuant to which a public contract must be concluded. The rules enable public authorities to give preference to goods, which, for example, contribute to the better preservation of the environment, which decrease harmful emissions, or which are manufactured by socially responsible companies.

The Directives provide that environmental protection requirements must be integrated into the definition and implementation of each EU member states' policies and public procurement activities and that they should promote sustainable development. Other requirements include:

- Contracting authorities may lay down special conditions relating to the performance of a contract, provided that these are compatible with Community and are indicated in the contract notice or in the specifications. The conditions governing the performance of a contract may, in particular, concern social and environmental considerations.[11]
- Contracting authorities may lay down environmental characteristics in terms of performance or functional requirements for the contract.[12]
- A contracting authority which supplies the information about a contract must request that all tenderers (or candidates in the contract award procedure) indicate in their tender documentation whether they have taken into account any relevant obligations relating to employment protection.[13]

Both Directives provide that the contract award criteria include the environmental characteristics of goods and services to be procured.[14]

In addition, both Directives allow the contracting authorities to set out their own environmental requirements, such as production methods and environment effect indicators. The contracting authorities may (but are not obligated to) request compliance with certain conditions or specifications like the ones defined in eco-labels, such as the European Eco-label.[15] In order to further the greening of public procurement, the European Commission issued the Communication 'Public procurement for a better environment'.[16] The document's general objective is to provide guidance on how to reduce the environmental impact caused by public sector consumption and to use GPP to

[10] See Official Journal L 134, 30.4.2004, pp. 1-113. Both Directives entered into force on 30 April 2004 and became effective on 30 January 2006.

[11] See Article 38 of Directive 2004/17/EC and Article 26 of Directive 2004/18/EC

[12] See Article 34 of Directive 2004/17/EC and Article 23 of Directive 2004/18/EC.

[13] See Article 39 of Directive 2004/17/EC and Article 27 of Directive 2004/18/EC.

[14] See Article 55 of Directive 2004/18/EC and Article 43 of Directive 2004/17/EC.

[15] See Preamble, Section 29 of both Directives.

[16] COM(2008) 400 final.

stimulate innovation in environmental technologies, products and services. Also of importance in this respect is the Handbook on green public procurement 'Buying Green!'[17] that explains in concrete terms how environmental considerations can be integrated into public procurement procedures, and clarifies the legal possibilities, as introduced and/or clarified by the Directives, of integrating environmental considerations in a tender's technical specifications, selection and award criteria and contract performance clauses.

As already noted, the system of green public procurement exists in many member states of the European Union, with policies in this sphere adopted not only on the national level but on the level of various governmental authorities. For example, the British Department for International Development ('DFID') has published its environmental public procurement policy,[18] which sets out the following priority areas:

- sustainable consumption and production
- climate change and energy
- natural resource protection and environmental enhancement
- sustainable communities.

Another case in point is the Netherlands, where public procurement by state level authorities is to be 100% sustainable in 2010.[19]

Thus, in the EU a multilevel legislative and policy system exists to promote green public procurement though the use of European directives – national laws – regional policies and standards – and through policies of separate governmental and regional authorities.

Russian laws and regulatory framework on public procurement

The Russian public procurement laws and regulations are based on the Civil Code of the Russian Federation (the 'Civil Code') and the Budgetary Code of the Russian Federation (the 'Budgetary Code'), and are regulated by the following two main regulatory acts:[20]

- *Federal Law No. 60-FZ* 'On Supplies for Federal Government Needs' of 13 December 1994 (as amended) (the Federal Public Procurement Law)[21]

---

[17] See <http://ec.europa.eu/environment/gpp/guideline_en.htm>.

[18] See <http://www.dfid.gov.uk/procurement/environmental-proc-policy.pdf>.

[19] See <http://www.senternovem.nl/sustainableprocurement/index.asp>.

[20] See also the following article which reviews public procurement laws in Russia: H.-J. Priess, ed., *Getting the Deal Through: Public Procurement, An Overview of Regulation in 43 Jurisdictions Worldwide'* (2009), see chapter on Russia by Falk Tischendorf, Kamil Karibov and Ekaterina Soboleva of Beiten Burkhardt, pp. 199-207, see also: <http://www.bblaw.com/uploads/media/PP2009_Russia_chapter.pdf>.

[21] It is noted that this federal law is a specific law (others include the Federal Law 'On State Order for Defence' of 27 December 1995 No. 213-FZ, the Federal Law 'On State Material Reserves' of 29 December

- *Federal Law No. 94-FZ* 'On Placement of Orders for Procurement of Goods, Works and Provide Services for State and Municipal Needs' of 21 July 2005 (as amended) (the Public Procurement Law)

Neither of these laws provide for 'green' procurement policies or processes. As the texts of these laws do not overlap and contain no references to one another, it seems appropriate to analyse them separately as regards how to improve their provisions from the perspective of integrating green public procurement policies.

Federal public procurement

The Federal Public Procurement Law establishes the main principals of public procurement of goods and services by the federal authorities. Article 3 of the Federal Public Procurement Law states as follows:

> '4. *Products to be supplied under a government contract must correspond to the mandatory requirements set forth by governmental standards and special terms and conditions established by such contract.*
> *Mandatory requirements include requirements relating to the quality of products, which ensure their safety for human life and health, environmental protection, compatibility and exchangeability of products.*
> *Products supplied under government contracts and subject to mandatory certification under the laws of the Russian Federation shall be accompanied with a certificate and conformity mark issued or recognized by the relevant authorized body.*'

Thus, the Federal Public Procurement Law acknowledges that federal public procurement contracts must comply with Russian environmental law. Such requirements can be complied with, either by observing the established mandatory standards (including mandatory certification) or through compliance with the special (environmental) conditions established by a specific public procurement contract.

Furthermore, the provisions of the Federal Public Procurement Law could be refered to in order to develop specific green public procurement policies at the federal level and to create the regulatory framework.

Public procurement on federal, regional and municipal levels

The federal state's, Russian Constituent Entities' and municipal authorities' public procurement activities are governed by the Public Procurement Law. It establishes the unified public procurement process on the entire territory of the Russian Federation, which is aimed, *inter alia*, at ensuring transparency within the public contracts award procedure and at preventing corruption.

1994 No. 79-FZ, the Federal Law 'On Procurement and Delivery of Agricultural Goods, Raw Materials and Commodities for the State Needs' of 2 December 1994 No. 53-FZ, among others) which governs procurement of specific types of goods , ibid., n. 20.

The Public Procurement Law regulates, in detail, the procedure relating to the awards of government contracts at the all levels and relating to the holding of tenders and auctions (including the selection of a winner). These requirements are binding on the governmental authorities of all levels and on local self-government bodies. Therefore, the inclusion in the Public Procurement Law of requirements regarding environmentally responsible public procurement would seem to be most efficient.

It is noted that this law establishes no general requirements with respect to the criteria, which is applicable to goods and services being supplied for public needs. The Public Procurement Law does, however, state that such requirements must be incorporated in the tender documentation (auction documentation), i.e., they must be case specific. The criteria which are stated to be taken into account when a contracting authority is reviewing the results of a tender do not include any criteria relating to environmental protection.[22] Moreover, the Public Procurement Law does not allow an authority to consider any criteria other than those listed in Article 28.

Thus, in order to promote the environmentally responsible public procurement within the framework of this law, it would need to be amended or supplemented to include a further provision, similar to above-mentioned Article 3 of the Federal Public Procurement Law.

Options for Green Public Procurement Laws in Russia

As mentioned above, the Russian laws on public procurement do not address environmental protection. However, the requirement of conducting 'green' public procurement is set out indirectly within the Russian Federal law 'On Environmental Protection' (the 'Environmental Protection Law').

The Environmental Protection Law provides that economic regulation in the 'environmental' sector shall serve as a mechanism to protect the environment. Article 14 of the Environmental Protection Law sets out, without limitation, the following methods of such economic regulation to protect the environment:

> '… *granting tax and other benefits upon implementation of the best available technologies, use of alternative power sources, recycled resources and waste processing, as well as [the] implementation of other efficient measures aimed at environmental protection in accordance with the laws of the Russian Federation*;
> *support of entrepreneurial, innovative and other activities (including environmental insurance) aimed at environmental protection*;'

Origin of goods and environmental sustainability

Presently, there is a need for legally binding requirements in public procurement with respect to the 'lawful' origin and environmental sustainability of goods, which apply to public and municipal procurement activity. In the opinion of the author, the absence of

[22] As set out under Article 28 of the Public Procurement Law.

such requirements is one of the obstacles for the development of a market advantage of environmentally responsible companies over companies, which continue to deliberately buy cheaper goods of illegal origin (e.g., illegally cut/produced wood). For example, in the forestry sector, this issue has become even more pressing after the enactment of the new Russian Forestry Code, which, in the opinion of the author, has contributed to an increase in the production and use of illegally cut wood.[23] As a result, the companies processing legally cut timber have started to loose ground to companies, which use illegally cut wood in their production processes.

But, in practice, if any production of forest and fishery sectors is designated for export, the market advantages of legal and environmentally sustainable goods is ensured by greater consumer demand (overseas) for such goods. In addition, if the legality and environmental sustainability of goods is confirmed by voluntary certificates (FSC for forest products, MSC for marine resources), including certification, eco-labeling or any other evidence of a good's legal origin and sustainable production, and is a requirement in the public procurement process, this would encourage Russian companies to be environmentally responsible when producing goods for the domestic market, too.

Energy efficiency

Environmental considerations within public procurement activity is also indirectly addressed in context of increasing the energy efficiency of the Russian economy – as required by the Presidential Decree No. 889 'On Measures to Increase Energy and Environmental Efficiency of the Russian Economy', dated 4 June 2008 (the Energy Efficiency Decree). In order to promote market advantages for modern and energy efficient products in Russia, the Energy Efficiency Decree requires the application of the highest energy efficiency standards when public authorities procure goods and services.

The need for publicly procured goods to meet the highest energy efficiency standards is reflected, too, in the '*2020 Concept of Long-Term Social and Economic Development of the Russian Federation*'.[24] This Concept is aimed at decreasing the energy usage per unit of GDP by at least 40% by 2020 (as compared with 2007 levels).

Suggestions for reform of Russian Public Procurement Law and Policies

The need to reform Russia's public procurement legislation is acknowledged at the highest governmental level. On 11 November 2008, President Dmitry Medvedev de-

---

[23] However, it is noted that the Foresty Code of Russian Federation (of 4 December 2006 No. 200-FZ) does not contain a definition of the term 'illegal logging' and there are no legally binding tools for tracking the supply chain of timber products, which, in practice, means a lack of state control on legal origin of such a products.

[24] This concept was approved by Resolution No. 1662-p of the RF Government dated 17 November 2008.

clared in his speech at the meeting with the presidium of the management board of the Russian Chamber of Commerce and Industry that amending the Public Procurement Law was necessary.[25] Furthermore, as mentioned above, the '*2020 Concept of Long-Term Social and Economic Development of the Russian Federation*' incorporates clear wording that public procurement laws should include provisions which encourage the procurement of goods which meet the highest energy efficiency standards.

In order to create a system of 'green' public procurement requirements in Russia, various further steps would be needed, including the passing of regulatory acts of general application, as well as individual legal acts on the federal level, in the Russian Constituent Entities and on the municipal level. Thus, and drawing from the EU experiences of employing a multilevel legislative and policy system to promote green public procurement, in order to implement such 'green' procurement policies and laws in Russia, the following rulemaking initiatives would be needed:

(i) Amendment and supplementing of federal laws and regulatory acts
(ii) Formulation of public procurement policies (strategies)
(iii) Approval of regulatory acts of federal ministries and agencies
(iv) Approval of general regulatory acts

Amending and supplementing federal laws and regulatory acts

As discussed above in connection with the Public Procurement Law, it would be necessary to supplement that law in order to incorporate a new Article – which requires that applicable environmental requirements should be met when goods and services are being procured by a public authority. This new Article would need to provide that:

- general and specific requirements to observe mandatory environmental standards be met during the public procurement process;
- the relevant goods and services be subject to mandatory certification and such certificates to be made available during the tender process;
- when awarding contracts for the supply of certain types of goods (such as wood and pulp products, fisheries), the governmental and local self-government authorities may give preference to suppliers of goods and services, which are certified under voluntary certification procedures, evidencing their legal origin and environmental sustainability.

Approval of regulatory acts of federal ministries and agencies

Such change in the Public Procurement Law would require the drafting of regulatory acts, which ensure mechanisms for the implementation of its provisions and, possibly, the development of new governmental standards and certification systems relating to environment protection requirements.

---

[25] See <http://www.rosbalt.ru/2008/11/11/540541.html> and <http://www.vesti.ru/doc.html?id=222485>.

The regulatory framework to establish such a standard certification system has already been developed by way of the Federal Law No. 184-FZ 'On Technical Regulation' of 27 December 2002. Article 11 of this law states that one of the goals of standardization is ensuring that governmental contracts are fulfilled by way of applying voluntary certification of goods (works and services), which attest to their quality. Furthermore, Article 31 of the Environmental Protection Law sets forth the general principles of environmental certification, which can be voluntary or mandatory, and provides that the environmental certification procedure be determined by the Russian government. This procedure has not been developed yet, but when it is developed, it will be necessary to include provisions which allow the use and application of environmental certification in the public procurement process.

Formulation of public procurement policies (strategies)

In order to develop a more comprehensive green public procurement system in Russia, public procurement policies (or strategies) need to be developed and approved, which set out what are the main principles of green public procurement.

Although (and as has been confirmed in practice) such policy or strategy documents operate as formal declarations (i.e., are not legally binding), their approval will motivate public authorities to incorporate environmental requirements when awarding public contracts.

For example, a green public procurement policy or strategy could provide that, when public authorities consider candidates partners for public contracts relating to goods or services, the following factors should be taken into account:

- Use of energy-efficient and water-efficient processes in manufacturing goods and rendering services;
- Reduced packaging and use of recyclable packaging;
- Volumes of wastes generated in manufacturing goods and services, including harmful and toxic wastes;
- Use of recycled materials;
- Use of renewable resources;
- Reducing the content of harmful and toxic substances in goods.

Such a governmental policy or strategy could, in addition, be applied to public contracts in all sectors of the economy and governmental/municipal management, including, without limitation, the following types of goods:

- Wood and pulp / paper products
- Marine resource products
- Power supplies
- Gas supplies for municipalities
- Transport
- Water management measures and measures to protect water bodies
- Utilities

Approval of general regulatory acts

Lastly, in order to reform and incorporate environmental requirements within public procurement laws in Russia, the approval of further regulatory acts (at the level of federal ministries and agencies) and the approval of departmental (industry-specific) regulatory acts are needed, which specify environmental requirements relating to goods and services being supplied to a specific ministry or agency. Such regulatory acts should be binding and be applied to all public contracts being awarded.

At the level of the Russian Constituent Entities and each of the municipal units, regulatory acts containing the same above-mentioned provisions need to be enacted, and regional regulatory acts be made binding on the entities acting in the territory of relevant Regions. In addition, regional regulatory acts could be approved by the heads of the Russian Constituent Entities, which would enable the establishment of the main principles of green public procurement by the Regional governmental authorities and municipal agencies.

In addition, at the level of municipal units, local self-government bodies could develop green public procurement policies for municipal needs in accordance with the federal and regional regulatory framework. Local self-government bodies might also wish to establish rules which require compliance with governmental standards and mandatory certification standards, as well as with voluntary certification rules.

Concluding remarks

The analysis of green public procurement regulations in the EU, as well as of the Russian regulatory framework relating to public procurement, shows that the implementation of environmental requirements and applying environmental criteria to the process of awarding government contracts for goods/services can be achieved in Russia. It is noted that the EU funded a Euro 3 million public procurement project with the Russian Ministry for Economic Development and Trade (now referred to as Ministry of Economic Development, 'MED') and the Russian Federal Anti-Monopoly Service ('FAS'). The project aims to support the reform of public procurement by introducing efficient monitoring tools, by setting up an e-procurement system, and by training Russian procurement officers. It was formally launched in Moscow on 4 October 2006 at the European Commission Delegation to Russia and continued through 2008.[26]

However, in addition to filling the gaps within relevant sections of Russian law and policy, it is also necessary for Russian governmental authorities and their heads to recognise the importance and necessity of this process. In this respect, the inclusion of the declaration within Russia's economic development concept and strategic priori-

[26] See <http://www.delrus.ec.europa.eu/en/news_819.htm>. The 'Public Procurement Reform II' project was said to contribute to improving the effectiveness, efficiency and transparency of public procurement in Russia. Together with the main project partners, the then MEDT and the FAS, procurement specialists from the EU and Russia will work towards a better implementation of the Federal Law No. 94-FZ 'On the procurement of goods, works and services for state and municipal needs'.

ties,[27] that public procurement of goods and services are to meet the highest energy efficiency standards and use renewable energy sources can, in the view of the author, be seen as the first step towards establishing a green public procurement system within the Russian Federation.

[27] See the '2020 Concept of Long-Term Social and Economic Development of the Russian Federation' (approved by Resolution No. 1662-p of the RF Government dated 17 November 2008) and 'The Main Directions of the State Policy in the Sphere of Increasing the Energy Efficiency and Power Energy on the Basis of the Use of Renewable Energy Sources for the Period up to 2020' (approved by Resolution No. 1-p of the Russian Government dated 8 January 2009).

Chapter 11
CAN RUSSIA BECOME A 'GREEN INVESTOR' IN SOUTHERN AFRICA?

Ivetta Gerasimchuk[1]

This chapter addresses Russian environmental finance from a different angle than most of the chapters in this book. The discussion focuses on how Russian companies and Russia, generally, engage in outward investment activity within the Southern African countries.

Introduction – Russian FDI in Southern Africa

Post-Soviet Russia is the latecomer amongst large-scale investors into African natural resources, having been preceded by both 'traditional' stakeholders such as the USA and EU, and by other emerging market economies ('EMEs'), particularly China.[2] However, this is a race that has continued over many decades and resulted in a lot of scepticism about FDI in Africa.

Whereas countries such as Norway, Canada and Australia have gained significant upstream and downstream spillover benefits from inward FDI in extractive industries, in other resource-rich countries, especially in Africa, such FDI has generally benefited local economies and societies to a very limited extent.[3] This implies that, as soon as foreign investors deplete deposits of natural resources of African nations, they as a rule abandon these host countries, and leave them with severely degraded environments and virtually no operating mechanisms of economic growth.

In this regard, the obvious question to be answered is whether the emerging South-South[4] cooperation will differ substantially from the historic North-South relationship. EMEs have naturally not engaged in colonisation-type activities in the past, but this fact is not sufficient to ensure that their investments provide tangible benefits to

---

[1] Scholar-in-Residence/Lecturer in Environmental Economics and Management at the Cyprus International Institute of Management (CIIM) and Senior Advisor, Trade and Investment Programme, WWF Russia; Email: *ivetta@ciim.ac.cy*.

[2] Commerce and Industry Chamber of the Russian Federation / ТПП РФ, 28 February 2006. Russia and Africa in the Short- and Mid-term. / Россия и Африка в кратко- и среднесрочной перспективе. Moscow.

[3] T. Moran, E. Graham and M. Blomstrom, eds., *Does Foreign Direct Investment Promote Development?* (Peterson Institute for International Economics, Washingon DC 2005).

[4] Though Russia is geographically in the North, politicians and scholars sometimes include it into the notion of 'South' due to other common development features (Aykut and Goldstein 2006).

W.Th. Douma and F.M. Mucklow, eds., Environmental Finance and Socially Responsible Business in Russia
© 2010, T·M·C·Asser Press, The Hague, The Netherlands and the Authors

African economies. In particular, China's expansion in Africa has already given rise to a tendency for scholars and journalists, primarily in developed countries, but to some extent also in Africa, to voice their concerns regarding a potential form of 'neo-colonialism' being introduced by Beijing.[5]

In the case of Russia, the possibility exists to channel its investment into Africa into more sustainable avenues. Russia, itself a resource-rich nation and as an ecological creditor, faces an urgent need to diversify the sources of its economic growth, just as is the case with the majority of African countries.

Russia's interests in Southern Africa

The turning point in Russia-Southern Africa relations was (then) President Vladimir Putin's visit to South Africa in September 2006. Putin was accompanied by a sizeable business delegation. In March 2007, (then) Prime Minister Fradkov paid a follow-up visit to South Africa, Namibia and Angola. Several bilateral treaties and cooperation agreements were concluded, in particular the Treaty of Friendship and Partnership between South Africa and the Russian Federation.

It was also during 2006 and 2007, when the bulk of Russian investment into Southern Africa occurred. Russia's direct investment stock in the Southern African Development Community ('SADC')[6] is estimated at USD 3 billion as of June 2008, based upon the aggregation of publicly reported values of Russian acquisitions in the region. This puts Russia's FDI in Southern Africa at approximately 14% of the total value of Russia's outward direct investment stock, which was officially worth USD 21 billion at the end of June 2008.[7] The estimate is based on publicly available announcements of deals and includes assessments of large-scale Russian investments in the region that official statistics might have failed to take into account, because transactions are often effected via third countries. In particular, Russia's largest acquisition in Southern Africa to date resulted from Norilsk Nickel's takeover of LionOre, a Canada-based company with key operating assets in South Africa, Botswana and Australia. A further example is the acquisition of South Africa's Highveld Steel and Vanadium Corporation by Evraz, a Russian capital group registered in Lichtenstein.

There are two key factors explaining Russia's revived interest towards Southern Africa:

---

[5] F. Kornegay, Africa's Strategic Diplomatic Engagement with China // New Impulses from the South. China's Engagement of Africa. Centre for Chinese Studies, University of Stellenbosch, South Africa (2008).

[6] The fifteen countries of Southern Africa, united by the Southern African Development Community (SADC), include: Angola, Botswana, Democratic Republic of Congo (DRC), Lesotho, Madagascar, Malawi, Mauritius, Mozambique, Namibia, Seychelles, Republic of South Africa (RSA), Swaziland, Tanzania, Zambia and Zimbabwe. The terms 'Southern Africa' and 'SADC' are used as synonyms in this article.

[7] Rosstat, 15 August 2008. On Foreign Investment in Russia in the First Half of 2008 / Об иностранных инвестициях в I полугодии 2008 г. Russian Federal Service of State Statistics, Moscow.

- *Russia's strategic interest in gaining access to Southern African natural resources.* According to official governmental forecasts, within the next decade, Russia's economically viable reserves of zinc, manganese, chromites, diamonds, uranium, gold, copper, nickel, platinum metals and oil will be depleted (see Table 1 below). Although Russia possesses significant untapped resource deposits, these are often difficult to access and costly to develop. For this reason, and given the volatility of world commodity prices, it is in the economic interests of Russia to gain access to sources of supply of such strategic minerals in regions where costs are lower, in particular, such as in Southern Africa. This strategic interest of Russia is reinforced by the increasing competition for African natural resources, not only from 'traditional' stakeholders such and the USA and EU, but also from key EMEs, especially China.
- *Russia's strategy to act as a global guarantor of energy security,* as was outlined, for instance, during the G8 Summit in Saint Petersburg in 2006. In this context, Africa is regarded as both a rich resource base, especially for uranium, and a market for Russian energy-related value-added products, including nuclear reactors.[8]

Russian companies in Southern Africa

In the meantime, alongside these two strategic considerations of Russia's leadership, the realities of the bilateral relationships with African countries are to a great extent determined by the myopic interest of a number of Russian companies in the acquisition of undervalued assets that can help to boost the companies' share price. World prices for non-ferrous metals such as nickel, manganese, platinum group metals, gold, chromites, vanadium, and others are highly volatile. As a rule, during periods of high prices, the increased revenues earned by industry leaders enable them to undertake large-scale mergers and acquisitions. By contrast, during periods of falling prices, the assets producing these strategic metals are often disposed of. In accordance with this logic, significant acquisition activities were undertaken by Russian metal and mining companies between 2006 and 2008, based on high prices for both steel and non-ferrous metals. This acquisition drive began domestically and expanded to Europe, North America, Australasia and Africa. Against this background, it is probable that some of the existing Russian investors in Southern Africa may withdraw in the short to medium term.

The interaction of these competing factors has resulted in differing socio-environmental policies on the part of Russian businesses operating in Southern Africa. For example, Norilsk Nickel's investments in nickel mining companies in Southern Africa have generally been regarded as beneficial both by the companies and the local communities. The Nkomati project in South Africa is a 50/50 joint venture, in which Norilsk Nickel's local partner is the Black Economic Empowerment company, African Rainbow Minerals. This arrangement ensures that the joint venture complies with the South African government's priorities in terms of improving the economic status of previ-

---

[8] Geopolitics, 22 March 2007, <http://www.geopolitics.ru/news/afr-004.htm>.

Table 1. Russia's natural base complementarities with Southern Africa.

Mineral	Year in which Russia will have depleted		SADC countries that possess reserves of the mineral (in alphabetical order)	SADC share in the world's total resources of the mineral
	economically producible reserves	all reserves		
Lead	2007	Beyond 2025	Namibia, South Africa, Zambia	–
Manganese ores	2008	Beyond 2025	DRC, South Africa	83 %
Zinc	2011	Beyond 2025	DRC, Namibia, Zambia	–
Chromium ores	2013	Beyond 2025	Madagascar, South Africa, Zimbabwe	93 %
Diamonds	2013	Beyond 2025	Angola, Botswana, DRC, Lesotho, Namibia, South Africa, Tanzania	> 50 %
Quartz	2013	2013	Madagascar, Mozambique	–
Tin	2015	Beyond 2025	DRC, Namibia, South Africa, Tanzania, Zimbabwe	–
Uranium	2015	Beyond 2025	Angola, DRC, Namibia, South Africa, Zambia	12 %
Gold	2015	Beyond 2025	Angola, DRC, Namibia, South Africa, Tanzania, Zambia, Zimbabwe	53 %
Oil	2015	Beyond 2025	Angola, DRC	–
Copper	2016	Beyond 2025	Angola, Botswana, DRC, Namibia, South Africa, Zambia, Zimbabwe	8 %
Nickel	2016	Beyond 2025	Botswana, South Africa, Tanzania, Zimbabwe	10 %
Tungsten	2016	Beyond 2025	Namibia	–
Platinum-group metals	2018	Beyond 2025	South Africa, Zimbabwe	88 %
Graphite	2018	Beyond 2025	Madagascar, Mozambique	–
Coal	Beyond 2025	Beyond 2025	Botswana, DRC, Madagascar, South Africa, Swaziland, Tanzania, Zambia, Zimbabwe	12 %
Phosphate	Beyond 2025	Beyond 2025	Angola, South Africa, Tanzania	10 %
Potash	Beyond 2025	Beyond 2025	Botswana	–
Bauxite	Beyond 2025	Beyond 2025	Angola, Madagascar	–
Iron ores	Beyond 2025	Beyond 2025	Angola, Botswana, South Africa, Tanzania, Zimbabwe	5 %
Natural gas	Beyond 2025	Beyond 2025	Angola, DRC, Mozambique, South Africa, Tanzania	–
Vanadium	Beyond 2025	Beyond 2025	South Africa	–
Fluorspar	Beyond 2025	Beyond 2025	Angola	–
Salt	Beyond 2025	Beyond 2025	Botswana, Madagascar, Namibia, South Africa	–

Source: 1) Long-Term State Programme of the Research of Subsoils and Reproduction of the Mineral and Raw Materials Base (2005-2010 and through 2020) of the Russian Federation, pp. 27-29; 2) SADC 1997; 3) US Geological Survey 2007.

ously disadvantaged sections of the population. Furthermore, Norilsk Nickel reported its plans to commit USD 830 million to investment in 'greenfield' developments and the modernisation of its facilities in South Africa and Botswana by 2010,[9] though in the new circumstances of the financial crisis in early 2009 the company announced its plans to scale down its overseas investment programme first in Australia and then in Botswana and South Africa.[10] Finally, the company is in the process of introducing an environmentally friendly technology, known as Activox, into its operations in the region.[11]

Another privately-held Russian metals company, Evraz, has accorded less consideration to the environmental aspects of its Southern African acquisition activities. In 2006, driven by an interest in acquiring a larger share of the global vanadium market, Evraz acquired the Highveld Steel and Vanadium Corporation Ltd in South Africa for USD 678 million.[12] The assets acquired by Evraz were generally characterised by poor environmental performance. In October 2007, an inspection by the 'Green Scorpions' on Highveld Steel's Vanchem vanadium plant exposed a number of transgressions of environmental legislation, including the contamination of groundwater, a criminal offence in terms of the Water Act.[13] These transgressions undoubtedly contributed to the situation in which Evraz was unable to recoup its investment, when it was forced by European competition authorities to divest itself of some of Highveld's assets.[14] According to media reports,[15] Evraz is as a result currently reviewing Highveld's capital expenditure strategy, in order to place more emphasis on investments in environment- and safety-related equipment, systems and services.

The environmental aspects of 'greenfield' investment projects proposed by Russian companies also merit some attention. These proposals include the development of manganese and uranium fields, as well as construction of a manganese smelter in South Africa by Renova, the development of uranium fields by a consortium of Russian companies in Namibia, the construction of an aluminium smelter by RUSAL and a fertiliser plant by Azot in South Africa, a number of small mineral prospecting projects by Sintez in Angola and Namibia, and oil prospecting and development by ALROSA, Gazprom and LUKOIL in Angola.

Trends of Russian business in Southern Africa

Consideration of these and other cases provides scope for the following conclusions regarding Russian investments in SADC:

---

[9] Mineweb, 26 September 2007, <http://www.mineweb.net/mineweb/view/mineweb/en/page36?oid=37493&sn=Detail>.

[10] RIA Novosti, 17 January 2009, <http://www.rian.ru/economy/20090117/159469833.html>.

[11] *Sunday Standard*, 20 May 2007, <http://www.sundaystandard.info/news/news_item.php?GroupID=1&NewsID=1499>.

[12] Smart Money, 9 October 2006, <http://www.smoney.ru/article.shtml?2006/10/09/1476>.

[13] Business Day, 5 October 2007, <http://www.businessday.co.za/articles/article.aspx?ID=BD4A579828>.

[14] Business Report, 24 April 2007, <http://www.busrep.co.za/index.php?fSectionId=553&fArticleId=4370487>.

[15] Engineering News, 17 August 2007, <http://www.engineeringnews.co.za/article.php?a_id=114089>.

- Russian businesses that are active in Southern Africa have in general pursued investment projects only in the industries in which they have been operating in Russia, namely ferrous and non-ferrous metals, diamonds, petroleum and uranium. All these projects are energy intensive and possess a considerable environmental footprint.
- There is no evidence suggesting that the environmental profile of Russian investment projects in Southern Africa is any 'dirtier' than that of projects undertaken by Western or African investors in the same industries.
- At the same time, there exist no examples of leapfrogging development projects promoted by Russian investors in Southern Africa; in other words, breakthrough projects that would enable Southern African countries to bypass the environmentally damaging and unsustainable stages of industrial growth that have marred the past of the majority of developed countries.

These trends are not particularly surprising, since they would appear to mirror existing business practices within Russia.[16] Russia's own economy is energy- and resource-intensive, and until this situation changes, it will be premature to expect Russian investors to promote leapfrogging development patterns abroad. At the same time, however, given the considerable environmental footprint of the projects being undertaken by Russian companies in Southern Africa, any steps towards improving their environmental practices are certain to reduce the sustainability risks attached to these projects.

Socio-environmental potential of Russia–Southern African cooperation

The heavy reliance by the Russian and Southern African economies on the extraction of non-renewable resources is of course not sustainable in the long term. These resources will, by their very nature, be depleted at some point, even if the most efficient technologies are applied, while the growing urgency to address the issue of climate change is certain to significantly decrease the size of world markets for fossil-fuel resources. At the same time, volatile rises in world commodity prices make a number of previously uncompetitive renewable and resource-efficient solutions increasingly viable. As a result of these factors, the world's development path is sure to move onto a more sustainable footing in the medium to long term, potentially leaving resource-based economies, such as those of Russia and Southern Africa, outside the new mainstream. In the words of Sheik Ahmed Zaki Yamani, Saudi Arabia's oil minister in 1970s, 'the Stone Age did not end for lack of stone, and the oil age will end long before the world runs out of oil';[17] the same holds true for many of the world's other mineral resources.

---

[16] N. Poussenkova and E. Solntseva, *Russian Companies in the 21st Century – Towards Competitive Corporate Citizenship* (Moscow, WWF-Russia 2007).

[17] *The New York Times*, 25 August 2005, <http://www.nytimes.com/2005/08/21/magazine/21OIL.html?pagewanted=all>.

Over the past two decades, the integration of environmental considerations into business decision-making has become a significant factor in the international competitiveness of both nations and corporations.[18] Improved efficiency in the use of energy and other resources holds the potential for significant savings by the private sector. Such 'greening' also assists both states and companies to prevent environmental degradation, and consequent expenditures on rehabilitation activities, as well as to avoid the reputational risks associated with such environmental damage. The environmental footprint of a project is increasingly taken into consideration in investment and insurance decisions regarding virtually all types of projects. This process has materially contributed to technological progress and the development of new environmental practices and leapfrogging technologies, which provide significant competitive advantages to businesses and nations.

At the same time, however, it would be premature to expect Russian investors in Southern Africa to pursue better sustainability and corporate social responsibly practices in Africa than they do in Russia. In many cases, the primary focus of Russian businesses (as often is the case globally) is the maximisation of profits for their owners in the short term, often with a view to a future sale. Nevertheless, it would be unfair to apportion blame for this short-sightedness exclusively to the Russian private sector, since it remains the case that in both Russia and Southern Africa, the investment climates continues to be somewhat unpredictable, with significant uncertainty regarding possible changes in taxation regimes and various regulatory practices. Given this scenario, it would appear that an increased focus on long-term planning and sustainability issues on the part of Russian companies might be prompted by an improvement in the transparency and predictability of the domestic business environment by the governments of both Russia and Southern Africa.

Among other factors, the introduction of more stringent environmental regulations in Russia will serve to prepare Russian business for competition in international markets, including those in Southern Africa. In this regard, a positive, albeit somewhat delayed, measure is the Russian Presidential decree of 4 June 2008, '*On Measures to make the Russian Economy More Energy and Environment Efficient*'.

A 'greener' approach towards cooperation with Southern Africa does not necessarily imply or require a divergence by the Russian government from the high-level choices it has already made in terms of its interest in Southern African natural resources, its strategy to act as a global guarantor of energy security, and its efforts to develop an international assistance programme. What need to be re-examined in the sustainability context, however, are the instruments by which these high-level goals might be achieved.

---

[18] M. Porter and C. van der Linde, 'Green and Competitive: Ending the Stalemate', 73 *Harvard Business Review* (1995) pp. 120-155; D. Esty and M. Porter, 'Ranking National Environmental Regulation and Performance: A Leading Indicator of Future Competitiveness?', *The Global Competitiveness Report 2001-2002* (New York, 2001) p. 78.

Enhancing sustainability of Russian businesses

Practical measures which can be implemented by Russian companies, in order to enhance the sustainability of their businesses both domestically and in Southern African, include the following:

- Voluntary standards – the introduction of and compliance with more stringent international quality standards and HSE (health, safety and environment) certifications, environmental information disclosure, open 'green' dialogue with local communities and NGOs
- Corporate social investment – greater attention to local social and educational issues, with the aim of training employees and improving their living standards, as well as enabling local communities to find alternative income sources once resources deposits are exhausted
- Analysis – the integration of environmental and social performance indicators into 'due diligence' and investment approval procedures that would prevent 'surprises' and financial losses (as was the case in the purchase and sale of the Vanchem vanadium plant by Evraz)
- Research and development – aimed at sustainable technology solutions, including footprint mitigations, use of renewable energy sources, sustainable transportation, waste recycling and energy, water and other resources savings

As mentioned, Norilsk Nickel's application of 'Activox' technology at its Southern African mines can serve as an example of positive actions in this area (even though this technology was initially developed in Australia).

Russia's role in guaranteeing regional energy security

Russia's focus on energy cooperation with Southern Africa can assist in resolving one of the most pressing development issues in the region, namely access to energy. In this regard, potential exists for the following contributions by Russian companies:

- Efficiency measures – broader application of existing energy-, water-, and other resources saving solutions
- Investment – construction of new power generation facilities based on renewable resources, such as wind and solar radiation, with minimal environmental footprint, particularly with regard to CO_2 emissions
- Research and development – particularly in terms of leapfrogging technologies, both in the sphere of energy savings and the use of renewable energy sources

In this regard, two interesting examples exist of cooperation between Russia and South Africa. The first concerns the cooperation between South Africa's state-owned electrical utility, Eskom, and the Russian scientific community in the field of underground coal gasification technology. This technology allows the energy content of coal to be extracted without mining, and to provide liquid petroleum gas as an input for electricity generation. This method of energy extraction can produce significant conversion efficiency improvements over conventional coal-fired electricity generation.[19] At present, however, the majority of the interest, and potential investment, in this technology is from Eskom, rather than from Russian companies.

The second example of cooperation between Russia and South Africa concerns the 'gas-to-liquid' ('GTL') technology owned by South Africa's PetroSA and SASOL, which has reportedly attracted the attention of Russia's two state-owned petroleum majors, Gazprom and Rosneft. This solution holds the potential to significantly reduce gas flaring in Russia. According to the Russian Ministry of Natural Resources, Rosneft plans to build a GTL plant at its Sakhalin facility, with support from PetroSA and SASOL.[20]

Uranium and nuclear energy

Despite the current focus by Russian policy and business decision-makers on uranium mining and nuclear energy projects in South Africa and Namibia, there exists considerable doubt regarding the long-term financial viability and environmental sustainability of such projects. From an environmental perspective, while an increase in the contribution of nuclear energy to electricity generation capacity can assist in the reduction of CO_2 emissions in the short term, in the longer term, nuclear capacity creates serious challenges such as spent fuel and nuclear waste disposal, as well as increased risks in terms of the unsanctioned use of enriched uranium. In economic terms, although nuclear energy directly increases the global availability of electricity, it is similar to fossil fuels in the sense that it continues to rely on a finite resource. Furthermore, the experience of another African country, Niger, shows little evidence of any positive impact of a large uranium extraction industry on local development. Nuclear power generation requires considerable economies of scale and should ideally be distributed through a national or regional grid, and is therefore far more suited to industrial electricity supply than to meeting the energy needs of average Africans living outside areas of grid coverage.[21]

Renewable energy

Southern Africa possesses abundant renewable energy reserves; solar radiation levels in South Africa and Namibia are amongst the highest in the world,[22] and both countries contain significant areas of semi-desert, unsuitable for agriculture, which can be utilised

---

[19] *Reuters*, 23 April 2008, <http://factiva.com>.

[20] *RBC Daily*, 19 March 2007, <http://www.rbcdaily.ru/2007/03/19/tek/268514>.

[21] P. Du Plooy, *Re-Think Investment in (South) Africa* (WWF South Africa, Johannesburg 2008) pp. 36, 47-48.

[22] B. Cuamba, Sustainable Energy Pathways for the SADC Region. Presentation at the DFID-DSA policy forum, *International Development in the Face of Climate Change: Beyond Mainstreaming?*, Greenwich, 2 June 2008.

for solar energy accumulation facilities. In the DRC, construction of what is scheduled to be the world's largest hydroelectric scheme, is currently underway on the Congo river's Inga Falls. Once completed, this facility will have an electricity generation capacity of 40 GW.[23] Interestingly, this project does not involve the construction of large-scale dams and water reservoirs, and as a result, its potential environmental footprint is limited. Although no Russian companies are participating in the Inga Falls project, ALROSA has already invested in the construction of smaller hydropower stations in Angola, and another Russian company, Tekhnopromexport, has acted as a contractor to a number of African hydroelectricity projects, thereby accumulating considerable experience in the field.

South Africa possesses the world largest reserves of palladium, a platinum group metal used as a catalyst in hydrogen-powered engines – a solution for hybrid motor vehicles that is receiving increased attention in developed countries, particularly in Japan and the USA.

While renewable energy sources, particularly hydro-power, are capable of feeding into national grids, one of their most tangible benefits is that they are also particularly suitable for small-scale applications, and can therefore be owned and operated by the communities that they supply, rather than by governments or corporations located many miles away, or in some cases even on a different continent. In this context, bio-energy and in particular solar energy are increasingly viewed as a vital component of SADC's rural (off-grid) electrification programmes, which have in many instances been slowed by the high costs of grid extension services.

Climate change and Joint Implementation projects

In addition to plans for new energy generation capacity, the potential exists for Russia and South Africa to cooperate in the implementation of projects aimed at increased efficiency in the use of already available energy resources and the reduction of their CO_2 emissions. Under the Kyoto Protocol, Joint Implementation projects in Russia and CDM projects in Southern Africa offer significant opportunities to increase energy efficiency in a commercially beneficial manner. Although neither Russia nor the Southern African countries are required to reduce their GHG emissions during the Kyoto Protocol's implementation period, ending in 2012, a new negotiated international agreement that replaces the Protocol may impose such requirements. Under such circumstances, it is obvious that the energy-intensive facilities owned by Russian investors in Southern Africa, in particular non-ferrous metal smelters, will need to substantially improve their energy efficiency standards. In this regard, a Russian company, Factor Ltd., has already begun working with South Africa's Eskom on energy-saving projects.

---

[23] BBC, 21 April 2008, <http://news.bbc.co.uk/2/hi/business/7358542.stm>.

CONCLUSIONS

The measures identified in this chapter form the basis for hedging commodities revenues for the future, by investing them in sustainable business solutions. Rethinking Russia's investment in Southern Africa might involve additional costs in the short term, both in financial and administrative terms, but this expenditure should be viewed as capital investment rather than costs. Furthermore, the returns on this investment, in the form of increased environmental sustainability and national and corporate competitiveness, are certain to be far in excess of the initial expenditure for both Russia and Southern Africa, and in fact for the world as a whole.

Chapter 12
THE EBRD AND RUSSIA: STIMULATING EUROPEAN PRINCIPLES FOR THE ENVIRONMENT

Wybe Th. Douma[1]

The investments by the European Bank for Reconstruction and Development (the 'EBRD' or the 'Bank') in Russia are carried out on the condition that besides meeting the obligations set out in EBRD's Environmental and Social Policy of 2008, the 'European Principles for the Environment' ('EPE')[2] are applied to such investments. The EPE encompass not only those environmental principles set out in Article 174(2) of the EC Treaty (i.e., precaution, prevention, polluter pays and rectification at source), but in principle also the entire environmental *acquis* (notably secondary EC environmental law, such as regulations and directives) and international conventions. This means that where EBRD is involved in Russia, it must apply Russian law and, in principle, EU law.

In this chapter, it is investigated how this innovative "extra-territorial" application of European law takes shape in EBRD's practice in Russia and how it can stimulate a two-sided approximation of Russian and EU environmental law.

Introduction

The EBRD supports projects in 30 countries in central or Eastern Europe to Central Asia and so stimulates these countries' transition from being centrally planned economies to market economies. Russia is one of the main focal countries with respect to EBRD's investment activity. In 2008, €1.8 billion was committed to Russia, which represents 36 per cent of the total annual business volume of the EBRD – more than any other of the countries in which the EBRD is active. Since 1991, the year in which it was established, the EBRD has invested some €11.4 billion in Russia.

The EBRD's priorities in Russia include improving the country's infrastructure, encouraging economic diversification, boosting competitiveness and entrepreneurship, and promoting environmental and energy efficiency. EBRD also engages in policy dialogue with the Russian government in their efforts to improve the overall legal framework and investment environment in Russia.

---

[1] Senior Researcher European Law and International Trade Law, Asser Institute, The Hague, The Netherlands; Email: *W.T.Douma@asser.nl*.

[2] See <http://www.eib.org/infocentre/epe/>, and see further detail below. The EBRD is a signatory to the EPE.

W.Th. Douma and F.M. Mucklow, eds., Environmental Finance and Socially Responsible Business in Russia
© 2010, T·M·C·Asser press, The Hague, The Netherlands and the Authors

In this contribution, firstly, the general purpose and environmental provisions in the Agreement Establishing the European Bank for Reconstruction and Development (EBRD's Establishing Agreement)[3] will be briefly discussed. Thereafter, the environmental policy of the EBRD will be examined and a short look will be taken as to the way in which this policy developed since 1991. Attention will be paid to the role of the EPE initiative and the way in which the Bank requires its clients to observe national environmental law, European environmental law and international conventions (including public participation and environmental impact assessment requirements). Also, the mediation facility in case of complaints about EBRD projects is briefly discussed with the help of an example in which environmental aspects of a power plant were examined. Thereby, the meaning of some of the principles of environmental policy in EBRD's practice is illustrated, notably where the principles of sustainable development and precaution are concerned.

After a brief review of how the Bank perceives Russia and some other countries in the region in terms of democracy and multi-party pluralism, two examples of EBRD projects in Russian practice will be discussed. Attention will be paid to a relatively small energy efficiency project in Siberia, and to the multi-billion oil and gas project Sakhalin II. Finally, concluding remarks will be presented on the role of the EBRD in promoting EU environmental law in Russia.

The EBRD

Purpose and functions

The EBRD was founded in 1991 and is owned by 61 countries (including the USA, Canada and Japan), the European Union ('EU') and the European Investment Bank ('EIB'). According to Article 1 of the EBRD's Establishing Agreement:

> '[i]n contributing to economic progress and reconstruction, the purpose of the Bank shall be to foster the transition towards open market-oriented economies and to promote private and entrepreneurial initiative in the Central and Eastern European countries committed to and applying the principles of multiparty democracy, pluralism and market economics.'[4]

The EBRD, thus, not only focuses on economic, but also on the political aspects of the transition economies of their recipient member countries.

In order to fulfil its purpose, the Bank is required to assist its recipient member countries in order to implement economic reforms and to help their economies become fully integrated into the international economy, *inter alia*, by way of measures which

[3] Dated 29 May 1990 and entered into force 28 March 1991. See <http://www.ebrd.com/pubs/insti/basics.pdf>.

[4] Article 1 of the EBRD's Establishing Agreement; an amendment to Article 1 of the EBRD's Establishment Agreement was approved by the Resolution of the Board of Governors which was adopted on 30 January 2004 and which entered into force on 15 October 2006.

'... promote in the full range of its activities environmentally sound and sustainable development'.[5] This provision made EBRD the first multilateral development bank with an explicit environmental mandate built into its charter. It is noted, however, that the notion of sustainable development already encompasses an environmental dimension.[6] Strictly speaking, this makes it superfluous to add the words 'environmentally sound' within the text of the provision, but considering that many still regard 'sustainable' as a synonym for 'ongoing' it does not hurt to stress the environmental dimension of sustainable development. Another minor issue is the fact that the purpose of the EBRD is stimulating economic progress, and that 'environmentally sound and sustainable development' are described as one of the functions to reach this purpose. This forms a somewhat indirect subscription to the concept of sustainable development. In contrast, the European Union regards sustainable development as one of its objectives.[7] However, what is key, is the steps the relevant organisation takes in order to implement the concept of sustainable development through its environmental policy – as will be discussed below – and how such environmental policy works in practice. The latter will be touched upon several times below when discussing the development of the Bank's policy over time and when discussing various EBRD financing projects in Russia.

The EBRD Environmental and Social Policy taking shape

The EBRD adopted its first Environmental Policy in 1991 at the initial meeting of its Board of Directors. The EBRD's environmental policy is reviewed regularly to reflect changes in international standards and practice. The need for improvements was stressed by the Center for International Environmental Law ('CIEL'). In a 1995 report, CIEL claimed that the Bank often did not implement its own environmental procedures, and instead was sacrificing environmental protection for expedience and thus undermined its commitment to the rule of law.[8] Internal EBRD reviews of its practices took place in 1996[9] and in 2003.[10] In spite of the resulting improvement, many stakeholders, including EBRD members, NGOs and clients, demanded greater clarity on what the Bank requires *in practice* if its clients where environmental protection is concerned. The EBRD's own Evaluation Department also conceded that:

---

[5] Article 2(1)(vii) of the EBRD's Establishing Agreement.

[6] See for instance N.J. Schrijver, *The Evolution of Sustainable Development in International Law: Inception, Meaning and Status*, Leiden, Martinus Nijhoff Publishers, 2008.

[7] Article 2 Treaty on the European Union. After the entry into force of the Lisbon Treaty, the reference to sustainable development can be found in Article 3(3) TEU.

[8] CIEL, *The European Bank for Reconstruction and Development: An Environmental Progress Report*, Washington, November 1995 and Donald M. Goldberg and David B. Hunter (both from CIEL), *EBRD's Environmental promise: a bounced check?*, Washington, 1994.

[9] Notably raising environmental due diligence standards to EU levels, pledging to follow national and World Bank standards in contexts where no EU legislation existed yet, strengthening EIA procedures and public information policies and emphasizing a commitment to the implementation of multilateral environmental agreements such as the 1992 United Nations Framework Convention on Climate Change.

[10] <http://www.ebrd.com/about/policies/enviro/policy/policy.pdf>.

> '[t]here were several cases in the EBRD lessons-learned database related to the need for the EBRD to have sufficient leverage on clients to ensure that they implement namely: "agreed environmental actions". Agreements should have binding commitments for clients to environmental improvements. For example, the EBRD should consider more often including incentives and sanctions in the agreements related to achieving environmental obligations.'[11]

In July 2006, new IFC Performance Standards ('PS') were adopted by over 40 large commercial financing institutions for project finance and advisory services, being the 'Equator Principles'.[12] Many of EBRD's co-financing and syndication partners have a commitment to meet such requirements, and for many clients these IFC's requirements will become an important reference point. The EBRD considered the time ripe for the preparation and adoption of a new Environmental Policy that would offer at least the same level of environmental and social safeguards as those of the IFC and the banks following the Equator Principles (the 'Equator Banks'), whilst reflecting EBRD's region of operation with its specific characteristics, governance structure, mandate, operational policies and staff resources, and its commitments under the EPE. In line with its Stakeholder Consultation and Disclosure Plan,[13] these issues were set out in an indicative 'Environmental Policy Discussion Paper'.[14] After receiving comments,[15] convening a client survey and meetings with international organisations (e.g., the ILO, WHO, EU, UNECE, other IFIs, Equator Banks), the publication of a draft revised policy and convening public meetings in six locations,[16] the new Environmental and Social Policy was adopted in 2008. Before turning to the content of that new Policy, the EPE will be discussed below.

European Principles for the Environment

The EPE Declaration[17] of May 2006 aims at increased harmonisation of environmental principles, practices and standards associated with the financing of projects. The initiative is founded on the commitment of the 5 signatory European-based Multilateral Financing Institutions ('MFIs') to ensuring environmental protection and the promo-

[11] Summary of common lessons learned on environment in the EBRD lessons learned Database, from the 2003 Environmental Policy Review. Achieving the Bank's environmental mandate through direct investments, January 2008.

[12] See Annex II. See also <http://www.equator-principles.com/index.shtml>.

[13] <http://www.ebrd.com/about/policies/enviro/policy/review/scdp2.pdf>.

[14] <http://www.ebrd.com/about/policies/enviro/policy/review/issues.pdf>. The paper was published both in English and Russian.

[15] See for instance: the Bank Information Center, *Comments on the EBRD Environmental Policy Discussion Paper*, 29 June 2007 and CEE Bankwatch Network, *Comments on the EBRD Environmental Policy Discussion Paper*, June 2007. For a summary of received comments and EBRD responses, see <http://www.ebrd.com/about/policies/enviro/policy/review/consult.pdf>.

[16] Budapest, Belgrade, Moscow, Tbilisi, Bishkek, and London.

[17] <http://www.eib.org/infocentre/epe/index.htm>.

tion of sustainable development globally and across all sectors of their activities.[18] The EPE are defined as:

- the guiding environmental principles in the EC Treaty (in particular those codified in Article 174(2), i.e., the precautionary principle, the prevention principle, the principle that environmental damage should as a priority be rectified at source, and the polluter pays principle);
- the practices and standards incorporated in EU secondary environmental legislation.

Through the common EPE approach, the MFI's aim is to promote responsible stewardship and provide a consistent and visible mechanism for engaging with project sponsors in addressing environmental issues. It is hoped that the EPE will allow the MFIs to better manage credit and project risk related to the environment.

The EPE cover at least the respective regions of operations of each signatory institution.

In the Member States of the EU, the EEA countries, the EU Acceding, Candidate and potential Candidate Countries, the MFI's agree to provide financing to public or private sponsors of projects only where the projects comply with the principles and the relevant secondary EC legislation. Of the EC secondary legislation, particular emphasis is given to the EC directives related to environmental assessment, to industrial production, water and waste management, air and soil pollution, occupational health and safety, and the protection of nature, where these can be applied to specific projects. Projects in this region should also comply with any obligations and standards enshrined in relevant Multilateral Environmental Agreements ('MEAs').

In all other countries, projects financed by the MFI's are expected to comply with the appropriate EU environmental principles, practices and standards – and with regard to EU financing, due respect for the European Neighbourhood Policy and the EU policy towards Russia –, subject to local conditions. In such financing, the MFI's agree:

> '*to apply the EPE, with reference to such factors as the costs of application, the local conditions that prevail and the time frame for the phased application for implementing the EPE.*'

In other words, local conditions, including economic and other reasons, can serve as a reason not to apply the EPE in full. The MFI's are expected to build up the capacity in these countries to meet and enforce the requirements of the EPE.

In the case of co-financed projects, the MFI's promise to work together in order to agree a common approach to the project, where possible, based on or consistent with the EPE and its methods of implementation as outlined above. Finally, the MFI's stress in the EPE Declaration that, as part of the EPE, they aim to promote best EU practice in the fields of environmental management, transparency, public consultation and reporting.[19]

---

[18] The Council of Europe Development Bank ('CEB'), EBRD, the European Investment Bank ('EIB'), the Nordic Environment Finance Corporation ('NEFCO') and the Nordic Investment Bank ('NIB').

[19] The EC 'Environmental Management and Audit Scheme' ('EMAS') is specifically mentioned as an example of such best practice.

The European Principles for the Environment are also seen as the basis of further harmonisation of approaches among the signatory parties, a magnet for additional parties to endorse the same approach and a public affirmation of the credentials of the EU in the field of environmental management.

EBRD 2008 Environmental and Social Policy and Performance Requirements

The EBRD's 2008 Environmental and Social Policy and Performance Requirements[20] were approved on 12 May 2008 (the 'EBRD 2008 Environmental Policy') and apply to new projects appraised after 12 November 2008. The EBRD's earlier 2003 Environmental Policy still applies to projects that were appraised prior to that date. The EBRD 2008 Environmental Policy describes how the Bank is to manage the impact of its operations on both people and the environment, and, thus, covers both the environmental and social dimensions of sustainable development.[21]

The EBRD 2008 Environmental Policy details the environmental and social standards applicable to each EBRD project and describes the mechanisms used to appraise, monitor and report on a project's environmental and social performance. EBRD-financed projects need to meet 'good international practice' related to sustainable development. To help clients and/or their projects achieve this, performance requirements ('PR's') were designed, modelled after those adopted by the International Finance Corporation (IFC) in 2006. The PR's encompass:

PR 1: Environmental and Social Appraisal and Management
PR 2: Labour and Working Conditions
PR 3: Pollution Prevention and Abatement
PR 4: Community Health, Safety and Security
PR 5: Land Acquisition, Involuntary Resettlement and Economic Displacement
PR 6: Biodiversity Conservation and Sustainable Natural Resource Management
PR 7: Indigenous Peoples
PR 8: Cultural Heritage
PR 9: Financial Intermediaries
PR 10: Information Disclosure and Stakeholder Engagement.

In PR 1, it is set out that the client is to consider in an integrated manner the potential environmental and social issues and impacts associated with the proposed project through appraisal activities, such as risk assessment, auditing, or environmental and social impact assessment. PR1 states that this appraisal should:

---

[20] <http://www.ebrd.com/about/policies/enviro/policy/2008policy.pdf>.

[21] The social dimension of the EBRD 2008 Environmental Policy, which to a certain extend already was acknowledged under the 2003 Environmental Policy, now encompasses (i) labour standards and working conditions including occupational health and safety and (ii) community impacts such as public health, safety and security, gender equality, impacts on indigenous peoples and cultural heritage, involuntary resettlement, and affordability of basic services.

> 'identify applicable laws and regulations of the jurisdictions in which the project operates that pertain to environmental and social matters, including those laws implementing host country obligations under international law.'

Examples mentioned are the Aarhus Convention, the Espoo Convention and the biodiversity conventions (including the 1992 Convention on Biological Diversity). It is explained that projects located in EU Member States are required to comply with the EU EIA Directive.[22] It can be noted that here, contrary to the EPE Declaration, the application of EU law to non-EU projects is not mentioned. PR 1 does set out that the client, taking into account the findings of the appraisal and the result of consultation with stakeholders, is to develop and implement a programme of mitigation and performance improvement measures and actions that address the identified social and environmental issues, impacts and opportunities in the form of an Environmental and Social Action Plan ('ESAP').

In PR 3, it is explained that pollution prevention and abatement are key ingredients of a sustainable development agenda and that EBRD-financed projects must meet good international practice in this regard. Also, the fact that the EBRD is committed to the EPE is stressed, and that, therefore:

> 'projects will be designed to comply with relevant EU environmental requirements as well as with applicable national law, and will be operated in accordance with these laws and requirements.'

For EU (potential) candidate countries, the time-frame to bring about compliance with EU legislation can be used. For projects in non-EU members and (potential) candidate countries, the ESAP is to set the time frame for achieving compliance with EU environmental requirements consistent with any bilateral agreements or action plans agreed between the EU and the relevant country. It is explained that the cost of application and the local conditions that prevail may be taken into account. This implies that compliance with EU law can be put on the back burner or set aside, in line with the EPE Declaration discussed above.

PR 3 notes further that where EU environmental requirements do not exist, the client will apply other good international practice, such as the World Bank Group Environmental Health and Safety Guidelines. In such cases the Bank will agree the applicable requirements with the client on a project by project basis. When such requirements or EU environmental requirements differ from the levels and measures presented in host country regulations pursuant to projects, the client is expected to meet whichever is more stringent.

EBRD complaints mechanism and the 2003 Environmental Policy

Complaints about Bank-financed projects are dealt with by the Independent Recourse Mechanism ('IRM'). This gives local groups and individuals that may be directly and

[22] Environmental and Social Policy 2008, p. 17, footnote 4.

adversely affected by a Bank project a means of raising complaints or grievances with the Bank. In the next section, the Calypso case will be discussed in which the final decision to grant a project loan had not been taken yet, but already a complaint was dealt with under the IRM. Here the compliance review regarding the Albanian Vlorë thermal power generation project (the '*Vlorë* Case')[23] will be discussed as it brought up some interesting points regarding the application of the 2003 Environmental Policy.

According to the recommendations of the Compliance Review Report in the *Vlorë* Case, the fact that the EBRD's 2003 Environmental Policy states that sustainable development ranks among the highest priorities of its activities does not mean that EBRD projects can be tested directly against the concept. Instead, a determination as to whether a Bank action or failure to act in respect of the project amounts to a violation of the objective of sustainable development, is necessarily bound up with, and dependent upon, an examination of the adequacy of the EIA carried out.[24] A similar conclusion was reached where the alleged violation of the precautionary principle was concerned:

> 'The question (…) concerning violation of the precautionary approach required under the EBRD Environmental Policy can only be addressed by determining whether the EIA process was adequately carried out having regard to the requirements of the precautionary principle.'[25]

However, the only material violation found with respect to the *Vlorë* Case was the fact that before deciding whether to lend to the client, the Bank had failed to take appropriate steps in order to satisfy itself that the project sponsor had given people potentially affected by the project the opportunity to express views on the location of the project (in accordance with Bank policy and with national legal requirements).

The same Compliance Review Report recommended, *inter alia*:

> '(iii) Having regard to the requirement under the Environmental Policy (…) to meet applicable national environmental law, it is recommended that EBRD Country Strategies address the capacity of the particular country to meet the requirements of the Aarhus and Espoo Conventions and of relevant EU Environmental Law'

and:

> '(iv) When addressing legislative, regulatory and policy considerations during the preparation of an EIA; it is recommended that the EBRD should consider obtaining independent legal advice with respect to the applicability, and adequacy, of national compliance with international legal obligations.'

The EBRD, however, dismissed these recommendations.[26] Where point (iii) was concerned, it noted that the Country Strategies contain a list of environmental agreements

---

[23] Document of the EBRD, Compliance review report relating to the Vlorë thermal power generation project, 17 April 2008 (see <http://www.ebrd.com/about/integrity/irm/0701crr.pdf>).

[24] Ibid., point 13, p. 9.

[25] Ibid., point 53, p. 19.

[26] Ibid., p. 38.

to which the country is party, and for potential EU countries a general statement on the agreements with the EU and progress reports on implementation from EU agencies. Furthermore, they stated that:

> '[i]n order to independently assess the ability of a country to meet requirements under any international environmental law would be onerous and beyond the existing capacity of the Bank. In our opinion this is beyond the scope and objective of the Bank's country strategies.'

Also, significant resources would be needed to assess all countries' capacity to implement international legal agreements every three years. In the view of the management, '[t]his would not be practical.'[27] As for point (iv), it was noted that the 2003 Environmental Policy does not require EBRD to conduct extensive legal due diligence in each project whether or not a country has duly complied with its international legal obligations, but rather aims at ensuring that the Bank does not finance a project that on the face of it clearly contravenes any international legal obligations. It was stated that, in practice, the Bank focuses on sponsor compliance with requirements under relevant EBRD policies and with EU and national legal requirements, as identified during due diligence. On a case-by-case basis, depending on the complexity of the project issues and regulatory requirements, the Bank has in the past occasionally sought legal advice on specific issues. (An example of this is the Sakhalin II project discussed below.) Where possible gaps in process or requirements were identified during due diligence, these will also be brought to the management and EBRD Board's attention as potential risk factors associated with the project.

EBRD projects in Russia

Russian democracy and pluralism

Whereas most of the countries in which the Bank is active, are committed to and are applying the principles of multi-party democracy and pluralism, the 2008 EBRD Annual report noted that in the Commonwealth of Independent States some set-backs in democratic reforms could be noticed. One example mentioned was the fact that an OSCE mission to monitor national elections in Russia had been made impossible.[28] The outgoing Head of the European Commission Delegation to the Russian Federation, Mr. Marc Franco, was more frank about the general developments in Russia. During the last 5 years the relationship with Moscow had suffered '*standstill, if not a regression*'. He also explained:

> 'that you cannot have rule of law … without the basic elements of democracy, implying free elections and a vibrant civil society supported by a free press.'

---

[27] Ibid., under III, p. 39.
[28] Annual Report 2008, p. 24.

The diplomat felt comfortable making his observations because they were in line with what President Medvedev himself says.[29]

Indeed, President Medvedev has made the rule of law, the fight against corruption and the pursuit of a more open political system priorities for his presidency. One example of his good intentions came to light in April 2009, when he stated his preparedness to amend legislation regulating non-profit and non-governmental organizations ('NGOs'), because of 'a mass of cases where activities of NGOs have been restricted without sufficient reason.'[30] The law 'On Public Associations' as amended under Putin in 2006, did mean that work became more and more difficult for NGO's in Russia, or in some cases even impossible.

Meanwhile, the EBRD only expressed concern about the lack of commitment to multi-party democracy and political pluralism in Belarus, Turkmenistan and Uzbekistan and decided to exclude funding of public sector projects there.[31] The EBRD continues to work with Russia and in that way can keep up its efforts to influence developments in Russia. The below describes some examples of ways in which the environmental policy of the EBRD takes shape in Russian practice.

Forestry waste to heat remote Siberian settlements

A project set up under the EBRD 2008 Environmental Policy is the overhaul of the district heating system in a number of remote small municipalities in the Khanty-Mansi Autonomous District in Siberia. This district has a population of 1.5 million, a quarter of which lives in small settlements which are not connected to the centralised systems used to supply heat and electricity in large parts of Russia. Under this project, the EBRD is providing 1.3 billion roubles (approximately €29 million) under a 13-year loan. The project aims to remove inefficient diesel and oil-fired generators and to replace these with more environmentally sound energy sources. This will be achieved through the use of sawdust and other waste from the timber industry in order to provide communal heating, so reducing pollution in a cost-effective manner. Part of the generators will be replaced by combined heat and power plant, which use biomass fuel, others by combined heat and power boilers fired by both gas and associated gas (which is at present largely being flared in oilfields). The loan will also cover the cost of building a factory to produce pellets and briquettes to feed these biomass-fired heat and power plants, using forestry waste cast off by local timber companies, especially sawdust. This will reduce the costs of both electricity and heat production.

The plan, to be implemented by the Open Joint Stock Company 'Yugra Territorial Energy Company' ('YUTEK', wholly owned by the Khanty-Mansi Autonomous District), also foresees tapping heat recovery techniques to increase fuel efficiency. This includes feeding into the district heating system the waste heat produced by the engines and turbines generating electricity at YUTEK's power plants.

---

29 'EU Ambassador says Russia at Risk', *Moscow Times*, 27 July 2009.

30 <http://www.mosnews.com/world/2009/04/15/newngolegislation>.

31 Annual Report 2008, p. 24.

The project aims to provide quality heating at affordable prices for people living in remote areas. The EBRD sees this as an energy-efficient model which other local government bodies in Russia can copy in order to put district heating on a sustainable basis and at the same time contribute to a cleaner environment.

A feasibility study with the assistance of international consultants paid by the government of Japan worked out a priority investment programme to be implemented in a total of 14 settlements. This is the sixth municipal infrastructure project the EBRD has signed in the Khanty-Mansi Autonomous District.

The Project is to be designed, constructed and operated to meet national legal requirements and relevant EU standards (Best Available Techniques or 'BAT') and good practice on health and safety requirements.

Sakhalin II

The island Sakhalin is located off Russia's far eastern coast. It stretches almost 1,000 kilometres from North to South and covers an area of 90,000 km^2. It is inhabited by some 600,000 people and home to 170 ethnic groups, of which four are native to Sakhalin. The environment is particularly rich, making Sakhalin the third largest producer of fish products in the Russian Far East. A unique population of some 130 Western Pacific Grey Whales (of which only 25-30 are reproductive females) feeds on Sakhalin's north-eastern coast during the summer. This type of grey whale is critically endangered of becoming extinct according to the IUCN Red List.[32] The Convention on International Trade in Endangered Species of Wild Fauna and Flora ('CITES') also lists these whales as a species threatened with extinction.[33]

Sakhalin II is one of the world's largest oil and gas development projects, worth some USD 20 billion. The project encompasses the construction of two very large offshore platforms, both linked to the shore by pipelines in order to facilitate year-round production, an onshore processing facility for the treatment of gas from one of the platforms, two 800 kilometres long onshore pipelines crossing numerous rivers to transport the treated gas and the oil, a new liquefied natural gas (LNG) plant and an oil and LNG export terminal. The project also foresees developing public services on Sakhalin Island, e.g., the construction of a hospital and three schools, and improving public infrastructure like roads and railways.[34]

In 2001, the shareholders of Sakhalin Energy – British-Dutch Shell owning 55%, and the Japanese companies Mitsui and Mitsubishi controlling 25% and 20% respectively – asked the EBRD to partially finance Sakhalin II. Reportedly, a loan of several

---

[32] See for exact information the IUCN Red List at <http://www.iucnredlist.org/details/8099/0>.

[33] The CITES Convention is implemented in the EU through Council Regulation (EC) No. 338/97 of 9 December 1996 on the protection of species of wild fauna and flora by regulating trade therein, *OJ* 1997, L 61, p. 1 as amended most recently by Commission Regulation (EC) No. 407/2009 of 14 May 2009, *OJ* 2009, L 123, p. 5.

[34] Sakhalin Energy Investment Company, *Executive summary of the Phase 2 environmental and social impact assessment process*, November 2006, p. 9.

hundreds of million dollars was requested.[35] Elaborate collaborative work with Sakhalin Energy was initiated to ensure the project could meet the expectations and, especially, the applicable environmental standards of EBRD.

In 2003, an EIA was produced by the company that was approved by the then Russian Ministry of Natural Resources. At EBRD's request, addenda to the EIA were issued by the company in order to bring it up to standard with best international practice. In December 2005, a draft Health, Safety, Environment and Social Plan was issued by Sakhalin Energy, in which various commitments to action in terms of mitigation measures to minimise or prevent impact to the environment and also to monitor measures taken were laid down. The various standards are set out in an annex to the action plan. Some of them are derived from the World Bank group, some from EU legislation. According to a Bank's spokesperson:

> '[c]ollectively it is a body of standards, benchmarks, safeguards, that you would expect to see on an oil and gas project in the world today. It is what you would consider to be best international practice.'[36]

The document would become a part of the loan agreement and make the company contractually bound to comply with its provisions, if the EBRD would decide to approve its co-financing of the project.[37]

The document would, thus, become part of the leverage that the Bank ultimately has over the company. If a loan is approved by the credit committee of the EBRD Board, there is a mechanism in place for compliance. Furthermore, if the company does not comply with the terms of the environmental plan as incorporated in the loan agreement, the loan could be declared due and payable.

Furthermore, even before the Bank decides to grant a loan, the IRM (discussed in the previous section) can be available to stakeholders. This is the case in situations where EBRD involvement is outstanding or may reasonably be expected, and a clear indication of the Bank's interest in financing a project has been provided. It was in fact invoked in the *Calypso* case on 26 July 2005 by a group of local individuals and enterprises involved in fishing and fish processing – they claimed that the Sakhalin II project was adversely affecting them and that their claims made in May 2004 for compensation from the company were not being dealt with adequately.[38] The affected group had brought forward *prima facie* evidence that the Sakhalin II project has or is likely to have a direct adverse and material effect on their interests, and that all other criteria to bring the case to the EBRD's mechanism were met. On 20 September 2005, the recommendation to deem the complaints eligible for further processing was taken. Thereupon, the company engaged in settlement negotiations with three of the affected parties

[35] Figures cited in the press ranged between USD 300 (The Guardian, 16 August 2007) and USD 500 million (Dow Jones Newswires 20 September 2006).

[36] EBRD, Sakhalin public consultation in London transcript, 28 February 2006, p. 12.

[37] Ibid., p. 2.

[38] Independent Recourse Mechanism, Eligibility Assessment Report, Complaint Sakhalin 2 Phase 2 Calypso.

in such a way that the Chief Compliance Officer reported that an 'effective dialogue' existed. This meant that issuing a so-called 'Problem-solving Initiative Report' in the proceedings was deemed to be inappropriate and, therefore, delayed pending the outcome of settlement negotiations. On 23 March 2006, settlement was reached with one of the parties (i.e., with Calypso Ltd.), and negotiations with the remainder of the affected group continued until the end of 2006. Before revealing how the case ended, the below returns to EBRD's involvement with Sakhalin II.

After considering investing in the Sakhalin II project over a period of five years, in December 2005 the Bank declared that the project met sufficient requirements for it to seek the views of the public.[39] An intensive consultation process was convened in Russia, Japan and London. WWF brought forward that the project was not in compliance with EBRD policies and never could comply with those, and, therefore, was not eligible for public funding. Cases of non-implementation of mitigation measures in practice were mentioned, notably where the disturbance of whales was concerned.[40] Compliance with Russian laws and decisions of various regulatory bodies as confirmed by certain Russian organisations were criticised because of their biased position and/or corruption.[41] Moreover, it was argued that the company's documents and project implementation contained a large number of breaches of Russian law, and the absence of an analysis of Russian legislation and how the company intended to deal with the prohibitions and restrictions on its activities imposed by Russian law and international conventions was noted.[42] EBRD explained that it had hired independent lawyers from the Moscow office of White & Case to look into the Russian legal aspects of the project.

Over the course of 2006, pending a decision to invest or not, the EBRD continued to monitor construction and encouraged the adoption of long-term safeguards, especially related to environmental and social aspects of the project. Also, the *Calypso* case continued. The Sakhalin Energy Investment Company itself was of the opinion that it was complying with pertinent EU environmental standards, 'insofar as these can be applied to a specific project' and where these standards 'do not unduly compromise national legislation'. It also assured that the standards it applies are 'in general equivalent to applicable World Bank Group environmental safeguard policies and guidelines'.[43]

According to a number of NGO's, in violation of EBRD environmental policy, there was scant collection of baseline environmental data on many elements of the Sakhalin II project prior to the start of construction work. The absence of this baseline data precludes the possibility of subsequent monitoring to determine the full extent of harmful impacts over time.[44] While some claimed that EBRD involvement would mean that

---

[39] <http://www.ebrd.com/new/pressrel/2005/175dec14.htm>. Reportedly, this decision had been postponed in June 2005, due to environmental concerns. 'People familiar with the delay say it was worried by the predicament of the whales, the impact on local inhabitants and the risk of pollution in river crossings used by salmon for spawning.' <http://www.eca-watch.org/problems/eu_russ/russia/sakhalin/WSJ_Sakhalin_11oct05.htm>.

[40] EBRD, Sakhalin pubic consultation in Moscow transcript, 14 March 2006, p. 18 ff.

[41] Ibid., p. 22.

[42] Ibid., p. 29 and 30.

[43] Sakhalin Energy Investment Company, *Executive summary of the Phase 2 environmental and social impact assessment process*, November 2006, p. 29.

[44] <http://www.ebrd.com/country/sector/natural/projects/sakhalin/corres/060905.pdf>.

the protection of the environment and local inhabitants could be better protected, others claim that EBRD investments would mean that a fundamentally flawed project would get the blessing of the EBRD.[45]

Amidst this heated debate and against the background of allegations of violations of Russian environmental law and the announcement of a major increase in the costs of getting the project done,[46] on 21 December 2006, a major change in the ownership of the Sakhalin Energy Investment Company ('SEIC') came about. Shell was forced to sell its majority stake to Gazprom (now owning 51% in the project). Shell's share fell from 55% to 27.5%, Mitsui's holding dropped from 25% to 12.5% and Mitsubishi's from 20% to 10%. The original Production Sharing Agreement ('PSA') from 1994 was rather unfavourable to Russia. Under the PSA, Shell could recover all project costs before it began to share profits with the Russian government. Thus, in itself the Russian desire to renegotiate a new deal was understandable. The way in which Shell was forced to accept the new deal – notably under threats of having to stop work on the project because of violations of environmental law – violations that by March 2007 were not that serious any more[47] – certainly did not contribute to Russia's reputation as concerning respect for the rule of law.[48]

The result was that the EBRD announced on 11 January 2007 that following this significant change in ownership, it would no longer consider the financing package of the Sakhalin II project. In its official statement, environmental concerns are not mentioned as a reason for this step.[49] EBRD's statement does explain that pending a decision on whether to finance, the EBRD 'has continued to monitor construction and

[45] Richard G. Lugar, *European Bank for Reconstruction and Development (EBRD)*, London; A report to Members of the Committee on Foreign Relations, United States Senate, 109th Congress, Second Session, 16 June 2006.

[46] Mr. Ryjenko from the Banking Department of the EBRD had stated at the 28 February 2006 public consultation in London that 'the company is facilitating that process and providing information on the costs, how and where they were incurred; and the Russian authorities are looking into it with a view to taking a decision on what their opinion would be. (…) Of course there were expressed concerns and surprise and the need for consideration of this matter (...). In this context, obviously there are the concerns of President Putin, which were reported from his trip to the Netherlands, and the Russian authorities are considering that issue.' EBRD, *Transcript of public consultation meeting in London on the environmental and social issues associated with the Sakhalin II Phase 2 project*, p. 29. <http://www.ebrd.com/country/sector/natural/projects/sakhalin/consult/london.pdf>.

[47] According to the Russian newspaper *Kommersant* of 29 March 2007, the Deputy Head of the Federal Resource Management Agency Oleg Mitvol announced that his agency will not pursue legal actions any more, and that Sakhalin Energy is willing to solve the environmental problems. <http://www.kommersant.com/p-10421/Sakhalin_Energy_Gazprom/>.

[48] Mr. Ryjenko (see *supra* n. 46) had also explained on 28 February 2006 that renegotiating the PSO 'certainly would be very unfortunate for the stability of the investment climate and perception of Russia as a state and how it wants to present itself to the international community'. (ibid., p. 29). Ksenia Manyushis from the EBRD Finance department put it as follows: 'As for reviewing the terms, a country that claims to be ruled by law does not usually review documents signed 10 years previously and implemented in the form of a law. (…) [T]hat would damage the country's present reputation.' EBRD, *Transcript of public consultation meeting in Korsakov on the Sakhalin II Phase 2 project*, p. 4. <http://www.ebrd.com/country/sector/natural/projects/sakhalin/consult/korsakov.pdf>.

[49] EBRD Press release, *EBRD no longer considers current financing package f[illegible]halin II*, 11 January 2007, <http://www.ebrd.com/new/pressrel/2007/070111.htm>.

encouraged the adoption of long-term safeguards, especially related to environmental and social aspects of the USD 20 billion project.'

The above exemplifies that the Bank indeed played a positive role in terms of the protection of the environment and social aspects of the project, without being able to solve all problems. Notably, together with NGO campaigns, it influenced the decision to re-route pipelines away from a path which could affect Western grey whales. Also, the way in which river crossings were handled seems to have improved. The settlement in favour of Calypso Ltd.'s claim probably benefitted from EBRD's Independent Recourse Mechanism, though the claims of the other affected parties in that case remained unsolved. Following the decision not to finance, on 17 January 2007, the IRM announced that it no longer had jurisdiction to process the hitherto eligible complaint and finalised it. IRM did make known that the possibility remained to reinstate the complaint if EBRD decided at some future date to finance the project. That did not happen. Neither the latest annual reports from the company nor its website[50] reveal whether the remaining affected parties were able to reach a settlement after EBRD's involvement fell away.

Without further EBRD involvement, Sakhalin II became operational in December 2008, after the company withdrew its applications for financing from the public export credit agencies of the United States and the United Kingdom in March 2008,[51] but having received financial aid from the Japan Bank for International Cooperation ('JBIC') and a consortium of international banks in June 2008 worth USD 5.3 billion. The project is expected to produce around 9.6 million tonnes of LNG annually, of which more than 50% will be delivered to Japan. With regard to crude oil, around 150,000 barrels will be produced per day (equivalent to some 4% of Japan's total oil imports) under the project, of which a substantial percentage will be supplied to Japan. To quote the Hokkaido Fisheries Environmental Centre on this deal: 'We have high hopes that JBIC will pay sufficient attention to our concerns, provide guidance and supervision to Sakhalin Energy.'[52]

As for the whales, in spite of the instalment of the IUCN Western Gray Whale Advisory Panel ('WGWAP') in 2006,[53] consisting of marine scientists who provide expert analysis and advice concerning impacts on the whale population from the Sakhalin-I and II projects, the number of western grey whales observed in the near-shore (primary) feeding area significantly declined in the summer of 2008. The WGWAP issued an urgent call for a 'moratorium on all industrial activities, both maritime and terrestrial, that have the potential to disturb grey whales in summer and autumn on and

[50] <www.sakhalinenergy.com>. The website does mention a payment of USD 11 million to the authorities for potential estimated impacts on the fishing industry and the Calypso Ltd settlement.

[51] The company cited drawn-out and uncertain timetables for its decision. Environmental groups claimed that the decision was a success for their campaign against the project. However, a spokesman for the British ECGD, said that 'From ECGD's perspective, we had not finished our due diligence on the project. There were still issues we were looking to address and consequently we were not … in the position to make a judgement.' <http://www.environmental-finance.com/onlinews/0306sak.html>.

[52] <http://www.jbic.go.jp/en/about/press/2008/0616-01/>.

[53] <http://www.iucn.org/wgwap>.

near their main feeding areas.'[54] This call was repeated in April 2009 and lead to Sakhalin Energy to suspend its planned summer 2009 seismic testing.[55]

Concluding remarks

EBRD provides financing for numerous projects in the Russian Federation that contribute to the country's economic development while taking account of the environmental and social aspects of the projects. The Siberian district heating system forms a clear example of an EBRD project aimed directly at sustainable development and energy efficiency. While it has been pointed out by critics that financing non-renewable energy projects like Sakhalin II does not contribute to the country's sustainable development *per se*, the involvement of the Bank (as discussed in this chapter) does show that EBRD can contribute considerably to the environmental (and social) aspects of a project which was going to go ahead with or without EBRD's financing.

EBRD clients only become eligible for financing on the condition that they abide by national law and European and/or international standards after an extensive process involving the public concerned, the publication of an EIA (in cases where the environment is potentially seriously affected) and the setting up of an Environmental and Social Action Plan. In principle, not only Russian, but also EU environmental standards (including the concept of BAT) should be applied when EBRD finances projects in Russia. If there are differences in such standards, the highest standard is to prevail. This *ad hoc* 'bottom-up' approach to 'approximating' Russia's environmental standards to EU standards could bring about instant results, contrary to the 'long march' of one-sidedly adjusting Russia's environmental legislation to that of the EU.[56]

However, in non-EU, non-(potential) candidate countries, like Russia, an important caveat applies where the application of EU standards is concerned. EU standards can be left aside where this is considered necessary due to local applicable circumstances. Fortunately, it seems that this caveat is not invoked where procedural requirements such as the EIA are concerned. In this manner, the strict EU approach is followed also in Russia and potential weaknesses of the Russian EIA-system can be overcome.

Furthermore, EBRD practice shows that its Independent Recourse Mechanism offers opportunities – even before projects are actually financed – to influence and promote responsible behaviour of clients where affected parties have complaints about projects.

One interesting point that remains in context of EBRD's practice in Russia is how EBRD (and EBRD 2008 Environmental Policy) deals with situations where it is contested that a project abides with applicable national and international law norms, as

---

[54] See section 17 at <http://cmsdata.iucn.org/downloads/wgwap_5_report_final_040209.pdf>.

[55] <http://www.iucn.org/wgwap/news_archive/?3079/IUCN-welcomes-reprieve-for-whales>.

[56] See chapter 6 (*Hahn and Begak*) which discusses the prospects of harmonising Russian environmental standards with EU legislation. Their project follows up on the TACIS project 'Harmonisation of environmental standards, Russia' (July 2002-January 2004), which among other matters, resulted in a gap analysis of EU and Russian environmental legislation.

was the case in the Sakhalin II. This point – also raised when the 2008 Environmental Policy was taking shape – probably needs to be reviewed when the EBRD 2008 Environmental Policy is up for revision in a couple of years time.

In sum, the way in which the EBRD operates in Russia through its EBRD 2008 Environmental Policy and the more concrete performance requirements, coupled with the public information policy and recourse mechanism, and the EPE Declaration all contribute to introducing higher standards in Russian practice. Where high standards already exist in Russia and/or when these are higher than EU standards, the Russian standards can and must be applied. Where EU standards turn out to be higher, in principle these are to be applied. In other words, the system under the EBRD 2008 Environmental Policy allows for a two-sided approximation and optimal protection of the environment. Nevertheless, making the system operational in Russian practice remains a formidable challenge for the EBRD.

Part III

Practical Trends and Synergies

A. Business, NGOs and environmental projects

Chapter 13
INVOLVING RUSSIAN BUSINESS IN KYOTO

Michael Yulkin[1]

In this chapter, the importance of Russian (and foreign) business participation in the success of the 1997 Kyoto Protocol to the United Nations Framework Convention on Climate Change (the 'Kyoto Protocol' or 'Protocol') is discussed. The Russian regulations and remaining issues and questions are reviewed, also in context of the new global climate change regime which is to be negotiated later this year. Furthermore, possible ways in which to control greenhouse gas ('GHG') emissions in Russia are considered, including the introduction of technical performance standards and GHG emissions permits. Finally, the ways in which Russian businesses are participating in the climate change mitigation process are explored in more detail.

INTRODUCTION

The Kyoto Protocol entered into force on 16 February 2005. The Protocol establishes limits on GHG emissions for the period from 2008 through 2012, and provides flexible economic mechanisms for the minimization of compliance costs. Altogether, over 180 countries are now party to the Kyoto Protocol, including 23 developed countries and 12 countries with economies in transition listed in Annex B to the Kyoto Protocol who have accepted quantitative commitments on GHG emissions limitation and reduction. In 2006, Belarus submitted an official application to join Annex B of the Kyoto Protocol with a commitment to reduce GHG emission by 8% below 1990 level. Kazakhstan is said to do the same soon.

The success of the Kyoto Protocol, and the ability of the Parties to meet its goals to limit and reduce GHG emissions will crucially depend upon the involvement of and the position taken by the business community and, in particular, the largest company-emitters of GHGs. Business participation in the 'Kyoto' process is twofold. On the one hand, companies are expected to limit and reduce their own emissions. On the other hand, they are encouraged to participate in the flexibility mechanisms of the Kyoto Protocol: in emissions trading ('ET') and, as investors, in joint implementation ('JI') and Clean Development Mechanism ('CDM') projects.

---

[1] Director General, Climate Change Global Services (CCGS); Email: *yulkin.ma@gmail.com*.

W.Th. Douma and F.M. Mucklow, eds., Environmental Finance and Socially Responsible Business in Russia
© 2010, T·M·C·ASSER PRESS, The Hague, The Netherlands and the Authors

KYOTO – IMPLEMENTATION MODELS

EU and USA

EU countries have made considerable progress towards the implementation of the Kyoto Protocol. The first (pilot) phase of the European emission trading scheme ('EU ETS') entered into force on 1st January 2005. The second phase started on 1st January 2008, and the third phase, which is to be launched on 1st January 2013, is now being designed. Within the framework of EU ETS, the main company-emitters in the different sectors of the EU economy received quotas (overall permits) relating to their GHG emissions, in the form of tradable emission allowances ('EUAs'). These emitters are now allowed to sell and buy EUAs on the EU ETS market. They may also invest in emission reduction projects (either JI or CDM) in other countries, and count emission reductions achieved in such projects towards their obligations. In 2007, the turnover on EU ETS market achieved 3,100 MtCO_2e and EUR 67 billion.

Similar schemes are being implemented even in countries, which have not yet ratified the Kyoto Protocol. For example, in the USA, 10 Northeastern and Mid-Atlantic states (Connecticut, Delaware, Maine, New Hampshire, New Jersey, New York, Vermont, Massachusetts, Rhode Island and Maryland) launched the so called Regional Greenhouse Gases Initiative ('RGGI') and the first 'climate stock exchange' was opened in Chicago in 2003. RGGI sets a cap on emissions of carbon dioxide (CO_2) from power plants, and allows sources to trade emissions allowances. The programme will begin by capping emissions at current levels in 2009, and then reducing emissions by 10% by 2019. Now, with newly elected President Obama in office, the USA is seriously considering a nationwide cap-and-trade scheme as an instrument to control and reduce major GHG emissions in order to achieve its climate change mitigation targets set forth in the UNFCCC.

Russia

Unfortunately, Russia lags far behind in the implementation process relating to the Kyoto Protocol. One of the reasons why no special policy and measures are envisaged is that, according to all existing projections, Russia will most certainly not exceed its GHG emission limit (or Assigned Amount) established under Annex B of the Kyoto Protocol. Indeed, having an annual GHG emission quota of 3,323 MtCO_2e, Russia today actually emits about 2,200 MtCO_2e per year.

Based on this data, most analysts agree that Russian cumulative GHG emissions over 2008-2012 would not exceed 10,500-11,000 MtCO_2e, while the Russian 'emission budget' under the Kyoto Protocol for this period is 16,617 MtCO_2e. Thus, Russia will automatically fulfill its obligations, resulting in an emission surplus of at least 5,600-6,100 MtCO_2e (see Fig. 1). The addition of maximum allowed carbon removals by sinks due to the improvement of land-use and forestry practices increases this surplus up to 6,200-6,700 MtCO_2e.

Figure 1. GHG emissions in Russia under various scenarios

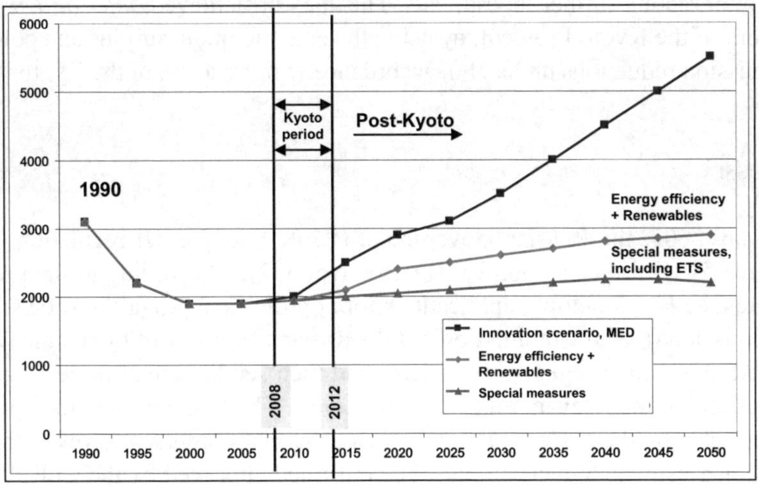

Source: Agency for energy balances forecasting, 2009

At the same time, CO_2 emissions per 1 dollar of GDP (when calculating GDP using the purchase parity power method) in Russia are 3.8 times as much as in European countries, 2.4 times as much as in USA, and 2 times as much as in Canada (see Fig. 2). In a carbon-constrained world, this may lead to material economic losses for Russia which needs to meet strong market requirements and standards in terms of carbon intensity and GHG emissions management.

Figure 2. Unit GHG emissions (emissions per 1 dollar of PPP GDP, kg CO_2e)

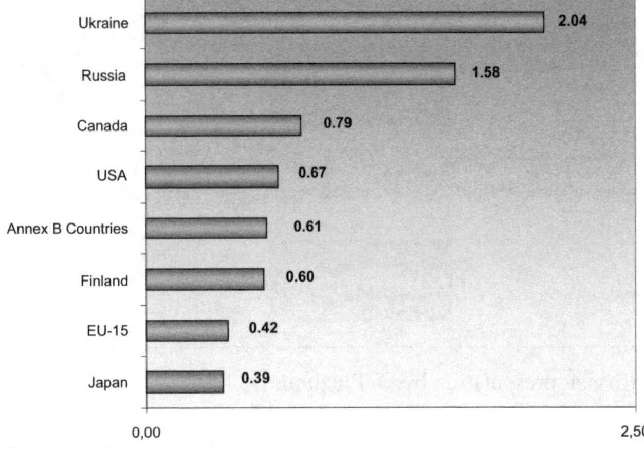

Source: UNFCCC Secretariat, 2004, World Bank, 2002.[2] Data for Russia refers to 1999, data for other countries relates to 2002

[2] These numbers were calculated by the author and his EIC colleagues, based on the UNFCCC data regarding GHG emissions and the World Bank data regarding PPP GDP.

This problem can be resolved through the modernization of the Russian economy and by way of attracting further investments. This may be achieved, *inter alia*, within the framework of the Kyoto Protocol, by using the existing quota surplus and potential of GHG emission reductions under JI in accordance with Article 6 of the Kyoto Protocol.

JI in Russia

On 28 May 2007, the Russian Government ('GoR') adopted JI regulations and the procedures for JI projects approval (Decree No. 332).[3] According to the prescribed procedure (see Fig. 3 below), applications for JI projects approval must be submitted to a commission (the 'JI Commission') at the Russian Ministry of Economic Development ('MED'), which registers JI projects and then sends each JI project 'specifications package' to the relevant ministries for review. The final decision to approve a JI project is made by the JI Commission, which consists of representatives of the 'interested' ministries, which decision must be confirmed/approved by the GoR. Whereas, emission reduction units ('ERUs') are to be issued in the Russian Carbon Registry annually on the basis of the monitoring reports that have to be submitted by JI project applicants on or before 15 February each year.

Figure 3. JI projects approval procedure in Russia

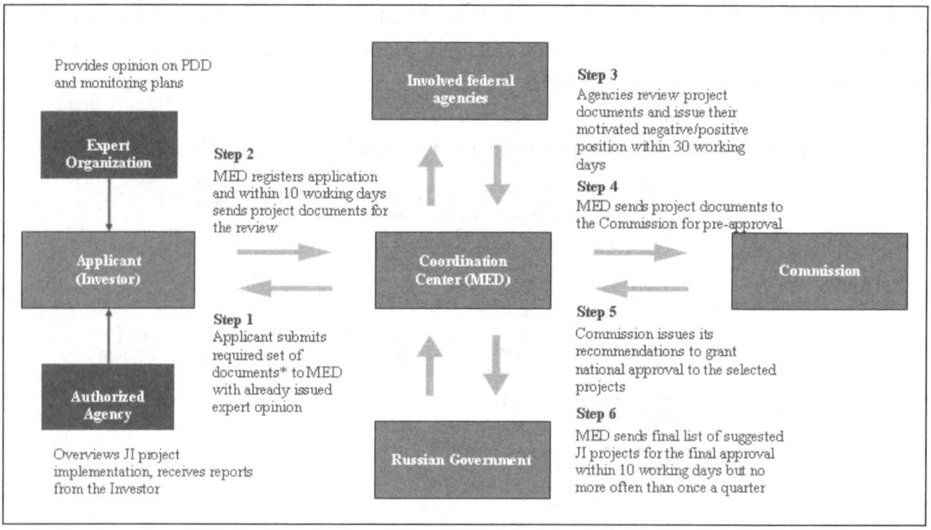

Source: JI Status in Russia, presentation by O. Pluzhnikov, 2008

The collection of JI project applications started on 10 March 2008. So far, some 40 applications have been submitted for approval to the JI Commission with a total pro-

[3] This chapter does not discuss the details of the new JI regulation signed by Prime Minister Putin on 28 October 2009.

jected volume of GHG emission reductions exceeding 100 $MtCO_2e$ over 2008-2012.[4] The biggest projects submitted for approval by Russian companies are: coal mine methane utilization in the Kuzbass region (17.9 $MtCO_2e$, by SUEK); associated petroleum gas utilization at the Komsomolskoye oil field (6.65 $MtCO_2e$, by Rosneft); HFC-23 and FS_6 co-destruction (5.25 $MtCO_2e$, by KCKK Polimer Plant); and resource and energy savings at steel production (3.2 $MtCO_2e$, by UralSteel). However, at the time of writing (November 2009) no JI projects were approved yet.

On 28 October 2009, the GoR re-approached the JI regulations once again and adopted a new regulation (Decree No. 843), which replaced the previous one. Under the new JI regulation a key role has been given to Sberbank. The procedure includes selection of JI projects through a bidding (contest) to be conducted by Sberbank by applying certain criteria and procedure which are still to be elaborated and approved by MED.

If one considers all the JI projects ever proposed under Track 1 or Track 2 by Annex B countries, Russia undoubtedly plays a key role in holding potentially as much as 58% of the market in terms of anticipated GHG emission reductions (see Fig. 4).

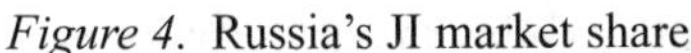

Figure 4. Russia's JI market share

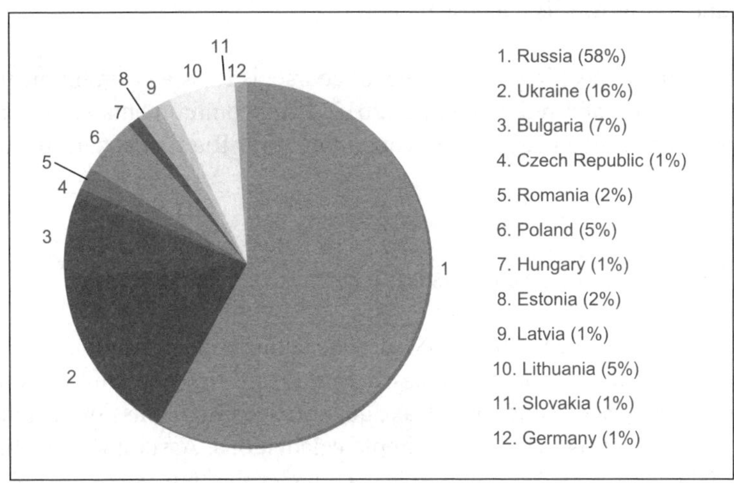

Source: FCCC Secretariat, 2009[5]

Thus, in order to become the real market leader in JI, the JI Commission should start approving JI projects without delay. In addition, the Russian JI procedures should be improved to avoid uncertainties and meet international standards

---

[4] As of May 2009.

[5] This graph was created by the author, based on the JI projects data provided on the UNFCCC website <http://ji.unfccc.int/JI_Projects/DeterAndVerif/Verification/PDD/index.html>.

In view of the post-2012 negotiations for a new international climate change regime (at COP 15) later this year, Russia should take advantage of its leading role in the world JI market in order to advocate and promote the JI mechanism in the new international treaty and to further improve it. In doing so, it should be recalled that, unlike with respect to CDM, JI is designed for countries with capped GHG emissions. Since more and more countries are expected to commit to GHG limitation and reductions in the future, the role of JI will gradually increase, and, thus, the competition in the JI market will become more pronounced, too.

In order to secure and strengthen its position in the JI market, Russia should, firstly, undertake to improve its domestic JI regulations. In particular, I would recommend the following steps to do so:

1) No caps on GHG emission reductions. Only emissions should be capped;
2) No '*efficiency criteria*' should be incorporated in JI projects approvals, except for such criteria that serves to prevent the use of old-fashioned technologies and equipment;
3) To provide for the issuance of ERUs into project owners' accounts with the Russian national registry;
4) To take into consideration the Chinese experience of applying taxes to carbon proceeds and to follow this Chinese model.

The biggest and the most important issue, of course, is Russia's commitment to GHG emissions limitation and reduction after 2012. If no commitments are taken, Russia will lose its market advantages and the opportunity for Russian businesses will dissipate.

Capping Russia's GHG emissions post 2012

At COP 14 in Poznan (December 2008), despite taking a rather cautious and uncertain position, the Russian delegation expressed with regard to GHG emissions limitation after 2012, that Russia can and should take quantitative GHG emissions commitments.

In this respect, let us make some simple calculations. Assuming that Russia will double its GDP from 2012 through to 2020,[6] under the 'business as usual' ('BaU') scenario this could lead to doubling of GHG emissions by 2020.[7]

However, the GoR has just introduced a long-term programme aimed at reducing the energy intensity of the Russian economy by 40% by 2020 and at developing carbon free renewable energy from less than 1% up to 4.5% of Russia's total energy generation and use. The doubling of GDP, while at the same time reducing its energy intensity by 40%, would mean that Russia's emissions would grow by $2 \times 0.6 = 1.2$, i.e., by only

---

[6] Which is equivalent to an average of 10.4% economic growth per year over such time period.

[7] Of course this is not quite so. Emissions are dependent upon many things that are not constant over time, including the structure of an economy, energy generation and use, etc. But for the simplicity of calculation let us assume that under the BaUscenario, emissions will grow in line with GDP.

20% from 2012 levels.[8] As renewables-based energy is carbon free, if Russia increases the proportion of energy derived from renewables to up to 4.5%, this would mean further reductions of Russian GHG emissions. With this in mind, Russia's GHG emissions would grow by only $2 \times 0.6 \times (1–0.045) = 1.146$, i.e., by only 14.6%.

Currently, Russian GHG emissions are at about 65% of their 1990 level. In 2009, they will be even less due to the global economic crisis and drop in production. Thus, let us assume that by 2012, Russia's GHG emissions will rise to 70% of their 1990 level (though I doubt this could happen). If one multiplies 70% by 1.146 growth rate, we arrive at only 80.2% of 1990 level to be achieved in the year 2020. This means that, on average, Russian GHG emissions in 2013-2020 would be less than 80% of 1990 levels.

In addition, Russia has a 'safety cushion' of between 5,600 and 6,700 $MtCO_2e$, which is the remaining Russian Assigned Amount that can potentially be transferred (from the Kyoto Protocol) over into 2013-2020 commitment period. Let us say this 'cushion' would be at least 5,600 $MtCO_2e$. Dividing this by 8 (the number of years between 2013 and 2020) we obtain an additional 700 $MtCO_2e$ for each year, which is another 21% of the 1990 GHG emissions levels.

Thus, if the GoR is really serious about its announced energy efficiency and renewables targets towards 2020, then Russia can relatively easily keep its GHG emissions at least 20% below 1990 levels. With the remaining part of the Assigned Amount recognized and transferred from the Kyoto Protocol 2008-2012 period into the new 2013-2020 period, Russia could even commit to keeping GHG emissions 40% below its 1990 levels. Whether Russia will do so, still remains to be seen.

How to best control GHG emissions?

So far, GHG emissions are not regulated or controlled in Russia at all. However, in order to stabilize GHG emissions in Russia and to decrease them (as the task towards 2050 requires), some special measures can be recommended as set out below, including the putting in place of technical performance standards and of an emissions permit (and trading) system.

Direct and indirect control

For purposes of effective GHG emissions regulation, GHG emission sources should be divided into two groups:

(i) the first group should include small and unorganized sources, such as transport and individual communal boilers; and

[8] It should be noted that there are, of course, GHG emissions not related to energy. But the share of such emissions is relatively small, i.e., less than 25%. Besides, we can assume, again for the sake of simplicity, that non-energy-related emissions will follow the dynamics.

(ii) the second group should consist of sources, which are controlled by companies and other enterprises.

GHG emissions of the first group should be regulated *indirectly*, through a system of technological performance standards applicable to individual technological units and pieces of equipment. These are considered to regulate fugitive emissions, fuel and energy consumption etc on the basis of best available technologies – and, thus, indirectly limiting GHG emissions.

The emissions sources of the second group, aside from such technical performance standards, should be *directly* capped (limited) under a kind of Russian emissions trading scheme ('RF ETS') similar to the EU ETS.

Creating RF ETS

The RF ETS could be structured in such a manner that absolute emission caps (i.e., quotas or permits as they are called in the EU ETS) are determined for each company participant and that each emission permit is then linked to a certain amount of tradable carbon units (in case of EU ETS these are called allowances or EUAs), whereby each carbon unit is accorded 1 tCO_2e nominal value. Under such a regime, carbon units can be bought, sold and/or banked (i.e., saved for use in future periods).

In context of setting up a Russian cap-and-trade regime, emissions allowances should be balanced against actual GHG emissions and ought to be cancelled out every year, after the submission of a participant's GHG emissions report. Such a report should be independently verified, and companies which submitted underestimated emission data should be fined. Such companies should cancel out additional permits, to balance their actual emissions against their emission allowances, and so comply with GHG regulation.

Within such a regime, aside from obtaining emissions permits, as mentioned above, companies could obtain additional emissions rights by way of their acquired carbon units (AAUs, RMUs, ERUs and CERs) issued in other countries. Thereby, when a company's annual emissions balance is calculated, these units are converted into local RF emission allowances and, then, immediately be cancelled out.

If a company exceeds its permitted emissions (i.e., if actual emissions exceed the number of emission allowances held by a company in its account with the carbon registry), it should be penalized. The violating company should then be obliged to provide 'double' compensation for this violation in the future commitment period, as follows: during the next commitment period, the number of emission permits issued in favour of the violating company will be automatically reduced by the exceeded emissions amount in the previous period, multiplied by two.

Therefore, the initial distribution of emission allowances/permits plays a central role in such the proposed RF ETS. The permit distribution schemes used in EU member countries (e.g., in Slovakia) can be used as an example, as follows:

1. The calculation of baseline emissions. Baseline emissions are calculated as average GHG emissions during several years, instead of just relating to 1990 or any other year;
2. The baseline level is further corrected taking into account expected growth of output of the enterprise. Thus, GHG emission limits are calculated separately for each year of the budget period;
3. The allowed emissions during each year are summed-up, on the basis that this sum equals the emission quota which is established for the entire budget period.

This scheme could be complemented by two further conditions, namely:

4. Emissions limits should be corrected in the context of sectoral 'scale factor'. This factor would reflect the relative differences which exist among unit emissions reductions in each sector of the economy, arising from economies-of-scale and as a result of investments, and
5. If actual output is slower than was projected, then the emissions quota for the next period should be corrected by way of subtracting the surplus which the enterprise acquired in the current period by overestimating its economic growth.

These two latter conditions should facilitate investments in Russia and the implementation of best available technologies, as a necessary prerequisite for economic expansion. An alternative solution for a company would always be the purchase of additional emissions permits or carbon units issued by other countries.

The RF ETS should be open to foreign companies, as this would stimulate market demand and create tangible incentives for Russian emitters. Foreign companies should be allowed to trade, and legally purchased emission permits should be freely convertable into exported carbon units (i.e., AAU and RMUs).

Other mechanisms

Aside from technical norm-setting and quotation, the GoR could provide fiscal and tariff incentives for GHG emissions reductions, the improvement of land use and forestry practices, and for investments in the framework of the flexibility mechanisms of the Kyoto Protocol.

Table 1 summarises possible methods for state regulation of GHG emissions in Russia.

Russian business taking the lead

Russian businesses are now starting to embrace the Kyoto Protocol and climate change mitigation issues. Russian companies invest in energy saving, energy efficiency, utilization of recyclable and renewable sources of energy and secondary biofuels. This results not only in the more than 100 JI projects set out on the UNFCCC website

Table 1. Methods of state regulation of GHG emissions[9]

Policies and measures	Objects of regulation and emission sources
Cap-and-trade (combination of emission permits and emission trading)	Large and medium emission sources, groups of small emission sources, which are controlled by enterprises and companies.
Technological norm-setting (performance standards)	Single units and installations, such as means of transportation (automobiles, water transport, trains, aircraft, pipelines, agricultural machinery, construction and other machines, energy-consuming and energy-generating installations).
Tariffs and taxes	Aimed at saving fuels and energy by small businesses and in the communal sector.
Land-use and forestry policies	Protection and enhancement of natural carbon sinks and reservoirs. Prevention of illegal logging, introduction of timber labeling, improvement of technological norms in land use and forestry, reform of land laws and forestry legislation, early detection and extinguishing of forest fires, afforestation, reforestation, revegetation.
Investment policy	Supporting investments in the communal sector for its technological modernization, investments in reduction of fuel consumption and losses during production and distribution of energy, investments in household waste management, recovery of associated gas and coalmine methane, investments in improved land use and forestry practices.

(which is an absolute world record!), but also in the more fundamental fact that GHG emissions in Russia are growing much slower than its GDP.

Tackling climate and GHG emissions at corporate level

The most advanced Russian companies now voluntarily count and report their annual GHG emissions using the recognized international standards. Some of them have proclaimed climate strategies and introduced GHG management systems as a part of their overall environmental policy and environmental management plan ('EMP').

The first to mention is RAO UES. This company no longer exists now (as a result of a major power sector reform undertaken recently in Russia), but before it was reorganised, as much as 77 regional energy companies embraced by RAO UES elaborated and submitted their GHG inventory reports, as well as their integral company's inventory report, which covered all sources of emissions, independently verified by

[9] Table compiled by the author in 2005.

the international auditors. RAO UES was also a Russian pioneer in implementing an overall GHG monitoring system and in submitting first JI projects.

The RAO UES example was followed by Lukoil, one of the biggest Russian oil producers. Not only has it produced a GHG emissions inventory covering 1990-2006, but it has also adopted a corporate strategy and an action plan aimed at meeting climate change mitigation requirements. In addition, corporate standards on GHG management and JI projects elaboration and submission have been introduced.

Norilsk Nickel is another good example. It produced a GHG emissions inventory for 1990-2005 relating to its polar affiliate and independently verified such emissions. This inventory was then used to estimate future GHG emissions (up to the year 2015). Now the company can calculate its GHG emissions by itself by using software specially elaborated to cover the company's specific emission sources.

Gazprom is also measuring its GHG emissions. All sources of its emissions have been identified, described and included in an overall inventory module which can be used for monitoring purposes, as well.

In 2008, Rusal was about to undertake an overall GHG inventory study that was supposed to cover all its companies and affiliates around the globe. Unfortunately, due to the economic crisis, this titanic work had to be postponed. In any event, this company also participates in some climate change mitigation activity. For example, Rusal is currently working on two huge JI projects dealing with perflourocarbons (PFCs) reduction at its Krasnoyarsk and Bratsk facilities in Russia.

A lot more Russian companies from different sectors are engaged in JI projects activity now, including Rosneft (oil and gas sector), SUEK (coal mines), Euraz (steel mills), Eurocement (construction materials), Ilim Group, Archangelsk PPM, Volga, Svetogorsk PPM, Segezha PPM (all – pulp and paper), Sawmill 25 (timber industry), TGK-4, TGK-6 (all – power sector) and many others.

It appears very much so as if the economic crisis has stimulated a whole new way of thinking among Russian managers and businessmen. Instead of using new resources and industrial assets in Russia and beyond its borders, Russian companies are now switching to more efficient uses of resources which they already have at their disposal. Thus, creating, *inter alia*, many GHG emission reductions, which are potentially convertible into cash and other economic benefits.

More than that, it also looks very much as if Russian business is taking the lead in climate change related activity in Russia and that they are leading the way. For example, a Climate Change Working Group was established at the Russian Union of Industrialists and Entrepreneurs ('RSPP') in 2008 with the aim to elaborate a reasonable strategy on how to deal with and benefit from climate change mitigation.

Archangelsk Pulp and Paper Mill

The most encouraging example in this regard is demonstrated by Archangelsk Pulp and Paper Mill ('APPM'). APPM started controlling its GHG emissions back in 2002 when the mill ran its first comprehensive inventory of GHG emissions covering the period from 1990. Since then, APPM publishes its GHG emission data annually using the model and software developed by the Environmental Investment Center.

In 2003, APPM became the first and, so far, the only Russian company which has committed to quantitative emission targets. The company pledged to keep its annual GHG emissions under 2.6 $MtCO_2e$ until 2012. This is 12% lower than its emissions in 1990. At the same time, pulping was expected to increase to 1 million tonnes per year, which is 8.5% more than in 1990 (see Fig. 5). This decision was announced by the General Director of the mill, Mr. Vladimir Beloglazov at COP 9 in Milan (Italy) on 10 December 2003.

Figure 5. Voluntary GHG Emission target of Archangelsk Pulp and Paper Mill towards 2012 (in per cent. of 1990 level)

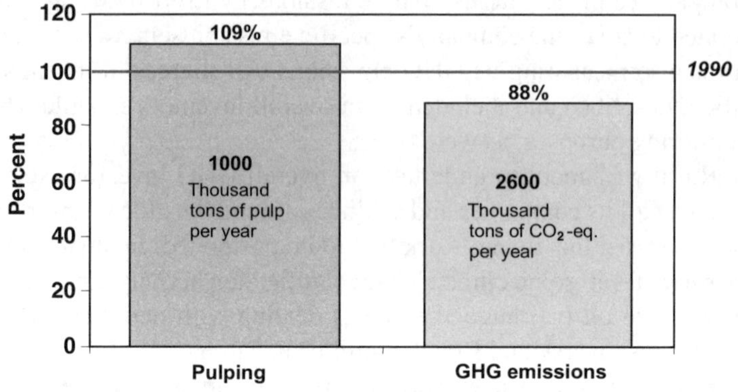

Source: JSC Archangelsk pulp and paper mill

To achieve this goal, APPM developed its climate strategy towards 2012, which includes the implementation of energy-saving technologies at the main production lines, increase of efficiency of energy generation at APPM's own power stations, the prevention of heat and power losses, and the further increase of the use of biofuels.

Due to this strategy, specific GHG emissions per ton of pulp have decreased by 45% against the 1994 year level; from almost 4.5 tCO_2e to 2.5 tCO_2e (see Fig. 6). Thus, production at APPM has grown almost twofold against 1994 level, while GHG emissions remain practically the same.

In 2007-2008, APPM received its first 'carbon money', EUR 1.5 million, for early (pre-Kyoto) GHG emission reductions under the waste biomass utilization project which has been implemented at the mill since 2000.[10]

From the moment the project was first put into operations in late 2000 until 31 December 2007, it generated some 807 thousand tCO_2e of GHG emission reductions, which were sold in the market under the *Voluntary Carbon Standard.*[11] Now the mill is looking forward to receiving additional money for its GHG emission reductions under

[10] Note: JI Commission approval is only needed for JI projects that generate ERUs during 2008-2012. With respect to the APPM situation, a PDD was prepared and validated against JI criteria, and then "early reductions" generated before 2008 were verified against voluntary carbon standards and sold for non-Kyoto (non-compliance) purposes to a company, which was seeking to be carbon neutral.

[11] See <http://www.v-c-s.org>.

Figure 6. Specific GHG Emissions at JSC Archangelsk Pulp and Paper Mill

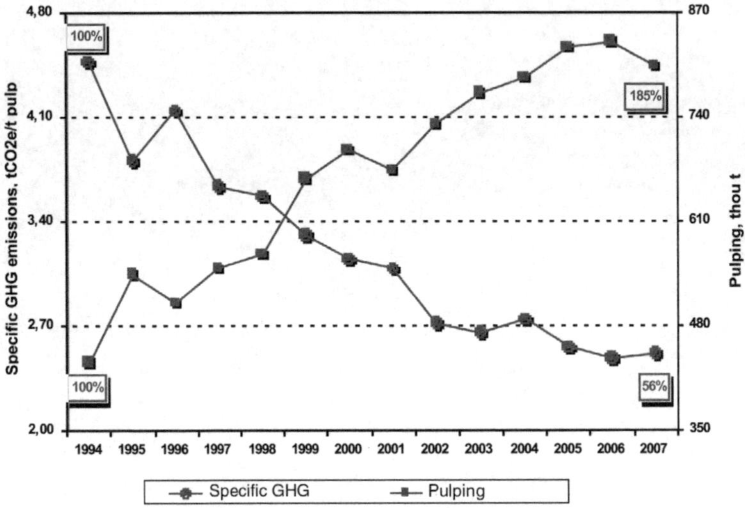

Source: JSC Archangelsk pulp and paper mill

the JI project during the Kyoto period (2008-2012) while at the same time working on another potential JI project that could double its emission reductions and, thus, increase the mill's total benefits both in terms of energy savings and in terms of carbon credits to be further supplied to the market in return for cash.

Final remarks

Examples of APPM and other Russian pioneer companies demonstrate now that implementation of the GHG limitation and reduction measures at the corporate level does not prevent business development and production growth. On the contrary, they help improve operations, strengthen company's competitiveness and sustainability. They also provide excellent opportunities for companies to earn money by selling carbon credits (GHG emission reductions) in the international carbon market and, thus, to profit from additional economic benefits which are especially valuable in the crisis times we now live in.

Chapter 14
THE CHANGING CONTEXT FOR INVESTMENT IN GREEN PROJECTS IN RUSSIA

Jonti Small[1], Trevor Barton[2] and James Lanman[3]

Environmentalists have argued since Soviet times that Russia's economic growth has been pursued at the expense of ecological concerns. This criticism intensified during Vladimir Putin's presidency when annual growth rates surpassed 7% and environmental degradation accelerated. However, at a meeting on 30 January 2008 in the Kremlin, President Putin and his now successor, Dmitry Medvedev, articulated the message that Russia's economic growth has been achieved at a cost to the environment. Furthermore, President Putin pronounced that Russia will only be economically competitive if an environmental clean-up takes place. The chronic need for green investment projects in Russia has long been evident; the new rhetoric employed by Russia's political leaders suggested that the conditions for green investment projects in Russia may never have been more promising.

This chapter will provide a wide-ranging survey that is intended to contextualise the opportunities for green projects in Russia. The policy agenda of the new president, Dimtry Medvedev, is still being formulated; only time will tell whether environmental concerns are taken seriously. Therefore, to assess the likely success of green projects, we will first review the significance of the recent political statements and will look at how environmental projects need to be considered in the context both of the domestic political agenda (specifically, Medvedev's stated objective of ending 'legal nihilism') and Russia's ambitions (specifically, Russia's desire to be a leading participant in the 'international arena'). We will then briefly review notable environmental projects to date before assessing Russia's ratification of the Kyoto Protocol. Kyoto, in theory, opens up a range of green opportunities for Western companies but the lack of progress at the time of writing also demonstrates that implementation of green projects can be challenging. In conclusion, this paper will seek to outline the current situation with regard to environmental projects in Russia and will look ahead to future developments, particularly in the run up to a possible new global deal to replace the Kyoto Protocol.

[1] Senior Account Manager, The PBN Company. The PBN Company is a consultancy specialising in advising Western investors into Russia and the former Soviet Union on market entry and expansion strategies, public affairs, government relations, communications and crisis management. The company has offices in Moscow, Kyiv, Almaty, London and Washington DC, <www.pbnco.com>; Email: *trevor.barton@pbnco.com*.

[2] Senior Vice President and Managing Director, Europe, The PBN Company; Email: *jonti.small@pbnco.com*.

[3] The PBN Company.

W.Th. Douma and F.M. Mucklow, eds., Environmental Finance and Socially Responsible Business in Russia
© 2010, T·M·C·ASSER PRESS, The Hague, The Netherlands and the Authors

The political environment

Until relatively recently, very little public attention was paid in Russia to environmental issues. In contrast, the international community has long recognised that Russia is home to some of the world's most significant environmental problems, arising from the Soviet legacy of outdated heavy industry, inefficient, under-funded municipal infrastructures, a decaying nuclear complex, and compounded by the rollercoaster economy of the 1990s.

The environment has been a very low political priority even though the United Nations estimates that more than 40% of people in Russia live in areas with poor environmental conditions.[4] Politicians have been criticised by environmentalists for exacerbating green problems. For example, in 2000, one of then President Putin's first acts was to dissolve the State Environmental Protection Committee. Responsibility for protecting the environment was transferred to the Natural Resources Ministry – the Ministry which is also charged with utilising resources.

As in other countries around the world, green issues have gradually gained greater political prominence in Russia. In the 2008 presidential election, Dmitry Medvedev made a point of criticising the environmental record of Russian companies and said that 'enterprise managers very calmly go about violating environmental legislation'. More stringent penalties were promised to ensure that businesses adhere to environmental regulations.

More significantly, as noted in the introduction to this paper, in early 2008, Vladimir Putin made a direct connection between economic growth and the environment. Addressing a meeting of the Security Council, then President Putin argued that government and industry must ensure that 'Russia's economic growth is based on high ecological standards … Working to protect nature must become the systematic, daily duty of state authorities at all levels.'[5]

'Legal nihilism'

The new recognition of environmental problems is an encouraging sign. In the long run, however, of greater significance may be President Medvedev's stated commitment to improve the 'legal environment'. In his inauguration on 6 May 2008, he said, 'I'm going to pay special attention to the fundamental role of the law. We must achieve a true respect of the law, and overcome the legal nihilism which is hampering modern development.' Improvements to the 'legal environment' and the broader 'investment climate' are fundamental to the success of green projects in Russia. Western investors and environmentalists alike should monitor improvements in the legal environment, as this will be a key contributing factor to the viability of green investment projects.

---

[4] As reported by Bloomberg, 30 April 2008.

[5] As reported by The Associated Press, 30 January 2008.

The 'international arena'

Russia's international ambitions are inextricably intertwined with the domestic political agenda. In recent years, Russia has played an increasingly active, even assertive, role in global affairs.

From an environmental perspective, there are many potential positives stemming from Russia's greater engagement in international affairs. The Russian government has a long-standing ambition to join the WTO. Part of this process will entail Russia redoubling its efforts to bring environmental, health and safety standards in line with international norms. Russia's ratification of Kyoto is another example of the country joining an international 'club'. When Western governments examine ways of encouraging and building relations with Russia, ecological concerns should be high on the agenda. The environment and climate change are areas where mutual interests converge, where engagement may be possible and which may encourage Russia's positive integration into the global economy.

However, Russia's integration into multilateral organisations has been far from smooth, and 'Russia's resurgence' is frequently seen as a threat in Western capitals. WTO accession is on hold. Indeed in August 2008, Prime Minister Vladimir Putin said Russia should abandon some of the commitments it made during accession talks. This followed the war between Russia and Georgia, which saw relations between Russia and the West become very strained. The way in which Russia conducts foreign policy (particularly the emphasis on restoring the nation state in a 'zero sum game' approach) will have consequences for the success of environmental projects: If Russia wants 'national interests' to prevail and compete, then environmental concerns are not going to be on the agenda.

Viewed through this geopolitical prism, it is significant that the environment is presented by Russia's political leaders as an area where Russia needs to protect the country's interests effectively. As then President Putin said in the same address to the Security Council, 'We have to learn to effectively protect Russian interests in this sphere in the international arena, in the first place by rebuffing the threat to environmental safety caused by cross border pollution of the Russian territory.' Such rhetoric, in line with Russia's resurgence and the broader assertion of the country's interests, is often viewed unconstructive, even menacing, in the West.

President Medvedvev echoed Putin when he said, 'We cannot handle environmental issues within one country, even such a big one as Russia.' He then went on to speak of need for international legal procedures to address trans-border air and water pollution. These two themes run through the recent experiences of environmental projects in Russia: Russia's greater involvement in international organisations and Russia's reassertion of its interests. Both trends have implications for green projects in Russia, and it remains to be seen how the two – potentially contradictory – trends play out.

Environmental investment to date

Before examining the impact and implementation of the Kyoto Protocol and also the high-profile stories that have attracted negative headlines in recent years, it is worth briefly reviewing the many examples of Western companies successfully engaging in clean-up operations and efficiency improvement projects.

Traditionally, international governmental and non-governmental organisations have provided the most significant channel for 'green' investment. A variety of multilateral bodies have become increasingly involved in environmental activities in Russia over the past decade and a half, typically through part-funding projects involving a contract with Western environmental product/service providers. For example, the Northern Dimension Environmental Partnership ('NDEP'), an organisation set up under the auspices of the European Bank for Reconstruction and Development ('EBRD'), has provided funding for a variety of schemes in the northern European part of Russia. The NDEP's environmental programme has so far approved 12 environmental investment projects at a total cost of more than €2 billion.[6]

NDEP's projects include restarting the construction of the St Petersburg Southwest Wastewater Treatment Plant – a sewage treatment plant that was left half-finished in the 1990s after the break-up of the Soviet Union. NDEP also has a €49 million project to improve solid waste management in Kaliningrad. This project involved the construction of a EU-compliant landfill site and a station for the bulk transfer of waste, together with the gradual closure or rehabilitation of existing dumps. The NDEP also oversaw and part-funded a €29 million project to update Murmansk's energy-inefficient residential heating system.

The United Nations Development Programme ('UNDP') has overseen a number of environmentally-focused projects in Russia, mostly by implementing Global Environment Facility ('GEF') projects. The GEF is a 178 member-strong international financing body which provides grants for projects related to biodiversity, climate change, international waters, land degradation, the ozone layer and persistent organic pollutants. In April 2008, the Council of the GEF approved two new climate change-related projects, which are being implemented by the EBRD and are part of a larger GEF-funded energy efficiency program in the Russian Federation.[7] With a combined $170.8 million in grants, credits and loans, these two projects will contribute to efficient use of energy in public buildings such as kindergartens, schools, hospitals and public offices, as well as municipal and mixed ownership urban housing in selected Russian regions. It is estimated that the refurbishment of housing stock covered by the project will generate an emission reduction of around 30% relative to the situation at the start of the project.

From 2002 until 2007, the UK Government's Department for International Development ('DFID') ran a bilateral programme which spent over £77 million in Russia. One of the programme's focuses was to improve public services, and it secured funding for a series of projects in 14 cities throughout Russia to improve municipal water

[6] See <www.ndep.org/projects>.

[7] See <http://www.gefweb.org/interior.aspx?id=21354>.

networks, including wastewater treatment facilities. One key partner has been and continues to be the UK engineering consultancy Halcrow Group, which has been working on projects related to rational planning and water resource development in Russia since 1996. After successfully cooperating on a project with the water providers of Rostov-on Don, Halcrow is currently the lead consultant of a team that has been appointed to undertake the design for the completion of the St Petersburg Navigation and Flood Protection Barrier.[8]

Specific opportunities have also opened to companies specialising in nuclear clean-up products and services. In the wake of the break-up of the Soviet Union, funding to Russia's huge nuclear complex all but dried up, producing a myriad of safety and pollution challenges. Additionally, under START I[9] and START II[10] the Russian government has specific obligations to reduce and destroy a large quota of strategic delivery vehicles, and it must find a way to reduce the total amount of deployed strategic warheads securely by storing or dismantling them.

Various countries have developed assistance programmes to aid Russia's nuclear industry and non-proliferation efforts, and several dedicated multilateral initiatives have also been established. A prominent example is the EBRD Nuclear Safety Account ('NSA'), the goal of which is to fund short-term safety improvements to older nuclear power plants in Eastern Europe and the former Soviet Union. It has provided $95.7 million for upgrades at the Leningrad, Novovoronezh and Kola nuclear plants. Work has included non-destructive examination, provision of fire protection systems and supply of components for emergency core cooling system upgrades. The US-coordinated Nunn-Lugar Cooperative Threat Reduction Program has overseen similar projects and claims to have helped to deactivate almost 6,000 nuclear warheads, while the 2002 G8 initiative, the Global Partnership Against the Spread of Weapons and Materials of Mass Destruction committed the G7 to raising up to $20 billion over 10 years to fund non-proliferation projects, principally in Russia.

Kyoto: 'Greening' Russia or an Opportunity Being Lost?

Russia's ratification of the Kyoto Protocol[11] in 2004 was by far the most promising development in the field of environmental investment in recent years. Kyoto, in theory,

[8] See <http://www.halcrow.com/html/our_projects/projects/st_petersburg_barrier.htm>.

[9] 1991 Treaty between the United States and the Soviet Socialist Republics on the Reduction and Limitation of Strategic Offensive Arms (START I), entered into force 5 December 1994, all parties met their implementation deadline (5 December 2001); Russia became a Party as one of the legal successors to the Soviet Union.

[10] The 1993 Strategic Arms Reduction Treaty (START II) was ratified by the US in 1996 and by Russia in April 2000. On 14 June 2002, Russia declared itself no longer bound by START II – largely in response to the US withdrawing from the Anti-Ballistic Missile (ABM) Treaty on 13 June 2002. However, a month before this, Russia and the US concluded the SORT treaty (Treaty on Strategic Offensive Reductions), which also limits deployed strategic warheads and so supercedes the START II caps on warheads. It is now unlikely that START III will be negotiated (see the commentary by Arms Control Association, <www.armscontrol.org>). SORT does not call for nuclear warhead destruction or tactical nuclear weapons limits.

[11] Here also referred to as 'Kyoto'.

offers major opportunities for both Russia and for Western investors to profit from the trading of greenhouse gas emissions as a new commodity. As the world's largest exporter of natural gas and the second largest oil exporter, Russia has vast potential to reduce energy consumption by the application of modern technology and efficiency gains. And yet, the opportunity to utilise Kyoto and to 'green' Russia in the process is slipping. To date, Russia's ratification of Kyoto was followed largely by delays and a lack of implementation.

The potential opportunities for Russia to benefit from Kyoto are two-fold. First, as a relatively low carbon emitter Russia stood to gain significantly under the emissions trading scheme, a financial mechanism whereby countries can earn revenue through the sale of portions of their emissions quotas to overproducing countries (who can then offset them against their own emissions surpluses).

Second, joint implementation ('JI') projects are conceived as a way of stimulating green investment, reducing greenhouse gas emissions and actually providing an incentive for Russian business to profit from JI initiatives. Because of the great potential to reduce the emissions of Russia's inefficient industry, Russia is potentially the largest seller of emissions credits on the international market and the biggest source of JI deals.

JI projects: theory and reality

During the ratification debate, some of Kyoto's strongest supporters were major Russian companies that were most likely to benefit from JI projects. Powerful Russian business groups remain enthusiastic about JI projects. In 2005, RAO UES, the power monopoly, signed one of the first JI agreements with the Danish government to replace coal-fired boilers at the Amurskaya power plant in Khabarovsk and to reduce greenhouse gas emissions of 1.2 million tons to be purchased by Danish Carbon.[12]

Before ratification, however, proposed JI projects faced difficulties. In 2003, two potential JI projects proposed by the Dutch government's ERUPT programme were also withdrawn after failing to secure Letters of Endorsement from the Russian government confirming that emissions reductions units would be transferred from the national registry if and when Kyoto comes into force. Likewise, a series of Activities Implemented Jointly projects, which were encouraged as part of a pilot phase of Kyoto preparation in the 1990s, ran into similar problems. Of nine projects started in Russia, five were either cancelled or delayed.[13]

Since ratification, potential JI projects have been held up by the failure to swiftly put in place an institutional and legal framework to facilitate implementation and monitoring. Investors expect the ground rules for a country's JI projects to be clear, consistent and transparent, and for there to be a codified, dedicated source of financing. Progress has been slow in Russia in each of these respects. In September 2004, the

---

[12] New York Times, 'In Russia, Pollution Is Good for Business', 28 December, 2005.

[13] See <http://www.climate-strategies.org/gisfinalreport.pdf> A study by 'Climate Strategies' 'A Russian Green Investment Scheme – Securing Environmental Benefits from International Emissions Trading'.

Russian government initiated the National Action Plan, which was designed as a way of getting different Federal Ministries and Agencies to work together to implement Kyoto. Only in early 2008 were the final legal statutes put in place. On 10 March, 2008, the newly formed government commission overseeing carbon credit trading began receiving applications for JI projects.[14] As the JI programme runs from 2008-2012, there is already criticism that an opportunity has been missed.

Why has Kyoto disappointed?

Russia's ratification of Kyoto has the potential to produce successful environmental projects, however, it is illuminating to outline why it has disappointed to date. The first factor is that Russia is able to considerably *increase* carbon emissions without exceeding its Kyoto targets. Russia's binding target for carbon emissions is set at the 1990 level. As Russia's emissions are far below this level, there is no imperative for Russia to act quickly. This is demonstrated by recent comments by Russian government representatives. Vsevolod Gavrilov, Deputy Head of the Ministry of Economic Development and Trade, outlined on 4 May 2008 that Russia's main goal is to reduce emissions to 1990 levels and that 'Russia has completely fulfilled this obligation'.

It appears that the Russian government has lost any appetite to implement Kyoto quickly and to use the protocol as a catalyst for green projects. This dovetails into a second factor, which is that the slow pace of reform is an all too familiar story. For a long time, there was uncertainty about JI projects and confusion as to the division of responsibilities between different Russian ministries. Interest from Russian businesses has been smothered by the lack of clarity and transparency in the processes.

Thirdly, until recently, high oil and other commodity prices created a situation where Russia's incentive to profit from Kyoto was greatly diminished. The economic benefits of Kyoto appear small when weighed against revenue from other sources, especially the natural resource sector.

At the end of 2008, it was reported that Russia was projected to stockpile $58 billion of carbon credits.[15] Rather than sell the carbon-emission credits, Russia appears to be looking to strengthen its position after 2012 with the successor treaty to Kyoto. This has led to accusations from environmentalists that Russia is working on ways to soften future targets and promote economic growth rather than on curbing emissions.

Kyoto going forward

Despite the delays and the negative factors outlined above, there is still hope, now that the government commission is finally accepting projects, that Russia will work within Kyoto and make use of the JI opportunities to raise money, modernise industry infrastructure and efficiency, and ultimately ensure stable long-term growth and inward investment.

---

[14] See also chapter 13 (*Yulkin*) for more detail.

[15] Bloomberg, 3 December 2008.

Going forward, there are two factors that investors will be monitoring. First is the appetite of Russian state companies to use JI programmes. In recent years, the Russian state has increased its control of certain economic sectors, particularly in strategic industries such as energy – the areas where JI projects are a natural participant. Recent nationalisations could scare foreign investors, but it is also theoretically possible that investors will have greater confidence in projects which are negotiated and approved directly by the state.

The second question is what replaces Kyoto in 2012, and specifically how Russia will be involved in any new scheme. Taken at face value, recent statements by government officials do not offer positive signs. In fact, the rhetoric suggests that Russia's 'national interest' is being asserted at the expense of the environment. Vsevolod Gavrilov, the official in charge of Russia's Kyoto obligations, said in May 2008 that Russia will not accept binding greenhouse gas emissions caps under a new climate regime being negotiated to replace the Kyoto Protocol. 'Energy must not be a barrier to our comfort', he said. 'Our emerging middle class ... demands a lot of energy, and it is our job to ensure comfortable supply.'[16]

Key considerations and conclusions

With a deteriorating investment climate and the impact of the global financial crisis, it is a time of low expectations for foreign investors into Russia. Western companies – both existing and potential foreign investors – have re-evaluated and scaled back their ambitions for the Russian market as hopes for the emergence of a market-friendly, 'normal' Russia have proved misplaced. All too often the 'rules of the game' for foreign investors appear to have been re-written by an increasingly assertive Kremlin while there has been little progress in addressing the enduring problems of corruption and bureaucracy. A series of high-profile disputes, such as those involving BP and Shell, have illustrated that doing business in Russia remains a challenge for even the largest of companies.

However, these well-publicised cases should not dominate the picture of business in Russia. The many Western investors that have operated very successfully and profitably in Russia seldom receive media coverage. This is certainly true of environmental projects – Western companies have brought much-needed skills and expertise to make clear and tangible contributions to the country. (This, in fact, is an excellent illustration of how Western companies need to approach investment in Russia.)

As noted in the introduction to this chapter, Vladimir Putin acknowledged that Russia's economic growth has been achieved at a cost to the environment. Now, however, the global financial crisis is hitting Russia particularly hard – but no one knows how hard. Since 1999, the Russian economy expanded at 7% a year. Oleg Vyugin, chairman of MDM Bank and a former central banker, has predicted that the economy may contract by as much as 4% in 2009 if export prices for raw materials remain low.[17]

---

[16] Reuters, 28 April 2008.

[17] Reported in Bloomberg, 3 December 2008.

What are the prospects for more environmentally-friendly policies during a period of slower growth or recession? In November 2008, industrial production fell by 8.7%[18] year-on-year, and manufacturing fell an annual 10.3%. Russia's idle factories will produce fewer carbon emissions, but recession is clearly not the recipe for investment in lower-carbon forms of energy in the long run.

The encouraging political rhetoric at the start of 2008 has not (yet) been followed by action. The economic crisis is limiting Russia's capabilities to act, and Russia's politicians are reluctant to impose extra costs on business. The Russian government is funding a massive bailout, but a 'Green New Deal', whereby government fiscal stimulus programmes invest in green technologies, is not on the agenda.

At the domestic level, opportunities for green projects will be determined by two great unknowns: how Russia emerges from the current financial crisis and whether the liberal agenda articulated by President Medvedev to end 'legal nihilism' gains traction. All investors – including those in green projects – need a prosperous and stable Russia that is increasingly less corrupt and which has a strong rule of law.

Internationally, the United Nations Climate Change conference in Copenhagen at the end of 2009 is intended to put in place a deal to replace Kyoto. The prospects for a global deal at Copenhagen will depend on the ability and willingness of nations to pool their interests and sovereignty for the common good. These liberal notions of 'the common good' and the need to pool sovereignty often appear to be anathema to Russia's current political leadership. How Russia chooses to engage with Copenhagen will be fascinating to watch. The reported stockpiling of carbon credits suggests that Russia is planning to participate actively in the successor to Kyoto – if only to use it to its own advantage.

One certainty is that Russia's political leaders will continue during the economic crisis to advance Russia's interests abroad. This will have a significant impact on the opportunities for green projects in the next decade.

---

[18] Bloomberg, 16 December 2008.

Chapter 15
RUSSIAN NATIONAL POLLUTION ABATEMENT FACILITY: PRACTICE AND RESULTS

Yurij L. Maksimenko[1] **and Irina D. Gorkina**[2]

The National Pollution Abatement Facility ('NPAF') is the sole organization of its kind in the Russian Federation. It acts as a practical tool for the financing of repayable investment projects ('IPs'), which are focused on the reduction of environmental pollution and the rehabilitation of natural resources. The main NPAF objective, alongside the effective utilization of its funds (amounting to about US$64 million) for the implementation of environmental protection projects, is the perfection of the financial mechanism according to international eligibility criteria and procedures taking into consideration the specific Russian economy conditions.

This chapter describes the work and operations of the NPAF, including its structure, operations and sample projects. It also provides an example of the way in which environmental finance projects can be conducted and achieved in Russia.

National Pollution Abatement Facility

In 1994, the government of the Russian Federation, with participation of the World Bank, the Global Environment Facility, the European Union, and the governments of Canada, Norway, USA, Finland, and Switzerland, developed the Framework Environmental Programme, to support and improve the implementation of Russian state environmental policy on the federal, regional, and local levels. As a part of this Framework Programme, the Russian Environmental Management Project ('EMP') was launched. For purposes of financing the EMP, a Loan Agreement (No. 38060-RU) was signed between the Russian Federation and International Bank of Reconstruction and Development ('IBRD' or 'World Bank'). The agreement provided a loan facility of US$110 million to the Russian Federation and entered into force on 30 August 1995.

The NPAF became one of the EMP components. The mission of the NPAF was to develop and strengthen the mechanisms of the financing of environmental investment

[1] Until 2008, Director General, Russian National Pollution Abatement Facility, <http://www.npaf.ru>. Now at Basic Element Company, Moscow, Russia; Email: *npaf@npaf.ru*.

[2] Until 2008, Deputy Director General, Russian National Pollution Abatement Facility, <http://www.npaf.ru>. Now at Basic Element Company, Moscow, Russia; Email: *gorkina@npaf.ru*.

W.Th. Douma and F.M. Mucklow, eds., Environmental Finance and Socially Responsible Business in Russia
© 2010, T·M·C·Asser Press, The Hague, The Netherlands and the Authors

projects ('IPs').The NPAF legally based its activities on a loan agreement (signed by the government of the Russian Federation and the World Bank, agreement No. 38060-RU) (the 'Loan Agreement'); the Grant Agreement between the government of Switzerland and the government of the Russian Federation (No. TF 3810 of 10 December 1996) (the 'Grant Agreement'); the Decision of the government of the Russian Federation 'On measures for implementation of Loan Agreement between the Russian Federation and International Bank of Reconstruction and Development, for financing of Environmental Management Project'; and, lastly, on the Resolution of the Government on the formal establishment of the NPAF (No. 808 of 11 August 1995).

The NPAF Supervisory Board provides general guidance on NPAF activities, and consists of officials from the Russian Ministry of Economic Development ('MED'), and the Russian Ministry of Finance ('MOF'). NPAF is managed by the NPAF Executive Directorate ('ED'), which has the following responsibilities: the selection and evaluation (including environmental assessment) of IPs, preparation of these projects for financing, drafting of legal documents, facilitation of procurement of goods and services, control and monitoring of project implementation. All activities of the NPAF ED comply with, and are guided by the 'Operational Manual', which was prepared by the World Bank, with the participation of Midlands Bank and the Know-How Foundation of the British government.

NPAF funds are used for the financing of environmental IPs and are comprised of monies received by way of the Loan Agreement and the Grant Agreement mentioned above. According to the Loan Agreement, the NPAF may cover up to 70% of total project costs, however, NPAF funding shall not exceed US$ 7 million for each individual project. Furthermore, the maturity period of a loan shall not exceed eight years, with a three year grace period. At the same time, the World Bank may agree to changing the financing terms on an *ad-hoc* basis (e.g., extend the maturity period and total amount of financing). Initially, the Loan Agreement envisaged two schemes of financing IPs:

(1) '*direct*' *scheme*: NPAF funds to be directly loaned to a company-applicant (as a sub-loan). In this case, the MOF and sub-borrower sign a sub-loan agreement ('SLA') under guarantees of a commercial bank which is acceptable to the MOF.
(2) '*through financial intermediary*': NPAF funds are loaned through a financial intermediary, such as a subject (i.e., Region) of the Russian Federation, or a participating bank. In this case, the MOF and financial intermediary sign the SLA, and then the financial intermediary and a beneficiary sign the final loan agreement ('FLA').

Environmental investment projects

The selection of projects eligible for NPAF financing is guided by a list of technical, environmental, economic and financial criteria. Each applicant should demonstrate the following:

(a) That the proposed investment project will (i) result in a sizeable reduction of adverse environmental impact, as well as the restoration (conservation) of natural resources; (ii) improve the environmental situation; and (iii) have an overall positive effect;
(b) Utilization of progressive technologies, internationally used methods and products; and
(c) Good financial standing of the applicant (sub-borrower or beneficiary), including demonstration of:
    - the absence of outstanding debts to the Federal budget and extra-budgetary funds, including the absence of debts under earlier loan agreements with Federal budget;
    - the capability to finance at least 30% of total IP's implementation costs, and to cover 20% of these costs out of its (the applicant's) own funds;
    - the ability to service debts, which may be incurred during the IP's implementation, in a timely and accurate manner, including repayment of loans under the SLA and FLA.

According to Russian legislation, all activities carried out in the territory of the Russian Federation should meet the requirements of the rational utilization of natural resources and environmental management. The Ministry of Natural Resources of the Russian Federation, and Russian Federal Agency for Technical Surveillance (*Rostechnadzor*) control compliance with these requirements.

In line with these requirements, the following types of IPs are eligible for NPAF financing:

(i) the construction of a new facility; and
(ii) the technical modernization, reconstruction, expansion or relocation of an existing facility,

providing such activities facilitate the restoration of natural resources, or pollution abatement.

NPAF project cycle and implementation

The NPAF project cycle includes the following procedures:

(1) *Project selection*: the selection and identification of environmental projects, their initial environmental and financial assessment (screening) to ensure that the IP meets NPAF criteria;
(2) *Documentation and approvals*: the development of supporting project documentation, including mandatory environmental impact assessment (EIA), drafting of a list of environmental conditions for project implementation, obtaining of the required approvals, including the approval of the State Environmental Review (*Государственной экологической экспертизы*);[3]

[3] See chapter 16 (*Dayman*) for more detail regarding SERs.

(3) *Evaluation*: in-depth evaluation of the IP, including financial, technical, economic and environmental assessments, selection of organizational and financial schemes, drafting of the project implementation plan;
(4) *Report*: the submission of the report on the results of an in-depth project evaluation to the NPAF Supervisory Board and the World Bank;
(5) *Negotiations*: negotiations with IP participants, signing of legal agreements (i.e., an SLA/ budgetary credit agreement, FLA, Project Implementation Agreement).

The steps of project implementation include the following: (1) the preparation of the IP for realization and transfer of NPAF funds, including the organization, facilitation and monitoring of tender for procurement of goods and services, the selection of consultants, signing and enforcement of contracts; (2) the disbursement of NPAF funds, including review of applications for withdrawal of NPAF funds (to ensure that NPAF funds is spent only on the intended purposes), the approval of withdrawal applications, transfer operations with NPAF account; control over the disbursement of NPAF funds, the calculation of loan repayment obligations and monitoring of the enforcement of financial obligations under the IP arrangements.

During the almost 14-year history of the NPAF, over 600 IP applications were considered and evaluated. To some extent, all these project applications attempted to solve environmental problems. For 70 projects, the in-depth pre-investment project reviews (i.e., feasibility studies) were organized and carried out in accordance with World Bank procedures, of which 45 project reviews were financed from monies obtained through a Swiss grant (No. TF 020217), and which is outsourced to the following consulting firms: Ecoling/PEG/KPMG Fides, COWI Consulting, Electrowatt Eng., Frisa Engineering, Arthur D. Little, ERM Ltd, Crown Agents, Dames & Moore, Man-West Environmental Group. Forty project evaluation missions have been organized, and upon their results, 40 IPs were approved by the NPAF Supervisory Board and IBRD (the World Bank 'no objections' visa was obtained).

The geography of the implementation of NPAF projects is rather diverse and take place in the following locations: Moscow and St. Petersburg, Northern and Central Russia regions, the Upper and Middle Volga, the Urals, Western and Eastern Siberia.

NPAF projects are aimed at the:

- abatement of air, groundwater and surface water pollution;
- utilization of industrial and solid household waste;
- development of infrastructure;
- creation of new jobs and sustenance of employment;
- improvement of work conditions and competitiveness of domestic production;
- production of additional goods; and
- saving of materials, energy, pollution fees and charges imposed on companies-applicants.

The implementation of the IPs is supported by governmental authorities and authorized environmental protection agencies in the Regions.

Thus, any businesses which are interested in undertaking environmental finance projects in Russia, and which wish to apply for NPAF funds, will need to take into consideration the requirements described above. Detailed requirements can be obtained from the NPAF ED. By way of example, the below describes how the first NPAF project was implemented.

Arkhangelsk Pulp and Paper Mill Project

From 1996 to 1999, the first NPAF project termed 'Technical modernization of production of cardboard at DAO "Karton" aimed to achieve the saving of energy and reductions of air emissions' was successfully implemented at the JSC 'Archangelsk Pulp and Paper Mill'. It is during the preparation and implementation of this project that the NPAF procedures were tested and fine-tuned.

Let us say a few words about this project. In 1995, Arkhangelsk Pulp and Paper Mill ('APPM') was the main source of air emissions containing sulfurous substances (i.e., methylmercaptan, dimethylsulfide, hydrogen sulfide, and methanol) in the North-Western region of Russia. The zones of influence of such air emissions from the APPM facility encompassed the cities of Novodvinsk and Arkhangelsk (situated some 25 km away from the mill), the mouth of Severnaya Dvina river, and coastal area of the White Sea. APPM management developed a programme of environmental investments, with the goal to reduce its noxious air emissions and limit their negative effects. This programme claimed a 98% reduction of the APPM's emissions of sulfurous substances by 2000, compared to its baseline 1990 emissions level. One element of this programme was the abovementioned IP, prepared by the APPM management on the basis of a pre-investment feasibility study made by COWI (Denmark) in March of 1996. After the NPAF ED conducted an in-depth project evaluation, this project was recommended for financing by way of a joint decision of the NPAF ED and the IBRD, as it fully meet the environmental, economic and financial criteria established by NPAF.

Initially, the project participants included the APPM subsidiary DAO 'Karton', which acted as the project applicant, the MOF, the NPAF ED and the Ministry of Natural Resources of the Russian Federation, the latter of which administered general oversight in the course of the project implementation. After the reorganization of the APPM, its subsidiary DAO 'Karton' merged with its 'parent', thus delegating all its rights and obligations with respect to the IP to APPM. Thus, APPM become responsible for the implementation of the SLA which was concluded with NPAF and entered in force on 15 December 1996. The total project cost was US$11.4 million, of which US$6.44 million was financed by way of a NPAF sub-loan. The NPAF SLA included the following financial conditions:

1. Payback period (in years) 5
2. Grace period (in years) 2
3. Annual interest rate (in %) 9.4
4. Payment for loan maintenance:
    - During the first five years (in % p.a.) 0.25
    - During the following two years (in % p.a.) 0.125
5. Payment for project evaluation (lump-sum) (in % of the sub-loan amount) 1

The SLA provided for the following two main options to pay-off the loan (repayment of the loan and interest, plus servicing commission), i.e., by way of:

(1) A special account ('SA') regime; or
(2) Guarantees of the APPM, as the parent company of the sub-borrower DAO 'Karton' (after the reorganization of DAO 'Karton', this second option was removed).

The IP goals included the:

(a) Production of more environmentally friendly cardboard in order to improve the company's standing and competitiveness on the domestic and international markets. This was achieved by way of technological modernization of the cardboard production.
(b) Decrease of the environmental burden caused by the mill in general. This was achieved through a technical overhaul and re-equipment of the cardboard production process.

The project did not envisage an increase in the facility's production capacity. Its profitability was achieved through saving of energy and pollution fees. The project consisted of two main elements:

(1) Modernization of two lines of regeneration of black liquor,[4] involving the liquidation of cascade evaporators and the replacement of water economizers, with an almost two-fold increase of the heating surface (from 5,600 m^2 to 9,004 m^2); reconstruction of three head installations at the evaporation plants, their conversion to less energy-intensive 'falling film' technology, and the installation of an automated control system for boilers of soda recovery units (SRK-1 and SRK-2) to increase their efficiency.
(2) Reconstruction of the pressing part or cardboard machine (KDM-2).

The replacement of the pressing part of KDM-2 involved the installation of rollers with a bigger diameter, and the increase of steam pressure during pressing (from 50 kg/cm to 350 kg/cm), while the pressing area increased from 40 mm to 120 mm. The increase of dryness of huckaback allowed to achieve heat economy during the drying of the cardboard, and to reduce emissions from the burning of coal and black oil at the power plant (HEPS-1), which supplied heat and steam for the KDM-2 machine.

The procurement plan for the implementation of this IP was financed through a subloan and was developed during an in-depth evaluation of the project. This plan became a constituent part of the Project Implementation Agreement between NPAF ED and the APPM. The APPM purchased goods and services for the implementation of the IP, in accordance with the procedures prescribed in the Procurement Manual for projects financed by IBRD loans and MAR credits, and as specified in the Loan Agreement.

[4] Lignosulfonates – substances derived from wood, which are produced during the cooking process.

Because the APPM had sufficient experience in foreign trading, it independently drafted its procurement contracts, using the IBRD contract templates and guidelines. All stages of the contact preparation and fulfillment were coordinated with the NPAF ED and the IBRD. The procurement stages included the preparation of the terms of reference, sending out requests, technical and financial evaluation of submitted proposals, the selection of suppliers and signing of contracts. In the course of the IP implementation, there have been no breaches of contract, and the actually disbursed procurement funds have not exceeded the amount budgeted in the SLA.

The procurement of goods and services in the framework of this IP was conducted using the APPM's own funds, at reasonable prices, taking into consideration such factors as the terms of delivery, reliability of goods, servicing arrangements, and the availability of spare parts. In the case of the purchasing of goods with accompanying services, the quality of these services and the professional qualities of service providers were taken into account. Besides this, the APPM management considered the quality and reliability of construction technologies and materials, experience and credentials of the contractors, in accordance with existing commercial practice. The IP tender selected the following contractors for the project implementation: Valmet Corporation Board Machines, Finland (equipment for KDM-2); SMAO 'Consul', Russia (assembling of pressing equipment for KDM-2); Clyde Bergemann GmbH, Germany (production and assembling of soot blowers); and JSC 'Sevzapenergomontazh', Russia (upgrading of evaporation plants with the equipment produced by the 'Tampella' firm, including the purchase and assembly of supplementary equipment).

The *IP preparation period* is defined as the time which has elapsed since the facility submitted its application for the sub-loan, and until the facility is authorized (by way of letter from the MOF, which notifies that the sub-loan agreement has entered into force) to drawdown the sub-loan monies from the relevant bank account. In the case of the APPM IP, this period extended over about one year (from 29 December 1995 to 15 December 1996). It took about six months to evaluate and approve the IP, after which the next six months were devoted to negotiations, signing of the legal documents, and entry-into-force of the SLA and the Project Implementation Agreement. The NPAF ED and the company-applicant played key roles in this stage.

The *IP implementation period* is defined as the time which has elapsed since the IP entered into force until the time of the commissioning of the last IP construction project (installation, etc.). During the IP preparation and evaluation, its implementation period was defined as 1.75 years. The parties planned to complete all IP activities before mid-1998. However, the last IP project was commissioned in December 1999, thus extending the IP implementation period by a further 1.25 years. This delay was caused, amongst other factors, by unscheduled repairs, when the equipment was put out of operation, as well as by seasonal factors during the start-and-adjustment works, and due to the faulty operation of the reconstructed evaporation plant. To some extent, the divergence between the planned and actual IP implementation periods was caused by an overly optimistic project implementation schedule, which was prepared during the in-depth project evaluation phase. The project implementation schedule was drafted on the basis of available information with respect to the implementation of similar projects, and failed to take into account the difficult procurement procedures established by the IBRD.

In accordance with the SLA, the IBRD authorised the withdrawal of the sub-loan funds, on the basis of the initial invoices which were submitted by the APPM to NPAF ED, the latter of which then filled out the standard IBRD applications for funding and sent them, together with all supporting documents, to the IBRD's Department of Disbursement. The sub-loan financing was also leveraged by way of a Swiss grant (in the proportion of 77/23).

Despite the numerous difficulties, which threatened this IP during the five years of its implementation (e.g., caused by default and reorganization of the APPM, which caused changes in financial and legal arrangements of the IP implementation, including the rewriting of the SLA), the APPM successfully fulfilled its obligations under the SLA.

The construction of the main project items within the framework of the IP was completed by September 1999. After the completion of the start-and-adjustment works at the reconstructed evaporation plants, and after the repair of the soda recovery unit (SRK) in December 1999, all stages of the IP were completed in accordance with the original project design. The total profits from the implementation of this IP were estimated as US$3,818,000, and the annual payments to the budgets of all levels were estimated at US$700,000 (see Table 1).

Table 1. IP returns

Income item	Energy savings, GCal	Economic effect, US$
Energy savings due to modernization of evaporation plants	13,989	341,200
Energy savings due to replacement of economizer at soda recovery unit SRK-2 of power plant HEPS-2	71,535	1,746,000
Steam economy at KDM-2	45,300	1,105,000
Savings of environmental payments	–	625,000
Total		*3,818,000*

The implementation of the IP measures resulted in the reductions of hazardous air emissions and the considerable decrease of air pollution in the city of Novodvinsk. Total air emission reductions, due to the reconstruction of the soda recovery units, were 3033.8 tonnes per year, estimated for planned production capacity. The direct results of the IP implementation included the total phasing out of emissions of hydrogen sulfide, methylmercaptane, and dimethylsulfide (see Table 2).

Total fuel savings at the power plant HEPS-1 were calculated at 20,460 tonnes, which in turn reduced air emissions by 1,193 tonnes (see Table 3).

Table 2. Reductions of air emissions with flue gases, due to reconstruction of the soda recovery units SRK-1 and SRK-2 (in tonnes per year)

Pollutant	Emissions before IP	Emission after IP	*Emission reduction*
Nitrous oxides	420	402	18
SO_2	160	171.6	-11.6
CO	1,814.6	775	1.039,6
TSP	2,254.6	2 254.6	0
H_2S	775.3	15.5	759.8
Methyl mercaptane	150	0	150
Dimethyldisulfide	628	0	628
Dimethylsulfide	450	0	450
Total:	*6,652.5*	*3,618.7*	*3,033.8*

Table 3. Reductions of air emissions from coal-fired power plant HEPS-1, due to fuel savings

Pollutant	Emission reduction
Coal ash	376.415
SO_2	552.97
SO_2 (above limit)	110.064
CO	54.22
NO	99.42

Thus, the main environmental benefits of the implementation of this IP included the reduction of the environmental burden, in terms of total air emissions and emissions of odour-nuisance substances. The economic benefits included the implementation of new modern technologies in the management of production processes, improved sanitary conditions at work, and advanced technological education of APPM personnel. Furthermore, it is noted that the environmental improvements in Novodvinsk resulted in the increased the role of the APPM in the implementation of environmental policy, carried out by the administration of the Arkhangelsk region.

Important factors, which contributed to the successful implementation of the IP included the experience of the APPM in the implementation of environmental programmes and projects, availability of qualified personnel, necessary infrastructure and working areas for the assembly of the new equipment. The APPM management claimed that the most important results of the IP implementation were the increases of the reliability of equipment and the quality of products. It is noted that the final report on implementation of the IP was submitted to IBRD in 2003.

APPM Second Project

After the successful completion of the above described IP, the APPM filed a further application to NPAF ED, after a new environmental investment project was prepared for financing, i.e., for the 'Reconstruction of pressing part of cardboard plant KDM-1, bark boiler KM-75-40 at power plant HEPS-3, and bark preparation unit DPC-3 at JSC APPM, with the goal to decrease environmental load of the mill'. The total amount of project investments was set at US$17.1 million, of which the NPAF's share was US$6.9 million. The project duration was 1.5 years. This project was to be implemented through a 'financial intermediary' – Vneshtorgbank of Russia. The project provided for technological improvements at a cardboard plant (KDM-1), a bark-fired boiler at a power plant (HEPS-3), and of a bark preparation unit (DPC-3). The environmental effects of this IP included: annual savings of 10,900 tonnes of coal and 38,200 tonnes of black oil; the utilization of 300,000 tonnes of bark waste per year; reduction of air emissions by 2866.6 tonnes per year, including the reduction of greenhouse gas emissions (CO_2) by 132,773 tonnes per year.

The capital expenditures of this IP included the procurement of equipment (production, shipment at CIF Arkhangelsk conditions, assembling), VAT and export duties. Several alternative suppliers (domestic and foreign) were considered at the project development stage. The profits of the IP were mainly to be sourced from savings of black oil and coal, mostly due to the reconstruction of the bark-fired boiler at power plant HEPS-3, a relatively insignificant increase of production of commercial cardboard after the reconstruction of the cardboard plant KDM-1, and the saving of costs of disposal (dumping) of bark waste. The more important results, in the view of the APPM management included the increased reliability of the facility's equipment, the maintenance and repair cost savings, and the improved quality of cardboard produced at cardboard plant KDM-1. Preliminary estimates showed that the budgets of all levels would receive about US$10.2 million in taxes over a 10-year period, as a direct result of the implementation of this IP. Having completed all the project preparation stages (preparation of supporting documents, in-depth review, obtaining of approvals, procurement and shipment of equipment), the APPM successfully completed this IP in August 2005.

Volgograd Project and other projects

As stated above, within the NPAF has seen the realization of 5 investment projects so far. Other projects, such as the Volgograd Project, are briefly described to provide the reader with a flavour of other types of NPAF projects.

In 2003, the NPAF completed the IP 'Improvement of environmental situation in Volgograd through essential reduction of air emissions from chemical enterprises "Plastcard" and "Volgogradneftepererabotka"'. The total project cost of this IP was US$9.5 million, of which the NPAF sub-loan was US$6.5 million. This project was implemented under the 'direct' financing scheme, including guarantees of a commercial bank. The environmental effects of this IP included the reduction of air emissions of JSC 'Plastcard' from 82,902 tonnes per year to 17,452 tonnes per year, and the reduction of emissions of fat gas by 'LukOil-Volgogradneftepererabotka Ltd.' by 60,000 to 62,000 tonnes per year. The project was completed in 2005.

During 2005-2007, the investment projects at Solombala and Kotlas pulp and paper mills were concluded. These two projects were implemented through a financial intermediary – i.e., Vneshtorgbank. At the time of writing, the total volume of used resources of the NPAF is US$ 27.5 million.

Currently, the implementation of two projects is underway, i.e., the 'Abatement of pollution in the Upper Volga River Basin by modernization of primary production and waste water treatment facilities at CJSC "Ostashkov Tannery" (Tverskaya Oblast)" and "Environmental enhancement of Podolsk city by modernization of gas purification system of the melting shop of the factory of secondary metals and alloys" (Podolsk, Moscow Oblast), both through another financial intermediary, i.e., CJSC "RosEximBank".'

NPAF Funds – applications and transparency

The NPAF repayment proceeds are such proceeds, which are paid to the NPAF by the Sub-borrowers to the MOF under the sub-loan/budgetary credit agreements. The repayment proceeds are transferred to the NPAF Account, which was opened by NPAF ED at CJSC 'RosEximBank'. The NPAF Account accumulates proceeds in Russian roubles as received from the Sub-borrowers or third parties acting as guarantors of Sub-borrower's liabilities to the MOF, namely: payments of principal; commitment fees; interest; appraisal fees; charged penalties and fines.

The disbursement of the NPAF Account funds is carried out according to the disbursement procedures in context of the following purposes: repayment of interest under the IBRD Loan and the commitment fee of the IBRD Loan; covering of the NPAF ED operating expenses for preparation and supervision of projects; financing priority nature protection projects (i.e., reinvestment).

The NPAF ED shall account receipts to the NPAF Account with break down by each Sub-borrower for purposes of monitoring the execution of their payment liabilities under Sub-loan agreements, and in order to provide separate accounting of the IBRD Loan and Swiss grant repayment proceeds. The NPAF ED shall inform the MOF in writing at a monthly basis, prior to the fifteenth day of the month following the accounting one, on receipts in the NPAF Account in the breakdown by sources of receipt with submission of bank extracts from the NPAF Account.

The reinvestment of NPAF repayment funds into new priority environmental projects is carried out under the following main principles:

- Repayment proceeds shall be first directed for financing the participation of the Russian Federation in international environmental projects and programmes implemented or prepared for implementation in Russia (for example, projects of the Global Environmental Facility, Joint Implementation Projects under the Kyoto Protocol on restriction of greenhouse gases air emissions, etc.), which will allow for the more active involvement of international donor funds and investors in these projects;
- Funds will be allocated to Russian enterprises and organizations as grants or loans on the basis of agreements made with the participation of the MOED, MOF, and Ministry of Natural Resources;

- The NPAF Supervisory Board shall carry out the general management of the repayment funds, including the approval of projects for financing and of conditions of granting funds (amount, form of granting – grant or loan, terms, and other conditions) in view of the conclusions and recommendations contained in the Appraisal Reports prepared by the NPAF ED. The NPAF Supervisory Board shall also define the programme of work and the NPAF ED operating budget for appraisal of projects;
- NPAF ED shall carry out the selection and appraisal of projects for the financing out of repayment proceeds, preparation of the necessary legal documents under loans and grants, supervision of projects and evaluation of results of their implementation.

Concluding remarks

The NPAF was initially envisaged as a revolving fund. In this scheme, the MOF receives funds from Russian enterprises, which implemented NPAF IPs (so-called revolving funds) and mobilized these funds for the financing of new environmental investment projects, while only a fraction of the revolving funds should have been spent on debt servicing under the Loan Agreement signed between the Russian Federation and IBRD. Currently, about US$25 million have been accumulated within the framework of the NPAF. As these funds are meant for reinvestment, NPAF ED is currently preparing a new portfolio of environmental IPs.

Thus, the almost 14-year history of the NPAF demonstrates that for the successful operation of this revolving fund, the environmental IPs should originate in such sectors of industry, which have caused serious environmental problems in the past. Any economic project aimed at the technological modernization or reconstruction, as well as restoration of natural resources and/or pollution abatement, is likely to meet the basic technical, environmental and economic criteria of the NPAF.

The mobilisation of special financial flows, such as are available through the NPAF to solve environmental problems at the level of enterprises and territories, allows companies to apply for less stringent credits and to implement projects with long payback periods, which are not very attractive from a general economic standpoint, but which have sizeable environmentally beneficial effects.

Chapter 16
ENVIRONMENTAL MANAGEMENT SYSTEMS IN RUSSIA: THE DELAYED WAVE

Sergey Dayman[1]

Over the last 30 years, the use quality management systems ('QMS') has become an internationally recognised way of doing any type of business. Environmental management systems ('EMS') were built on the success of QMS and many Russian businesses have, since the late 1990s, been implementing EMS on the basis that they wish to manage, and demonstrate that they manage, the environmental impacts of their operations effectively. In Russia, the history of management approaches differs significantly from that in the Europe, Japan and the USA. Yet these approaches, including the ISO 14001 EMS standard, have found their way into Russian businesses practices and have had positive effects on them.

Late introduction of EMS into Russia

The mid-1990s saw the beginning of wide-reaching discussions in Russia of the efficiency and potential of 'market instruments for environmental management'. Within these discussions, environmental audit was one of the most popular subjects. This was partially so because Russia entered into the sphere of interest of private Western businesses, as well as international development and investment banks. This resulted in a number of transactions being subjected to environmental audits. Shortly thereafter, information regarding EMS began to appear in Russia. It was mainly introduced from non-Russian sources and was based on the ISO 14000 series of standards already highly popular in the West.

At first, EMS information which was available in Russia was very vague and unspecific. This lead to many misinterpretations within Russian business community as to what EMS was, as well as to unreasonable expectations and fears as to its effects on business. One reason for this was the lack of Russia-based management experts working in environmental protection, the latter of which had long been considered to be a purely technical issue.[2] Furthermore, the Russian national standards body, *Gosstandart* at that time (which at present is the Federal Agency on Technical Regulation and Metrol-

[1] Senior consultant of the Sustainability Assurance and Advisory Services group in Ernst & Young (CIS) branch in Moscow; Email: *Sergey.Dayman@ru.ey.com*.

[2] See also chapter 3 (*Guseva*) with respect to understanding of sustainable development in Russia.

W.Th. Douma and F.M. Mucklow, eds., Environmental Finance and Socially Responsible Business in Russia
© 2010, T·M·C·Asser press, The Hague, The Netherlands and the Authors

ogy or *Rostekhregulirovanie*),[3] which could have participated in the preparation of the international standards mentioned, did not have enough knowledgeable experts for the task even to be recognised. Later, with information on ISO EMS standards having been brought into the country by international aid agencies and consultancies, *Gosstandart* ventured to translate the first few standards, namely the ISO 14001 and 14004 on management systems; ISO 14010, 14011, 14012 on environmental auditing, and ISO 14050 on terminology. In 1998, these standards were then introduced in Russia as voluntary national standards.

Only one representative of *Gosstandart* carried out all contacts with the ISO Technical Committee 207 responsible for the standard, as well as the translation of the standards. As a result, the Russian translations of the standards were not only published four years after the original English language version, but the Russian translation of the standards also used confused terminology, contained mistakes and inaccuracies – and the text was published without reference to the parallel original English text. These problems have not been resolved until quite recently.[4]

In the late 1990s, efforts to promote better environmental management in Russian enterprises were undertaken through many not-for-profit programmes supported by European or US governmental or private funds. At first the programmes were aimed at the governmental body responsible for environmental protection, *Goskomekologii*.[5] The training was lead by major international consultants, many of whom directly participated in the development of the EMS standards. These projects did not, however, succeed in their purpose. There have been many reasons for this: the informational materials did not take into account Russian management practices and background; they were frequently translated incorrectly; and not least, the Russian environmental authorities had little experience with working with businesses. As a result, the general impression of those being trained was that either EMS was completely 'new and extremely necessary', and that enterprises, therefore, should be obliged to apply EMS, or that EMS was 'already being applied' at every factory and that, therefore, nothing needed to be done to promote this concept in Russia. The potential opportunities of EMS were, thus, at that time not perceived adequately.

In any event, the Russian legal system allows the establishment of many voluntary certification systems, provided basic rules are observed.[6] Over a dozen certification

---

[3] Gosstandart (Rostekhregulirovanie since 2004) is the Russian specially authorised body in the field of standardisation and acts as the Russian representative to the International Organisation for Standardisation ('ISO').

[4] The 1998 version of the Russian ISO 14001 standard was only replaced with a new version (translation of ISO 14001:2004) in late 2007. This time, however, the translation was reviewed and corrected by a number of experts and, in particular, one of the Russian NGOs whose role is noted in the issued standard.

[5] State Committee for Environmental Protection was dismissed in 2000, with new structure of environmental authorities which has been evolving since then.

[6] Federal Law on Certification of Products and Services of 10 June 1993 (No. 5151-I); replaced in 2002 by the Law on Technical Regulation No. 184 which have not changed positions. The Law did not require but only allowed that voluntary certification systems are registered with the technical regulation body; it allowed for the establishment of many independent voluntary certification systems. The role of the certification system lies in the accreditation (including checking conformity to the system's criteria, appellation commission) of certification companies that wish to provide certification with respect to the voluntary standards.

systems relating to compliance with GOST R ISO 14001-98 have been registered with *Gosstandard*, with little supervision by it. Some such systems were established under the auspices of sector ministries, in particular the Ministry for Natural Resources ('MNR', recently re-named the Ministry for Natural Resources and Ecology). Many of the systems were created for the purposes of one certification body and not designed to control quality and independence of its services.

Goskomekologii also attempted to transform EMS, which is voluntary by its nature, into an obligatory activity, strictly regulated by state bodies. In December 1997, the Chairman of *Goskomekologii* approved the statute relating to the federal system of obligatory environmental certification. This system included certification of management systems in accordance with the ISO 14001 requirements. Such obligatory certification was, however, later removed on the basis that it contradicts pre-existing legislation. Formally, the structures of such certification systems are still maintained and certification companies which are registered within them exist. In the majority of cases, the establishing documents and procedures have not been reviewed since their initial development and do not refer to international norms supporting reliability of ISO 14001 certification.[7] And, to this date, some of the certification bodies have not yet issued any certificates.

Slow uptake in business

The situation in Russia, in general, has not facilitated wide implementation of EMS.[8] Regardless of the growing 'green' movement, only a few highly polluting enterprises experienced pressure from stakeholders concerned with environmental impacts. Most criticism by NGOs and academics fell instead on the environmental authorities. These stakeholders and the public were unaware of the potential of EMS applications.

Language difficulties, too, have lead to some degree of isolation of Russian business from current trends in this relatively new and narrow business area. This considered, it is perhaps not surprising that Russian enterprises did not at first express any notable interest towards EMS implementation and certification. It would of course be wrong to think that Russian enterprises did not undertake any 'environmental management' before the introduction of ISO 14001. Certain elements of such a system had already been established by GOST 24525-80 '*Management of production association and industrial enterprise. Management of environmental protection. Major requirements*'.[9] The Enterprises that effectively developed their quality management system in accordance with the ISO 9000 series of standards in some cases included environmental protection in the sphere of such system, too.

It is important to mention that, in Russia, standards were traditionally considered as technical requirements. However, EMS standards include none such. Both directors

---

[7] In particular, the requirements applicable to auditors and audits of ISO 19011:2002; requirements as to the certification bodies; and the process of certification in accordance with ISO/IEC Guide 66.

[8] See also chapter 8 (*Shvarts and Gerasimchuck*) for futher analysis of EMS in Russia.

[9] However, this document was in force only for 5 years until 1985.

and specialists had difficulties in understanding the approach adopted in environmental management standards and tried to extract and limit themselves to the 'technical' requirements on the basis that they had done so in earlier times in context of the requirements of the ISO 9000 series (Quality Management Standards) in the 1994 version.

Who was the first?

The first to introduce the ISO 14001 in Russian companies were, as one might expect, multinational corporations ('MNCs') operating in Russia. Many MNCs consider it necessary to have an ISO 14001-certified EMS at all their sites. In most cases, Western companies have tried to fully introduce their management approaches, and as these often included EMS, they developed fully functional systems within their Russian operations.

However, there were exceptions to this general rule: the first ISO 14001 certificate in Russia that we know of, obtained as early as in 1998, was awarded to a joint venture production site by an international certification body. As it emerged later, the site merely held 10 copies of their environmental policy stored in a safe box; it never communicated any information about the EMS to its employees nor to any external parties, even upon specific requests. Needless to say, this management system that formally was Russia's first, never actually existed.

Nevertheless, the requirements that MNCs applied to their oeprations and suppliers with regards to ISO 14001 have played and continue to play a very important role in the dissemination of EMS in Russia. For example, many corporations in the automobile sector require their suppliers to obtain ISO 14001 certification. As a result, in Russia (as has similarly been the case in other countries) the EMS approaches are being disseminated by way of cascade.

The path that Russians followed

With the MNCs leading the way, Russian companies followed suit soon. The managers and owners of Russian companies received information on EMS, and the benefits it proposes, from publications, national and international consultants, partners and colleagues, or from their own environmental staff who received such information from non-profit projects or motivated governmental officials.

Some of these Russian companies concluded that developing an EMS, or achieving certification, or both, could be useful from the viewpoint of business process, market position, or improving stakeholder relations. Some Russian companies single-handedly started to develop EMS practices, others joined various not-for-profit pilot projects, while yet others requested the assistance of various Russian or Western consultants. Again, MNCs continued their role as a source of experience and provided practical examples relating to the efficiency improvements and other opportunities associated with EMS. For example, an independent Russian company located adjacent to a production site of a MNC, followed that particular MNC's example and thereby started introducing EMS into its operations.

Top managers of Russian companies provided different reasons for the implementation and/or certification of their EMS based on their character and possible consequences, as follows:

- certification is requested by their target market or by particular consumers important to the company;
- certification could help eliminate conflicts with public or enforcement bodies;
- certification could be an added value and so is necessary due to strong competition (this affected both leaders in certification and then other players who struggled not to loose their share);
- EMS implementation could help in achieving compliance with environmental legislation;
- EMS could help to optimize the use of raw materials and resources, and so reduce production costs.

And finally, some of the managers considered reduction of companies' environmental and health impacts as the main reason for developing the EMS.

In the author's opinion, the majority of the points mentioned above are correct, but the last point relating to the use of resources was particularly important: many companies have not previously managed to reduce costs of production, which meant EMS was of great help to such companies. There were cases where, for example, energy saving measures reduced their electricity consumption by two thirds. Since then, many more companies, with or without the use of EMS or quality management systems, have implemented measures to optimise the use of energy, raw materials and process efficiency. Of course, this does not mean that there is nothing else that could be done to improve the current situation.

The approaches which Russian management has taken to achieve the expected benefits differ. When the directors of such companies expected some advantages to result from certification, they usually set their subordinates the task of obtaining such certification cheaply and quickly. These companies usually were successful in finding a certification body that did not pay much attention to the performance of the EMS system, and provided necessary documentation blueprints through an associated consulting company. This way of arranging certification of course led to some difficulties, among which were extra costs and troubles with passing inspection audits; lack of employee interest and involvement in the EMS; and more often than not, the absence of any improvements in the efficiency of the company's operations or in the reduction of its environmental impacts. However, this approach was partially successful: those Western partners, who were satisfied by the presence of the EMS certificate alone, continued their business with such companies.[10] The failure to implement an EMS effectively

[10] In this scenario, negative roles have been played by the certification and, in particular, Western companies that often sent just one or two auditors-in-training and used equally bad local technical experts,

means, however, that such companies face the risk of having no instruments available to control environmental impacts and related financial and image risks. Naturally, at a certain moment in time, this will become obvious.

Top managers who in their own, or someone else's, experience learnt the importance of improving their management systems, saw EMS as a way to improve their operations. These companies seriously selected an appropriate and knowledgeable consultant; allocated the resources and ensured that employees were involved. They developed company-specific procedures, approaches and organizational solutions. Even if a company decided not to obtain a certificate, its EMS started to bear fruit, including as follows:

- EMS helps strengthen the QMS and an organisational management system in general, supports development of an organisational culture;
- Often, even at the very first stages of EMS implementation, companies find opportunities for reducing unit raw materials, water and energy use;
- The 'third party' view in EMS implementation helps to reveal issues that were not accorded with the necessary attention, as well as find new solutions for difficult problems;
- EMS implementation positively affects the perception of the company's environmental activities as a whole, and helps personnel to recognise gradually that it is an integral part of their daily job;
- It helps to improve the production culture and workplaces;
- As a result of EMS, the introduction of cooperation and information exchanges amongst the company's functional units;
- Dissemination of information about the company's improved environmental performance creates a better image and marketing position.

Furthermore, and as certification experts frequently note, Russian companies which gradually increase their experience in operating the EMS, accordingly deepen their understanding of its role and possibilities, and thereby achieve greater positive results. This is evidenced by numerous Russian examples:

- According to the CEO of a mechanical engineering company, as a result of the implementation of certain EMS approaches due to developing management practices, the stability and controllability of the company improved notably.
- A not-for-profit project[11] has revealed that implementing EMS practices could result in water use reduction opportunities within a beverage enterprise. With the required capital investment of just US$1,500, the expected annual savings of such water use reductions were US$19,000.

---

or relied on very moderate language skills and legislation knowledge of such experts. As a result of such practices, for example, a company successfully passed its first certification audit and inspection audit, having had no such essential and required EMS elements as environmental aspects register or aspects identification procedure.

[11] Environmental Component of the Samara Regional Investment Initiative, Final Report (USA, Chemonics International Inc. December 1999).

- A company which produces diesel engines, while implementing an EMS, has reported to have reduced its production costs by 5% by minimising the use of copper and related waste.
- An oil-and-gas service company implemented a set of pollution prevention measures, spending US$200,000 on equipment, and achieved annual savings due to technology improvements of US$207,000 and due to reduction of 'polluter pays' fees and fines – additional savings of US$110,000.

Of course, such examples are not unique, and equivalents can be found at any company that successfully implements, or has implemented, an EMS.

Until recently, only industrial companies obtained EMS certification in Russia.[12] Similar to managers' experience in other sectors, managers and directors of industrial enterprises frequently found it difficult and time consuming to implement an EMS. This could not be sufficiently explained by the fact that it is difficult to transform a traditional administration model into a modern management model which is used by the ISO 14001 standard. In practice, while modernising the management of the facilities, many boards and CEOs have not unreasonably placed their first priority on financial output, and decided that it suffices to ensure there is no major problem which might affect the company's financial results. As pollution fees and fines have been, and still are, quite affordable, and as the market pressure for use of EMS in the majority of sectors is nonexistent, the environment seldom came to the attention of the very top management. This of course resulted in insufficient resources and employee motivation, even if a company took some first efforts to introduce new management approaches in the environmental sphere and higher environmental targets to achieve.

One of the largest Russian industries, having the lion's share of the market in its price range, can serve as an example of how long it can take to obtain certification. It first announced its readiness for EMS certification in 1998, and since then joined a not-for-profit project on EMS introduction, published numerous press releases and articles related to an almost-completed EMS introduction and, finally obtained the much-awaited certification near the end of 2005.

Current business practice

Nowadays, already many top Russian managers are increasingly watchful of the practices of their peers abroad and aim to improve the efficiency of their companies due to increased international competition, authorities' pressure, or resource costs – and now, therefore, consider implementing EMS into their operations. Regardless of the constraints described in this article, the interest and understanding in Russia of environmental management is increasing as experience grows.

---

[12] To the best of the author's knowledge, the first certificate in the agricultural sector in Russia was obtained in February 2005. As of March 2008, no certificate is known to be issued to organisations in education or in the public sector.

After a delayed start, the number of certified enterprises in Russia has begun to grow rapidly. In 2004, Russia passed the milestone of having issued 100 certificates (see Fig. 1). Those who have certified their EMS are just the tip of the iceberg: many more are developing their EMS gradually, or even implement the EMS in full, but are not interested in obtaining a certificate. For example there are oil service companies that pay certification bodies to conduct full audits, but then do not pay for the issuance of the relevant certificates – they do this because they were interested primarily in an independent assessment of their EMS, rather than in demonstrating publicly that they have implemented an EMS.

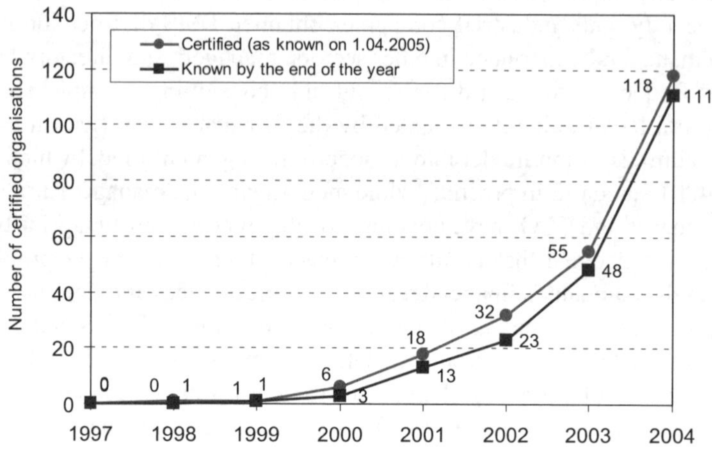

Figure 1. ISO 14001 certification statistics in Russia[13]

In this respect, the management of one of the largest Russian companies decided, not too long ago, to implement a full EMS in all of its assets. With plans coordinators, guidance and resources assigned, the original plans to do the work within one year are shifted again and again. The implementation of EMS transpired to be not as simple as originally perceived – various hampering factors included the lack of time, qualified staff and consultants, conflicting urgent business needs and many other reasons.

Concluding remarks

EMS in Russia has become an instrument which is of interest to many companies and it does not merely represent a fashionable trend followed by a few companies. The current situation with respect to the use of EMS in Russia, offers many opportunities both for foreign investors and Russian companies. Benefits can be derived from such entities working jointly with their partners, suppliers and contractors in order to im-

[13] *Source*: Open Register of EMS Certification in Russia, at <http://www.14000.ru/register/>; The ISO Surveys of ISO 9000 and ISO 14000 Certificates for 1998-2004.

prove their performance and reliability by using modern management instruments such as EMS. One point which should be remembered, however – Russians are reluctant to take general advice. Advice must be targeted, very practical, and developed jointly, which, in any event is the best approach to motivate and involve your partners.

Chapter 17
RENEWABLE ENERGY OPPORTUNITIES ON THE KOLA PENINSULA

Nina Lesikhina[1] **and Michele Hege Grønbech**[2]

There are vast opportunities for the development of renewable energy resources in Northwest Russia, as well as numerous reasons to do so. The Kola Peninsula, in particular, is one area which would benefit from the utilisation of clean renewable energy resources. Renewable resources have been analysed in great detail in this region due to the efforts of the international environmental organisation the Bellona Foundation and the Kola Science Centre of the Russian Academy of Sciences. In 2006, the Bellona Foundation commissioned scientists at the Kola Science Centre's Institute for Physical and Technological Problems of Energy in Northern Areas (hereinafter the Kola Science Centre) to investigate the prospects for renewable energy sources on the Kola Peninsula. The complete results of the scientific investigation were published in Bellona's report '*Prospects for Development of Non-conventional and Renewable Sources of Energy on the Kola Peninsula*' in 2007.[3] This article highlights some of the most important findings of the report. We begin with a brief summary of the significance of renewable energy for the region, recount the most important findings of the Kola Science Centre study, and then discuss the current regional initiatives of the Murmansk Oblast administration to devise a regional programme for the development of its renewable energy resources.

Introduction – clean energy and the Kola Peninsula

One of the regions in Russia where a transition to clean energy sources is most urgent is the Kola Peninsula. Today, nearly half of the energy on the Kola Peninsula is provided by nuclear power produced by the Kola Nuclear Power Plant ('KNPP') which poses an environmental risk to Northwest Russia as well as to its Nordic neighbours. The power plant is comprised of four reactors, two of which are operating with an

[1] Renewable Energy Adviser, Russian Department, The Bellona Foundation, Oslo, Norway; <www.bellona.no> and <www.bellona.org> ; Email: *nina@bellona.ru.*

[2] Coordinator of Energy Projects, Bellona-Murmansk, Murmansk, Russia; <www.bellona.ru> and www.bellona.org.; Email: *michele@bellona.no.*

[3] 'Prospects for Development of Non-conventional and Renewable Sources of Energy on the Kola Peninsula' is available at the Bellona website <http://www.bellona.no/rapporter/Clean_energy_Kola> in English and <http://www.bellona.ru/reports/Energy_Kola_Peninsula/Kola_Peninsula_report> in Russian.

W.Th. Douma and F.M. Mucklow, eds., Environmental Finance and Socially Responsible Business in Russia
© 2010, T·M·C·Asser press, The Hague, The Netherlands and the Authors

extended license to 2011-2014, nearly 10 years beyond their service life expiration date. Nevertheless, reliance on nuclear energy continues. There are plans to build an additional nuclear power plant, a KNPP-2, as well as a new power plant operating on coal. Currently, annual energy consumption is approximately 16 TWh but this is expected to increase due to industrial development plans. In addition to nuclear energy, power and heat is also provided by large hydro-electric stations, thermal power plants, boilers and diesel-electric stations running on mazut (heavy oil) in decentralized settlements.

The region suffers from environmental degradation. It has an unresolved problem with storage of radioactive waste left over after the Cold War. In fact, the Kola Peninsula was nominated among the top 25 most polluted places on earth in the Blacksmith Institute report '*The World's Worst Polluted Places – the Top Ten*' in 2006.[4] The area also suffers from high sulfur dioxide pollution due to emissions from the nickel industry.

It is against this background that the Bellona Foundation commissioned the Kola Science Centre to investigate the prospects for utilisation of renewable energy resources with a view to finding a solution to some of the environmental challenges in the region. Bellona's aim is to provide an impetus for renewable energy initiatives by providing scientific evidence of regional potential upon which decision-makers, as well as industry and investors, can base development of renewable energy on the Kola Peninsula. There are already regional signals that this is occurring. Bellona hopes that one day implementation of renewable energy will occur on a large-scale, sufficient to warrant decommissioning of the KNPP.

Below, we begin with a consideration of some of the reasons why renewable energy is important for the region. Next, we highlight the most important findings of renewable energy potential from the Kola Science Centre study. Thereafter, we discuss the regional initiative of the Murmansk Oblast that is currently underway to develop a long-sighted regional programme for renewable energy development. Finally, we make a number of recommendations that might facilitate a transition to clean renewable energy.

The significance of renewable energy for the Kola Peninsula

There are numerous reasons for developing renewable energy sources on the Kola Peninsula, foremost among them ecological, but also economic and socio-economic, political and practical reasons.

Above all, renewable energy should be developed for regional environmental safety and health considerations. Unlike nuclear energy, renewable energy sources impose no risk to human health; they are ecologically clean and produce no hazardous waste. The most pressing issue in the region is the condition of the two oldest reactors at the KNPP

[4] 'The World's Worst Polluted Places – the Top Ten', a project of the Blacksmith Institute (October 2006) Appendix 1, pp. 33-34, <http://www.blacksmithinstitute.org/top10/10worst2.pdf>.

which exceeded their service life in 2003 and 2004. These reactors are over 30 years old and have lower safety standards than Western reactors. In addition, the region is yet to recover from the legacy of radioactive pollution from the Cold War. Nuclear energy entails considerable risks – danger of nuclear accidents, leakages from radioactive waste storage and spent nuclear fuel assemblies, transportation, reprocessing, nuclear weapon proliferation and security issues. According to a safety analysis made by the International Atomic Energy Agency in 1991, the chance of a reactor meltdown is 25% in the course of 23 years.[5] These risks have been reduced, but not eliminated, through many costly projects to improve nuclear safety at the KNPP via multilateral projects funded by the EU Technical Assistance to the CIS ('TACIS') programme and the EBRD's Nuclear Safety Account ('NSA'), as well as unilateral initiatives like the Norwegian government's Action Plan for Nuclear Safety in Northwest Russia.[6]

There are various good reasons for developing renewable energy on the Kola Peninsula:

- Renewable energy sources should be developed for *climate change mitigation reasons* because they emit no, or very little, CO_2 emissions. Russia is the third largest emitter of CO_2 gases, after China and the USA. Reducing emissions should be a priority in a region which is already suffering considerably from pollution. It should also be a priority to prevent the undesirable and negative consequences of climate change, many of which will be particularly detrimental for the regions of the north: erosion, loss of coastline, reduction of biological diversity and extinction of some polar species, collapsing infrastructure, houses and roads in permafrost areas, damage to pipelines, and more extreme, unpredictable weather. In fact, there has been a 7% reduction in permafrost in Russia over the last 100 years – whilst thawing permafrost releases methane, a greenhouse gas more detrimental than CO_2. According to the Russian Federal Agency on Hydrometeorology and Monitoring of the Environment, general climatic warming in Russia over the last 100 years is greater than the global average and there is an increased occurrence of extreme acts of nature.[7]
- Renewable energy should be developed for *economic reasons*. Renewable energy can become a profitable sector providing the basis for new employment opportunities and income. There is general trend toward population decrease and loss of skilled labourers in the region. Since its peak in 1991, the population has been decreasing at the rate of 2-3% per year.[8] From 1991-2002, the population declined 16.7% in the Murmansk Oblast and 50% of this decline was due to emigration; this population decline is more severe than in other regions of Northwest Russia.[9] Expanding

[5] 'The Arctic Nuclear Challenge', Bellona Report, Vol. 3 (2001) pp. 40-43.

[6] <http://www.nti.org/db/nisprofs/russia/forasst/intnatl/intnatl.htm#TACIS>; <http://www.nti.org/db/nisprofs/russia/reactor/power/kola.htm>; <http://www.nti.org/db/nisprofs/russia/forasst/intnatl/intnatl.htm>; <http://handlingsplan.nrpa.no/English/englishfrontpage.htm> .

[7] Doklad ob Ocobennoctyax Klimata ha Territoriy Rossiskoy Federatsiy za 2006 god.

[8] 'Environment and Living Conditions on the Kola Peninsula', Fafo-Report 260, FAFO Institute for Applied Social Science, Section 2.2 Overall Population Development, p. 28.

[9] N. Filatov et al., *White Sea: Its Marine Environment and Ecosystem Dynamics Influenced by Global Change* (New York, Springer 2005) pp. 250 and 252.

the industrial base, or utilising the existing one, for the production of renewable equipment and parts, for example, can boost economic development, as well as generate jobs and tax revenue in the region. Renewable energy can also help meet the increasing demands for energy in connection with the revival in industrial development in the region.

- Renewable energy sources should be developed for *energy security reasons*. On the Kola Peninsula energy is based to a significant degree on fuel imported to the region. In addition, traditionally, fuel costs and tariffs on electricity and heating have been high. Nearly 100 small decentralised settlements and outposts on the eastern portion of the peninsula are off-grid and entirely dependent on fuel import. Transportation is limited due to harsh climate and undeveloped infrastructure. In fact, many of the areas experiencing a lot of difficulty with fuel supply possess renewable resources. Developing the renewable sources of energy, which are locally available, can ensure a stable heating and electricity supply to remote settlements relieving them from unreliable fuel supply, unstable fuel prices and high costs associated with fuel transport over long distances. Renewable energy can also protect consumers who are dependent on centralised energy supply and experience electrical power shortages and cut-offs, by guaranteeing them a minimum supply of energy during periods of high demand and energy deficit.
- Renewable energy sources are *easily accessible, virtually unlimited and sustainable*, in contrast to fossil fuels. Despite large deposits of oil, gas and coal, Russia's supply of organic fossil fuels is limited. Fossil fuel reserves are already diminishing and the most easily accessible reserves are already exploited. Seventy-five percent of onshore oil and gas fields are already in production and have an average depletion of 50%.[10] The remaining reserves are primarily offshore in areas characterised by extreme climatic conditions and whose development necessitates high investment and special technology. The Kola Peninsula is dependent on fossil fuel import as it has no large fossil fuel reserves of its own, yet there are no commercial fields in operation nearby in the Barents region. At the same time the Kola Peninsula possesses an enormous untapped potential of renewable energy sources, especially wind and small river hydro, which is sustainable and easily accessible.

Another reason to develop renewable energy in the region is its popular appeal. Neither the local population on the Kola Peninsula, nor the neighbouring region wants the risks associated with another nuclear power plant. According to a survey conducted in the Murmansk region by the Russian market research company Romir in December 2007, 87% of respondents were against construction of a new nuclear station, and 85% were against prolonging the life of old reactors at the KNPP.[11] Other results from polling at public hearings where the region's energy policy was discussed in December

---

[10] Y.P. Trutnev, RF Minister of Natural Resources, 'Concerning the Enhancement of Efficiency of Hydrocarbon Resources Development on the Continental Shelf of the Russian Federation', <http://www.geoinform.ru/?an=trutnev_en>

[11] <http://www.romir.ru/en/news/res_results/184.html>, and <http://www.bellona.org/articles/articles_2008/murmansk_poll>.

2006 showed that 93% of Murmansk citizens surveyed were in favour of prioritising wind energy development. Mobilising support for renewable energy will be easy at the regional level.

The potential for renewable energy on the Kola Peninsula

The prospects for six renewable energy resources – solar, wind, hydro, tidal, wave and bioenergy resources were evaluated by the Kola Science Centre. The complete results are published in the Bellona's report 'Prospects for Development of Non-conventional and Renewable Sources of Energy on the Kola Peninsula'.[12] A short summary of the potential for each renewable resource is provided below.

Solar energy

Solar resources are the most abundant of alternative energy sources. However, given the characteristically cloudy conditions on the Kola Peninsula, direct radiation is reduced by 60-70%. The biggest obstacle to developing this energy source is the absence of solar radiation in the winter, when the demand for energy is highest. Seasonally, there is great monthly variation in sunshine, from zero hours in December to 200-300 hours in June and July when the most intensive solar radiation is observed. Fortunately, solar and wind energy resources peak in opposite seasons making it possible for these two resources to complement one another when used jointly.

Scandinavian experience shows that the application of solar energy can be effective for the entire heat supply system. The challenge is to store solar energy in sufficiently large quantities during the summer months. Reservoirs serving as heat accumulators can be either above ground or underground. The main elements of a simple system are a solar collector, an accumulator for storing energy until it is needed to meet heating requirements of the consumer, and a reserve energy source for use when the sun is absent for prolonged periods of time. An advanced circuit can include an auxiliary energy source such as a wind installation. Although solar energy is not the most feasible renewable energy source to develop given the Kola Peninsula's latitude as well as high costs of solar panels, it can be prospective for remote isolated consumers with expensive heating costs and difficult fuel supply, and southern locations with a high technical potential. Solar radiation data for the region is comparable to that for Sweden. According to the Kola Science Centre's data, radiation in the vicinity of Umba (a remote settlement with a population of 6,500) is similar to that from Ingelstad, Sweden where a solar heating station successfully warms 52 houses. This makes Umba a promising site for a solar project. In the northern latitudes, solar energy potential is inferior and should be considered as a complement to hydro or wind power, the latter of which peaks in winter.

---

[12] *Supra* n. 3 for details.

In recent years, solar energy was used successfully in one project in the region. A joint Norwegian-Russian project gradually replaced decaying, radioactive strontium batteries in lighthouses along the northern coast of the Kola Peninsula with solar cells.

Wind energy

The opportunities for large-scale development of a wind energy system in the Murmansk region are just as large as they are in Denmark, Germany, Spain or the United States where wind energy is already a source of revenue. Russia possesses the necessary scientific and industrial capabilities, as well as experience from pilot wind electrical stations in Vorkuta, Kalmyka and Kaliningrad.

Wind resources on the Kola Peninsula are enormous and estimated at 360 billion kWh. The greatest wind speed is observed in the coastal regions of the Barents Sea. Along the northern coast of the Kola Peninsula, which is the windiest place in the European part of Northern Russia, average wind speed reaches 7-9 meters per second. This average annual wind speed is actually higher than in coastal areas in Denmark, Germany and the Netherlands where wind energy predominates and is profitable. On the Kola Peninsula the variation in average annual wind speed is only 5-8%, which is low in comparison with the 15-20% variation for river flow in the region.

Maximum wind speeds occur in the winter and coincide with the seasonal peak in heat and energy consumption. This winter wind maximum is in a counter phase to the annual river flow creating an advantageous opportunity for wind energy and hydro power to successfully complement one another when used jointly. During the summer when wind speed is reduced, the maximum wind speed occurs during the daytime when an increased consumption of energy on the part of consumers is usually observed, creating another favourable condition for efficient use of wind energy.

The high wind potential on the Kola Peninsula, the correspondence between winter maximum wind intensity and maximum energy consumption, and the presence of 17 hydroelectric stations with reservoirs create unique conditions for utilizing wind energy on a large scale. The region is also especially favourable for multi-turbine windparks due to the presence of prevailing winds which have the greatest energy concentration and allow a compact, and less expensive, location of wind installations with minimal interference and energy loss. Along the northern shore, south-western winds predominate approximately 50-60% of the year. Particularly prosperous areas for construction of multiple windparks are Dalniye Zelentsy and Teriberka located near the Serebryansky and Teribersky hydro-electric stations, which are connected to the Kola energy system and capable of facilitating large-scale use of wind energy in the region. A system that links windparks to hydroelectric stations will produce the greatest economical effectiveness.

There are also several favourable conditions for utilising wind energy to supply electricity or heat to small isolated settlements, meteorological stations, lighthouses, frontier posts and Northern Fleet establishments, located in windy areas. These dispersed consumers receive electrical energy from autonomous diesel electric stations, rather than from the Kola energy system. They pay high costs for diesel supply due to poor transport arteries; expenses are 30-70% higher along the coast and 150-200% higher in inaccessible inland areas. Wind energy can enable economising of expensive diesel fuel which in the most extreme cases is transported a distance of up to 1,500-2,000 kilometres. In the windiest areas, wind electric converters can supplant up to 60-70% of fossil fuel. During periods of prolonged lulls in wind, special accumulating units or auxiliary heating systems can be switched on.

In selecting appropriate sites for location of windparks, it is necessary to consider areas with high wind potential, infrastructure (roads) and an entrance to the energy system (grid) or a high-power sub-station, location of base personnel and installation facilities. From a practical point of view, wide open spaces are recommended over hills because this simplifies construction of underground paths and erection of the wind converters. Sites which meet these requirements are the area surrounding Serebryansk and Teribersk hydroelectric stations, the village of Lodeinoye, Verkhny Teribersky water reservoir, and the village of Tumanny.

There is one pilot windmill in the region, located at the Murmansk hotel 'Ogny Murmanska'. This is a recycled windmill imported from Denmark. Including transport, construction of the foundation and erection the windmill cost about 4 million roubles (approximately US$120,440), or $750 per kilowatt. Annual operation costs are 3 cents/kWh, less that the hotel would pay to buy energy from the grid. The windmill was connected to the grid in 2008 and now supplies energy to the Kola energy system.

There are two approved projects to build windmill parks on the Kola Peninsula: a 100 MW windmill park scheduled to be constructed in Liinahamary (by 'Russkiy Veter'), and a 200 MW windmill park, which is scheduled to be constructed in Teriberka (by Windlife Arctic Power). Both of these companies are in the process of obtaining land permits. The larger of these two parks will be financed through an EBRD loan. There are also plans to build a 5 MW windmill park near Ostrovnoy. In addition, there are ideas for the construction of a 1.5 MW park near Kislogubskaya Tidal Station (by a daughter company of RusHydro), as well as for erecting windmills along the Terskiy coast, a remote area of the Kola Peninsula.

The energy of small rivers

On the Kola Peninsula, the economically feasible hydro resources are already exploited on large and medium-sized rivers by 17 hydropower stations which produce 6 billion kWh and provide 42% of the total energy production in the Murmansk region. Ecological and financial limitations make construction of hydro-electric stations on the remaining undeveloped large rivers problematic. 'Kolenergo' is working on improving utilisation of the existing hydro-electric stations, but small rivers remain undeveloped. Compared to traditional large hydropower stations, development of small hydro resources will have less environmental impact.

The technical potential of small hydro resources is approximately 4.4 billion kWh, one-third of which is economically feasible to exploit. A confluence of several factors make development of small hydro power advantageous – periodic fuel crises, increases in tariffs on electrical energy, restrictions on construction of large complexes due to their negative ecological consequences, and progress in automation and remote control of hydro-generators. Small hydro-electric stations are defined as stations with an installed capacity under 20-30 megawatts. In the Russian north, the majority of small hydro-electric stations do not exceed 3-5 megawatts. China is the leader in the construction of small and micro hydro-electric stations with over 100,000 plants in operation and produces small turbines based on Russian technology. Two basic trends observed in the use of small hydro-electric stations in industrialised countries, namely the use of dams and water reservoirs created for water supply and the use of small channels and traditional lay-outs, are virtually non-existent in Russia at the present time.

Excluding from consideration the rivers which are significant for the fishing economy of the Kola Peninsula such as the Nota, Varzina, majority of the Umba, Varzyga and Strelna, leaves many prospective rivers for small hydro power. The following rivers have promising sites for small system integrated hydro-electric stations: Pirenga (1 site), Bolshoy Olenka (2), Ura (2), Zapadnaya Litsa (1), Titovka (1), Tumcha (3), Umba (1).[13] Kolenergo and Territorial Generating Company Nr. 1 (TGC-1) are planning to utilise all these rivers (with the exception of Umba) in accordance with their development strategy.[14] Construction of small hydro power plants on the Pirenga and the Tumcha are probably the most feasible and effective.

Small hydro power can provide an inexpensive, independent source of electrical and thermal power to remote areas with de-centralised heating supply, where diesel electric stations are the primary source of electric power and the cost of electrical energy is largely dependent on the cost of diesel. The Murmansk region has previous experience from the 1950s in using small hydro power for the benefit of isolated consumers. Currently, 80-100 settlements exist which are not encompassed by the centralised electric power supply. Their energy requirements vary from 5-10 to 500-800 kilowatts. Three isolated villages are prime candidates for small hydro power: Krasnoschele, Chavanga and Chapoma, as well as the military border settlement Svetly. Fuel supply to these areas is extremely difficult due to the absence of roadways. Small hydro power can be used as a supplement to diesel electric stations which can cover electric demands during periods of little water and act as a reserve for emergencies. The Ponoy tributary on the Yelreka River, the Chavanga River and the Lotta River are the best candidates for pilot small hydro electric stations that can demonstrate the advantages and efficiency of small hydro power for decentralised consumers.

Tidal energy

Tidal electric power stations are also a source of ecologically clean energy which do not pollute habitat with harmful wastes (unavoidable during operation of thermal electric power stations) and do not require any kind of flooding (unavoidable during construction of large hydro electric power stations on level rivers). Tidal energy is constant throughout the month and is independent of annual and yearly water content, despite

[13] See *supra* n. 3, pp. 53-54.
[14] See *supra* n. 3, p. 59.

intervals in the 24-hour cycle and irregularities during the lunar month. These qualities make it a rather powerful source of energy which can be combined with river hydro electric stations containing water reservoirs.

The area of the tidal basin and the tidal height are the most important indicators of theoretical potential of tidal energy. Along the coast of the Kola Peninsula, the relatively small size of the tidal range (2-3 metres) and the limitations of an aquatory allowing a cut-off from the dam, make construction of some possible tidal electric stations economically unfeasible. Economic and technical calculations show that it is most economically expedient to construct medium- and large-size tidal power plants (TPP) and utilise water reservoirs used by pumped storage power plants, or hydroelectric stations which can transform the cyclical unevenness of tidal energy into controlled dependable energy. Additional costs for construction of a regulating reservoir and for transmission lines connecting the hydroelectric station with the TPP are compensated for by the exchange of a portion of energy from the thermal station to the tidal.

There are several promising sites for development of TPP. The Mezen TPP, currently under development at the Abramov-Michaelovsky Cape where the tide reaches 10 metres has the greatest technical potential. This project requires construction of a pumped storage power station on the Ribachy Cape of the Kola Peninsula, which could guarantee an additional 3 million kilowatts of electrical power to the central northwest provided a high-voltage connection exists to the Mezen station. The expected total annual output for Mezen TPP is 50 billion kWh. Another site is a remote area at the Lumbovsky gulf where the average size of the tide is 4.2 metres and the aquatory is 70-90 km^2. A third site, the Kola TPP, is a pilot project under development in the Dolga Bay and will provide a prototype design for the Mezen TPP. Lastly, there is the Kislaya Bay TPP. Constructed as a pilot project in the 1960s to provide scientific and technical experience necessary for larger, powerful tidal plants like Mezen, its operation is now terminated. A ship-building factory in Severodvinsk, Archangelsk region, is currently preparing a prototype working wheel for the Mezen TPP which will be tested at Kislaya Bay.

Russian and foreign research conclude the importance of preliminary small pilot tidal electric stations for experimenting with hydraulic turbine equipment when constructing tidal electric power stations with unique output capacities like that of the Mezen TPP.

Wave energy

Wave energy possesses a higher energy density than wind and solar energy. Two merits of ocean waves along the Kola Peninsula are their ability to accumulate wind energy over a significant distance and their availability to a large group of consumers along 1,000 km of coastline. The disadvantages of wave energy are its periodic instability, dependence on ice conditions, and difficulties of conversion and transmission to the consumer. The possibilities of exploiting wave energy have been explored for over 200 years. However, the practical utilisation of wave energy is very complex necessitating:

reception and conversion of energy; a strong reinforcing system capable of withstanding a large load, especially in extreme conditions; evaluation of parameters and behaviour of windy seas; and study of the impacts of a wave installation on the environment.

An important peculiarity of ocean waves in the northern zone of the Atlantic Ocean is their seasonal irregularity. During the winter the waves are approximately twice as high as during the summer. The Barents Sea is estimated to have a wave energy potential of 22-29 kWh per meter which is close to data on wave potential in the neighbouring coastal region of Norway. Generalising from data, scientists conclude that the specific average annual wave capacity in the Barents Sea consists of approximately 20-25 kWh per meter. As far as the White Sea is concerned, the average annual potential of wave energy is considerably lower, 9-10 kWh per meter, due to the comparatively small size of the sea and the presence of ice cover in winter.

Wave energy has one of the highest efficiency rates of renewable energy. The total net efficiency of wave electric stations in converting energy to electricity is 30-80%. If one calculates wave energy in the Barents and the White Seas using a 60% efficiency rate as a basis then total technical resources of approximately 450 billion kWh per year is estimated. Along the coast of the Kola Peninsula the technical resources of wave energy for a 10 kilometre strip consist of 1.2 billion kWh for the Barents Sea coast and .4 billion kWh hours for the White Sea coast. The expected capacities of wave power plants in these locations are 230 MW and 100 MW. But to date no models have been tested in these areas.

Bioenergy (biofuels) resources

In comparison with other types of renewable energy on the Kola Peninsula, bioenergy resources are relatively small. In the Murmansk oblast, the potential for bio resources from agricultural waste is approximately 1 billion kWh annually. Processing agricultural waste by the application of anaerobic fermentation provides a solution to three problems: ecological – disinfection of animal wastes and removal of disease-causing substances, food supply – production of high-quality organic fertiliser which increases agricultural productivity by at least 10%, and energy – a partial replacement of liquid and gas fuel by biogas. Russia has conducted research on optimisation of technical methane fertilization of organic animal and bird waste, and began construction of experimental pilot-industrial installations in the 1980's.

Biogas can be produced from the anaerobic treatment of organic wastes. In its purified form biogas is compressed into cylinders and used for operating cars and tractors, or burned for heating. One cubic meter biogas equals .6 litres mazut/black fuel oil and is sufficient to power a motor with a 2-horse power capacity for 1 hour or yield 1.25 kWh electricity. Annually, the Murmansk oblast can produce: 2.3 million cubic meters biogas or 1,700 tons fuel oil equivalent from pig manure; 5.6 million cubic meters biogas or 4,000 tons fuel oil equivalent from cattle manure; and 15.1 million cubic meters biogas or 9,800 tons fuel oil equivalent from poultry manure. The marketable yield of biogas is estimated at 12-17 million cubic meters biogas or 7,000-11,000 tons fuel oil equivalent.

In the Murmansk region, waste from the forest and wood-processing industry is used in an insignificant volume as fuel for electricity and heating. A main obstacle to utilising timber waste is undeveloped infrastructure. Murmansk possesses the least forest residues of all regions in Northwest Russia and its potential for bio resources from lumber waste is estimated at only 1.5 billion kWh. Therefore, small population points which receive electricity from local diesel electric stations and heat from community boilers are a potential sector for utilisation of lumber waste. There are 150 diesel electric stations and 355 small boilers in the Murmansk oblast, but no pilot projects to use bio wood waste. Introduction of technology for briquettes and pellets from sawdust, and combustible distillates from industry and domestic waste could contribute to the development of bioenergy as a resource.

The Kola Peninsula has two successful experiences with bioenergy. The boiler station at Verkny Tulomskiy has been converted and now operates on wood waste provided by a local sawmill. The Verkny Tulomskiy bioenergy station supplies heat to the small settlement of Verknytulomskiy. The Kovdorskiy agricultural complex at Kovdor uses manure for biogas production to provide the farm with energy. A similar project based on farm waste is being planned for the Tuloma agricultural complex in Tuloma. And the agricultural complex Industriya in Apatity is another potential site for biogas production.

Regional initiatives

Despite the region's current reliance on nuclear energy and imported fossil fuel, development of renewable energy on the Kola Peninsula is supported by the Murmansk Oblast administration. (However, it is seen as a supplement to, rather than a replacement for, nuclear energy.) Following up on NGOs initiative to put renewable energy on the agenda in 2006, the Murmansk Oblast administration created a Working Group on Renewable Energy in March 2007 under the Ministry of Natural Resources and Environmental Protection (hereinafter the Working Group). The mandate of the Working Group is to consider all types of renewable energy sources and work out a regional programme for renewable energy development.

The composition of the Working Group deserves some attention. It is comprised of a diverse group of 17 representatives from the region: the Ministry of Natural Resources and Environmental Protection, the Ministry of Industry and Transport, the Department of Construction and Housing, the Kola Science Centre, the Kola branch of TGK-1, the fishing cooperative 'Chapoma', the 'Cleaner Production Program', Murmansk State Technical University, the Murmansk Oblast Energy Efficiency Center, the regional dispatcher of the Murmansk Energy System 'Kolskoye RDY', the wind developer 'Vetroenergo' and NGOs (Nature and Youth, Gaia, and Bellona Murmansk). The inclusion of NGOs, and not just one, but three, is truly remarkable.

At an early stage, participants of the Working Group agreed to use the scientific findings of the Kola Science Centre study, that is, Bellona's report 'Prospects for Development of Non-Conventional and Renewable Sources of Energy on the Kola Peninsula', in development of the Murmansk Oblast regional programme. Governmental authorities, the Ministry of Industry and Transport and the Ministry of Natural Resources and Environmental Protection, ordered the Murmansk Oblast regional programme. The Kola Science Centre won the tender to develop it and is doing so in consultation with the Working Group.

A draft for a long-term target programme 'Development of Non-Conventional Renewable Sources of Energy in the Murmansk Oblast from 2009-2015' is already completed and undergoing revision. At the time of writing, the final document was expected to be finalised in the autumn of 2009 and then to be approved by the Murmansk regional government. The draft programme calls for increasing renewable energy sources in the regional energy balance from the current 0,01% to 7,4% by 2015. The programme is expected to create the necessary conditions for implementation of renewable energy and produce the following results: experience with renewable energy, improvement of energy supply particularly to decentralised areas, a decrease in expenditures on fossil fuels, a decrease in harmful emissions and an improvement in the ecological conditions of the region. The draft programme also specifies which renewable energy projects will get priority and extra support. Bellona has reviewed the draft legislation and offered recommendations to strengthen the content of the regional programme.

Bellona's initiative

Bellona advocates a cross-border, multidisciplinary approach that includes a wide variety of specialists as the best means of supporting rapid development and implementation of renewable energy sources on the Kola Peninsula. In April 2008, Bellona Murmansk and the Bellona Foundation established the *Northwest Russia Renewable Energy Forum* as a permanent arena for dialog, cooperation and problem-solving in developing and implementing renewable energy in Northwest Russia. Bellona conducts regular seminars and meetings for diverse stakeholders to discuss the possibilities, obstacles and mechanisms related to development of the region's clean, sustainable energy sources and to work towards implementation of renewable energy. The purpose of the forum is to stimulate information exchange, research, development, investment, implementation and commercialization of renewable energy resources in the region. Bellona is convinced that by pooling the resources from research, business and industry, government and NGOs it is possible to achieve environmentally sustainable development of the renewable energy resources in the region and attract investors.

Bellona intends used the Northwest Russia Renewable Energy Forum[15] to promote good environmental standards and favourable conditions for renewable energy and to facilitate the involvement of regional participation in development of renewable energy resources on the Kola Peninsula.

---

[15] See <http://www.bellona.org/activities/Northwest%20Russia%20Renewab[illegible]Energy%20Forumeng>.

Conclusion

Developing sustainable renewable energy sources will have positive environmental consequences, provide a solution to local, as well as regional, concerns about energy supply and security, economise on expenditures for fossil fuel import and contribute to regional business and industrial development. Many prerequisites exist for development of renewable energy on the Kola Peninsula: tremendous natural resources, public opposition to nuclear energy and support for renewable energy, the presence and involvement of a scientific community and regional government support (legislation pending). There is really no reason not to pursue an aggressive exploitation of renewable energy on a large scale. Renewable energy can benefit both centralised and de-centralised users, and resources are in sufficient quantity to meet both current and future energy demands.

According to the findings of the Kola Science Centre study the Kola Peninsula possesses a sufficient quantity of natural renewable energy sources to make renewable energy a feasible alternative to nuclear energy and fossil fuels. In particular, the region possesses one of the greatest wind resources in Europe estimated at 360 billion kWh. In addition to wind resources, hydro resources of small rivers, tidal power and biomass are prospective and should be developed. Clearly, even if only the most easily accessible resources are exploited, the combined use of all renewable energy resources available in the region could easily meet the current power demands of the region and, in Bellona's opinion, match the power capabilities of the reactors at the KNPP, thus permitting their retirement.

Development of renewable energy on a large scale will necessitate technical and economic feasibility studies, pilot projects and reinforcements in the region's infrastructure (roads, power lines and connecting stations). Bellona expects that the approval of the regional programme by the Murmansk administration will open opportunities for participation of a wide range of stakeholders, such as energy consultants, project developers, technology experts, suppliers, international financial institutions and investors, researchers, the construction industry, and so forth. Regional actors should make an effort to participate in development of renewable energy by providing capacity building on project design, technical training, scientific collaboration, cooperation on infrastructure projects, technology transfer, and environmentally responsible financial investments. Lastly, Nordic neighbours should refuse to import energy produced by the nuclear reactors of the KNPP and begin to work towards creating the necessary conditions for import of renewable energy from the Kola Peninsula as an additional incentive to develop clean renewable energy on a large scale.

Chapter 18

INVESTMENTS IN MURMANSK OBLAST AND POMOR REGION – SOME REFLECTIONS

Jeffrey Roberts[1]

Murmansk Oblast and the Pomor Region, both located in the far north of the North-West Federal District of Russia, are still a remote corner of Russia, not just with respect to environmental investments and socially responsible business activities, but generally, too, with respect to foreign direct investment ('FDI'). This chapter is a personal account of my experiences of doing business in Murmansk Oblast and the Pomor Region and is a first-hand account of how environmental investments are still few and far between in this far Northern region. However, there are signs that this part of Russia, too, is starting to attract environmental investments. There are many opportunities that might develop in view of possible legal developments to incentivize companies to engage in environmentally sensitive investments.[2] Whilst precedent projects in the region, such as the greening of JSC 'Archangelsk Pulp and Paper Mill' with the support of the Russian National Pollution Abatement Facility,[3] the clean production programme[4] and considering the renewable energy opportunities in the Kola Peninsula,[5] promise that environmental investments are starting to take a foothold in this remote region of Russia.

When discussing the investment trends and practices in Murmansk Oblast and the Pomor Region, I focus on the abundant natural resources (in particular, oil, gas and minerals) as possible investment opportunities. This stems perhaps from my geologist background, but also reflects the predominant current (and past) investment trends in this region. Although there are numerous projects underway (for example, the Shtokman project[6]), the geographical problems of Murmansk Oblast and the Pomor Region re-

[1] Director, Impivaara Securites/Pomor Petroleum Ltd <www.impivaara.com>, or <www.pomor.com> or <www.kovdor.com>. Email: *jrobertsa@cix.co.uk*.

[2] See chapter 4 (*Ratsiborinskaya*) and 7 (*Nystén-Haarala*).

[3] See chapter 15 (*Maksimenko and Gorkina*) and see further environmental projects being developed in connection with 'joint implementation' projects in Russia at chapter 13 (*Yulkin*).

[4] See chapter 4 (*Ratsiborinskaya*).

[5] See chapter 17 (*Lesikhina and Grønbech*).

[6] The Shtokman Development Company, owned by Gazprom with its key foreign partners, Total and StatoilHydro, is preparing for the implementation phase of the Shtokman gas project (see <www.barentsoberver.com>, newsfeed 'Credits to finance Shtokman development', 8 December 2008) and the project is reported to be scheduled to be in operation by 2013 (see <www.barentsobserver.com>, newsfeed 'Foreign partners might get bigger role in Shtokman, Nord Stream', 16 December 2008). The Barents Observer further reported that the investments for the first Shtokman development phase has earlier been estimated at US$ 15 billion (see newsfeed, 16 December 2008). The Shtokman field is one of the

W.Th. Douma and F.M. Mucklow, eds., Environmental Finance and Socially Responsible Business in Russia
© 2010, T·M·C·ASSER PRESS, The Hague, The Netherlands and the Authors

main and hamper full-scale investment activity. These problems include climatic conditions and environmental problems (e.g., oil pollution, nuclear contamination),[7] language barriers, cultural and historical context of this region of Russia, as well as accessibility and informational barriers. Nevertheless, I do not wish to paint a bleak picture. My personal experiences of successfully investing in this part of Russia promise that with the correct amount of enthusiasm, persistence and knowledge of the Region, green investments, too, could flourish in this part of Russia, promising returns and profits together with environmental benefits.

THE SHARE MARKET: THE ROAD NOT TAKEN

For the last twenty years, I have been much involved in natural resources and share markets bordering Northwestern Russia. Finland was my starting point and Murmansk Oblast the jewel in the crown of natural resources investment. Despite its extraordinary importance as the only ocean going (and all-year round ice-free) port in the west of Russia, the mineral wealth of the Kola Peninsula[8] and the enormous hydrocarbon potential of the Barents and Kara Seas, I have to report a history of disappointment from a major foreign portfolio investment point of view.

On the other hand, my personal investment in Murmansk Oblast, with a very small number of Russian and English friends, has been a remarkable success, bearing out old wisdom on modest, helpful business experience in both the Soviet Union and Tsarist Russia.

However, a stockbroker is supposed to make money for his clients, and my first mentor in the City of London in 1973 told me to look after 'other people's money' honestly through success and failure by reliable intelligence, and that such 'other people' would then turn into trusty clients who then paid you.

So it is disappointing to have done well for myself on a modest scale, whilst trying to lead 'greater horses' to water and failing, on the basis of reliable intelligence, to persuade them to 'drink' at the correct time. However, I have to note that there have been plenty potential 'wild horse' investors, whose activities did not come to fruition, by stampeding to drink at the wrong wells at the wrong time, but that is '*ein weites Feld*' as Effi Briest's father used to say.

---

world's largest natural gas fields and lies in the central part of Russian sector of the Barents Sea, 600 km north of the Kola Peninsula

[7] See for example, WWF Russia website <www.wwf.ru>; A. Bambulyak, B. Frantzen, *Oil Transport from the Russian Part of the Barents Region* (Status 2005), Svanhord Environmental Centre, at <http://www.wwf.ru/data/publ/oil_report_english_1.pdf> for more detail.

[8] The Kola Peninsula is in the far north of Russia and part of the Murmansk Oblast. It borders upon the Barents Sea on the North and the White Sea on the East and South. The west border of the Kola Peninsula stretches along a meridian from the Kola Gulf through the Imandra Lake, the Kola Lake, and the Niva River to the Kandalaksha Gulf.

CULTURAL AND GEOGRAPHICAL PROBLEMS

I have first to confess to a polar bias in my view of Europe and Russia: I look at it from 71 degrees north. I first came to Finland from Finnmark,[9] the northernmost county of Norway, in 1965 when I had done my undergraduate field mapping in the Tanafjord. British geologists very much favoured Finnmark because there were not too many British geologists there; in contrast to the Northwest Highlands of Scotland where they were like flies on a fresh pancake. As for the Caledonian geology of Finnmark, it was 'difficult enough to be interesting, but not so difficult as to be impossible' as put to me by Olaf Holtedahl, the grand old man of Norwegian geology, forty years ago. His student, Sven Føyn had made brilliant discoveries in the 'Gaissa' of Finnmarksvidda showing a clear analogy with the Scottish Highlands at a time when Finnmark seemed too far away to people in Oslo. And far away it was – in 1965, we travelled by coastal boat from Bergen over 6 nights (4 nights on the Express) or drove up winding roads for a similar time, with a whole day crossing 5 ferries in Lofoten. Today, Finnmark is better known and more accessible due to better roads (whose building has encouraged remote communities to leave).

Language

When I first arrived in Finnmark, nobody spoke English, and the local storekeeper in Leirpollskogen changed my 'literary' book Swedish into proper spoken Norwegian, so that Nordmenn know which fjord I learnt it in to this day. Here was my first language lesson, which I will offer below as one of the missing links in Russian investment. Business is verbal (oral) not written, and depends on personal contacts. Thus, as I will explore further, the need for foreign investors to learn (at least some) Russian language is vital, especially with respect to business dealings in the far North of Northwest Russia, in order to establish good personal contacts and to build trusting business relationships with local partners. It seems some investors overlook this vital point. For example, during all my years of visiting Murmansk and attending, or speaking at meetings there, I have never met anyone from the financial sector doing research on the greatest frontier resources imagineable. However, it should be said that at the 2006 Arctic Shelf Conference,[10] there was excellent attendance (300) of oil industry people who knew its importance. Sadly, their numbers had dwindled to a fraction of that at the 2008 conference.

Naturally, language skills are key to making a success of investments in this far northern part of Russia. In addition, however, access, local geography and the local administration's incentives to attract FDI are some of the other vital elements which could facilitate investments in a particular region.

---

[9] Finnmark is a county in the extreme northeast of Norway. By land it borders Troms county to the west, Finland (Lapland) to the south and Russia (Murmansk Oblast) to the east, and by water, the Norwegian Sea (Atlantic Ocean) to the northwest, and the Barents Sea (Arctic Ocean) to the north and northeast.

[10] See <www.impivaara.com or www.pomor.com> or <http://www.arcticshelf.ru/index.php?option=com_content&task=view&id=52&Itemid=61&lang=en>.

Some geographical pointers – North West Russia

During his tenancy in 2000, President Vladimir Putin introduced a new administrative level in 2000: the Federal District [*federalnyy okrug*], which is between the Regions and the Federal level. Seven such Federal Districts resulted, including the North-West Federal District ('NWFD') comprising 11 regions and with St. Petersburg as its administrative centre.[11]

The regions within the NWFD have a number of natural, geographical, climatic, historical and cultural particularities in common which define the socio-economic development of their populations. The region lies in the northwest of Russia; in the north it is surrounded by the Barents, White and Kara Seas, in the southwest, by the Baltic. It borders on Norway, Finland, the Baltic countries and Poland. The basic natural resources of the NWFD are vast deposits of apatite, bauxite, coal, copper and nickel ores, platinum metals, nepheline, manganese, gas, oil, oil shales and peat. North-West Russia contains 60% of the wood harvested in the European portion of the country.[12]

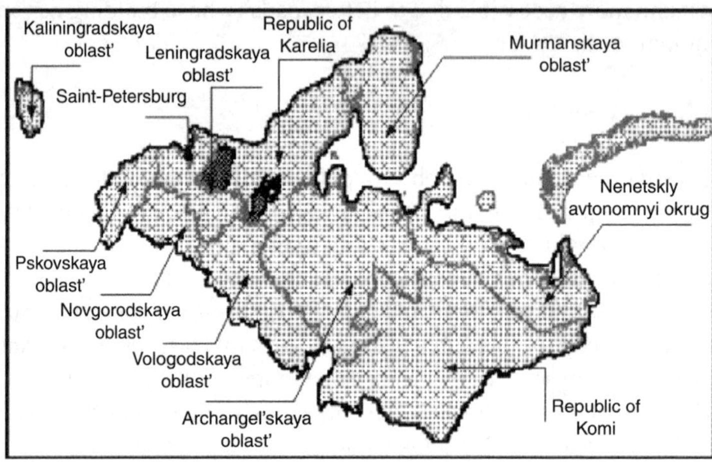

Map 1. North-West Federal District of Russian Federation.[13]

The basic regional industries in the Northwest of Russia are: production of machinery and instruments, metal processing, shipbuilding; metallurgy, pulp and paper industry, wood processing, chemistry, oil, gas and fuel extraction. There are some consumer goods and electronics production facilities which are developing, too. In the agricultural sector, farming, cattle breeding and fur-farming prevail.[14]

This said, although the fishing industry is well developed, further development of transport infrastructure, including the St. Petersburg, Archangel'sk, Murmansk and

---

[11] See the website of *Rosnedvizhimost* (*Федеральное Агентство Кадастра Объектов Нежвижимости*), the Federal Real Estate Cadastre Agency (or Federal Agency of Real Estate Objects' Cadastre): <http://eng.www.kadastr.ru/field_of_activity/lands_statisics/northwest/>.

[12] See <http://www.pdg.ru/news/2006/11/03/portalallnwrussia/en/>.

[13] Map taken from <http://www.wicnwr.ru/eng/nw/index.html>.

[14] See <http://www.pdg.ru/news/2006/11/03/portalallnwrussia/en/>.

Kaliningrad ports and building of new ports on the Baltic Sea is key to further economic growth in this Region.

As mentioned, this chapter will focus primarily on the far north of the NWFD, including the areas close to the Barents Sea, Murmansk Oblast and the Pomor Region. I do not venture to discuss the St. Petersburg hub, which has attracted much FDI since the 1990s, and is beyond the scope of this chapter. It is well known that St. Petersburg and the Leningrad Oblast are amongst the leaders in Russia with respect to attracting FDI. The Republic of Komi and Archangel'sk Oblast are among the top 20 in the Russian Federation; and by the level of FDI, the Northwest occupies second place, and is only behind the central federal district.[15]

Other institutions

In addition, various institutions exist in the NWFD, including those which promote small and medium-sized enterprises ('SMEs'). For example, the North-West Centre for SME Support[16] has been established in order to assist the formation of favourable conditions for entrepreneurship support and SME development in the NWFD. It also operates in Murmansk Oblast. Another organization promoting the development of the NWFD is the 'Association North-West', which was founded in 1991 with the main goal 'to promote effective interaction of the Northwest regions in creating conditions for interregional integration, social and economic development'.[17] These institutions are key in launching and promoting investment opportunities and facilitating cooperation in the far North of Northwest Russia.

The Barents Sea region and the Arctic of North America – frontier investment opportunities

I was working in Helsinki in 1989 when the Berlin Wall fell. But it was a conference in Tromsø in 1990 which ignited my passion for Murmansk Oblast.

Changing investment opportunities

When I did my PhD in Finnmark[18] we could only gaze eastwards toward the Fisherman's Peninsula as the start of the Timan-Pechora orogeny. It had only just been recognized in Varanger Pensinula in Finnmark, and again, by foreigners; two distinguished Polish geologists made the mental breakthrough in this very remote terrain.

---

[15] See <http://www.pdg.ru/news/2006/11/03/portalallnwrussia/en/>.

[16] See <http://nw.siora.ru/english/index.php>.

[17] See The Stockholm Region: Northwest Russia: Facts and Events (Newsletter by the Stockholm Region Office in St. Petersburg), Issue 91, 2008, week 6.

[18] J. Roberts, 'Late Precambrian Glaciation: an Anti-Greenhouse Effect', 234 *Nature* (1971) pp. 216-217; and J. Roberts, 'Late Precambrian Dolomites, Vendian Glaciation and Synchroneity of Vendian Glaciations', 84 *Journal of Geology* (1976) pp. 47-63.

It was in the 1970s that 'frontier resources' became increasingly of interest. After the Arab Oil Embargo and first oil shock of the early part of the decade, there was colossal stock market interest in Frontier oil and gas. With the Prudhoe Bay oil discovery on the Alaskan North Slope came huge gas reserves. Serious gas had been found nearby on the Canadian side in the Tuk Peninsula of the MacKenzie delta, as well as in the Canadian Arctic Islands. Publicity for these was given an enormous boost by a blowout; this is not normally welcomed in the oil world but it made great television. It put PanArctic on the map, and the then leading (but controversial) Canadian stock market actor, Dome Petroleum, grew enormously before it collapsed.

I stopped being an oil analyst in the early 1980s when the price of oil fell and I realized I had lost my touch for it. (It came back to me a few years ago). Regardless of geology, it is clear that the link between oil prices and the development of this entire region in Russia are inextricably linked.

Gas reserves and exploration

The oil and gas resources of the Barents Sea shelf became very attractive for such 'frontier' investment. Tromsø (in Norway), 1990 saw the first Soviet and Western geologists' conference on circumpolar petroleum. Just as the west had sought to deal with its excessive oil dependence on the Persian Gulf in the 1970s through developing the North Sea and other Frontier areas, so had the Soviet Union in their Arctic Seas. Absolutely enormous gas discoveries had been made in the Barents Sea in the 1980s. Maps of these discoveries were laid before an audience which had contact with Soviet geologists for the first time in open forum. It was blindingly obvious to me that the Barents Sea offered much greater gas supplies for Europe than the Arctic of North America could have done in the 1970s. There was not the problem of shifting ice over most of the prospects, and subsea completions offered the prospect of development. Pipeline distances to market, e.g., to Germany, were a half or a third those to Chicago from similar latitudes in the Canadian Arctic Islands.

But the 1990s saw the collapse of the state in Russia, and the transfer of natural resources to oligarchs. For ten years, there was no geological or other exploration work in the Barents or Kara Seas; besides the oil price was falling towards US$11/bl and even these great resources needed more than that. At that time, it appeared to me that Norwegian companies would be sure to play a major role in future geological exploration, as they had the deep water North Sea/Norwegian Sea experience to offer.

Nevertheless, since 2000, investments in gas exploration have increased, including the Shtokman project and this region is again 'on the map' in investment terms, including for environmental projects.

During a conference in 2005, which Impivaara/Pomor organized in London,[19] Murmansk Oblast was represented by sea and by land, with helpful presentations by Murmansk port and the environmental agency Bellona.[20] The Vice Chairman of the

[19] 'Natural resources on the Russian frontier' (London, 2005).

[20] See <http://www.bellona.org/>.

Duma Natural Resources Committee, Valentin Luntsevich, led the presentations on what were clearly strategic natural resources.

I note that although the investment opportunities in the oil and gas sector, including a new refinery, are vast, the investment opportunities in connection with energy efficiency, wind energy and other renewable energy sources, and investments relating to environmental services, are less so compared to much needed pollution abatement projects, although they exist and are on the rise.[21]

Training in Murmansk and Archangel'sk Oblasts

The issue of training is too big a subject for my chapter,[22] but I would like to highlight the important and generally unsung work done by Norwegian companies in training subcontractors in Murmansk and Archangel'sk Oblasts. Norway has done much to help train small businesses which would be needed to supply goods and services to Barents Sea development. Of course, this was done by Norsk Hydro and Statoil as part of enlightened self-interest. Nevertheless, they did it, though few outside conferences in Kirkenes (Norway) could have known much about this.

The Pomor Region

Sometimes, the most obvious things are a simple giveaway. I have always suspected that companies using the term 'Commonwealth of Independent States' or CIS for short, did not actually understand that it did not mean too much. But even *BBC Monitoring* has failed to find a better term than 'Former Soviet Union' for its wonderful daily summaries of XSU media, so perhaps CIS is just arose by any other name. It does not offend anyone, but has a sniff of ignorance about it, so it suits large companies.

Similarly, the term '*Barents Region*' which is often used to describe the Pomor Region is a term mostly used outside this region (in EU circles and others). There is the Barents Sea, but the term '*Barents Region*' means nothing to people who live in the Pomor Region.

The Pomor Region is named after the Pomor people who came from the shores of the White Sea to trade between Archangel'sk and Tromsø. There was no Murmansk in those days, though there has been a settlement at Kola for the best part of four centuries. The 'Pomori' were originally people who lived by the sea and traditional livelihoods were based upon animal hunting, whaling and fishing or reindeer herding. Just as the Saami could easily travel overland by sled, once the Tsar gave the Pomori permission to reach pastures new, distance was little object, so they could trade fish for wood with the Norwegians. Thus, the territory of practically the whole European Russian north, including the Murmansk region, Archangel'sk and Vologda regions, Karelia and Komi republics, started to be called Pomor'e.[23]

---

[21] See chapter 17 (*Lesikhina and Grønbech*).

[22] However, see chapter 9 (*Guseva*) for more detail on environmental education and capacity building in Russia.

[23] It is reported that this trade was so intensive that a kind of Russian-Norwegian pidgin/creole language *Moja på tvoja* (or 'Russenorsk') was created and used on the North Norwegian coast in 1750-1920. However, this language has now died out.

However, over the centuries, various settlements, including whaling stations in the extreme North were abandoned. It is interesting to note that there is no need to apply preservatives to pine from Archangel'sk – and at Hamningberg (Norway), an old whaling station at the extremity of Varanger Halvøya, there remain houses built centuries ago. The retreating Germans burnt everything in Finnmark, but could not get to Hamningberg as there was no road there. The same applies to the fishing community of Bugøynes on the south of Varangerfjord,[24] whose 300 inhabitants advertised themselves in the early 1990s in an effort to evacuate themselves (which the Germans army had not managed to do in 1944).

These cultural and historical circumstances explain the close links which people in Varangerfjord and other Norwegians (and northern Finns alike) feel for Russians and *vice versa*. Investors interested in doing business in this far Northern part of Northwest Russia will need to bear in mind these cultural and historical interlinkages shared by the countries which touch borders in the European far North, including the Russian Far North.

Investment opportunities – responsible minerals exploitation and mining

Returning to Kola which is the oldest town of the Kola Peninsula – it is notable that rocks do not stop at international boundaries though their colours on maps and the mentality of some people certainly do.

When I first went to Apatity (in Russia)[25] in 1992, I was astonished to see how much work had been done on the Soviet side. I knew there were mines along the railway from Leningrad to Murmansk, but did not realize how much excellent work had been done at the reconnaissance level further east.[26]

Just as the Barents and Kara Seas remained untouched by exploration and investment in the 1990s (despite my efforts!) the landward mining potential of the Kola Pensinsula was almost entirely forgotten, with honourable exceptions, confined to platinum metals exploration. BHP had played a key role in platinum group metals (PGM) exploration east of Apatity on Panskiy Tundra, but the company retreated almost totally in 1995 as the markets for iron ore and most other things it produced collapsed. At the same time, Barrick Gold of Canada, one of the world's largest gold mining companies, came to the conclusion that, after so many years' dull development in the gold price, it might not be a bad idea to participate in platinum metals in Russia.

---

[24] Varangerfjord is in the county of Finnmark and is the easternmost fjord in Norway.

[25] Apatity is a town in the Murmansk Oblast, located between Lake Imandra and Khibiny Massif some 23 km west of Kirovsk. The settlement of Apatity was founded in 1935 and granted town status in 1966. It was named after one of its most abundant natural resources in the area, apatite, the raw mineral used in the production of phosphate mineral fertilizers.

[26] See a paper by Akad. Felix Mitrofanov on the mineral resources of the Kola (presented at our 2005 London conference), at <www.impivaara.com>. During this conference, the environmental issues, especially for oil and gas pipelines were presented.

Apatite

The most important mining exploration company in Murmansk Oblast is O.A.O. Pana in Apatity.[27] It was the operator for licences held by major foreign companies, and Impivaara/Pomor helped with a little critical investment advice on how to deal with foreigners. In this space, it is perhaps not necessary to cover the history of Pana in detail, but my own belief in it has proved to be absolutely justified.[28] Back in the mid 1990s, the situation was indeed grim as BHP retreated, but Barrick saved the day, and O.A.O. Pana now has several drilling rigs, as well as a major Russian partner in its richest licence area, 'Little Pana'. As a minor shareholder, the little company, Pomor Petroleum Limited (which I put together in 1991), was amply rewarded for its efforts. Dividends from O.A.O. Pana helped us run our London conference in 2005 on Frontier Russian Resources. They also taught me the importance of having trusted friends in Russia, when it came to the complexities of transferring our dividend share to our English company abroad.

Platinum and possible multi-regional cooperation

Across the border, in northern Finland, there are similar rocks to Pana, and a few rich deposits have been found, as well as total resources equal to one year of world platinum metals production. I recognized Pana immediately as I had been correctly brought up on layered intrusions at Oxford, which Professor Larry Wager had 'discovered' in East Greenland and his own great student, Malcolm Brown, taught us proper respect for these extraordinary rocks: 'sedimentary' rocks laid down by currents in a white hot magma chamber two billion years ago. The Pana accumulations are unlike the best known model, the Bushveld Complex,[29] in that palladium is five times more important than platinum. At the time I 'discovered' Pana, the price of palladium was five times below that of platinum (an extremely rare metal), and as there was five times as much of it, we were approximately 25 times behind the Bushveld Complex, in value.

But in the later 1990s, palladium became inflated in price (to twice the platinum price and over $1,000 per oz) because of expected demand by General Motors and others for car exhaust catalysers.

Furthermore, until the late 1990s, many Western companies missed out on many opportunities in the Pomor Region due to Finland's strong state role in this Region. Finland farmed out its licences (regarding low-grade PGM deposits), but Finnish state companies had monopolised exploration of the northern regions for decades. The presence of a Finnish state monopoly in exploration, their neglect to actually explore these

---

[27] See <www.panapgm.com>.

[28] The results in exploration for platinum metals are beautifully presented on the Geological Survey of Finland website (<www.en.gtk.fi/>) as the result of a major EU TACIS project pulling together Russian, Swedish and Finnish experts.

[29] The Bushveld Igneous Complex is located in South Africa and contains some of the richest ore deposits, worldwide. The reserves of PGMs (as defined below), platinum, palladium, and others are considered the world's largest.

regions and falling (and then rising) prices in palladium and other circumstances (e.g., absence of Finnish investment banks and a Stock Exchange able to develop mining exploration) essentially led to no development in these northern parts.

My colleagues and I have proposed that PGM concentrates be transported from the low grade deposits of northern Finland and to refine them, jointly with Pana material, at a new state-of-the-art and environmentally compliant refinery in Apatity in Russia. This would be a major investment, which would transform the economy of the region. To give an idea of the value of such concentrates, if only half of the 1,100 tonnes proven by Gold Fields Limited in Finland were developed, that would mean transport of a little less than 1 million tonnes of platinum ore concentrates *per annum*, enough to make partial justification of a White Sea to Baltic railway by itself. The value of each tonne of platinum metals is currently a little over US$10 million, so that we would be thinking about 550 x US$10 million, i.e., some US$6 billion gross value of material. On the Kola side, there should be similar, or perhaps even greater quantities, but refining platinum jointly would bring in better economics for both Russia and Finland. Furthermore, by applying new technologies, using environmentally sound machinery and transportation modes, coupled with responsible business activity (e.g., through implementation of corporate social responsibility), this could be a major showcase in the Far North of the NWFD. Ironically, in such a case, Russia would be importing strategic materials from abroad, and exporting refined products.

Mining expertise is tremendously developed in Murmansk Oblast but, to my chagrin, there is little sign of that being recognized in the West, beyond the great companies mentioned above.

Renewable energy in the Far North

Although the issue of renewable energy opportunities is addressed in more detail in Chapter 17 by Nina Lesikhina and Michele Grønbech, I wish to point out that not merely do vast geological investment opportunities exist in the Pomor Region, but that many renewable energy projects could be developed, too. Natural resources investments are energy intensive and often bear numerous environmental problems. Thus, the integration of environmental aspects into such investments and other 'environmental' projects would be significantly beneficial, not only to the local populations, but also to the local mining industry in that they bring environmental expertise closer to and within reach of a traditionally 'non-green' sector. A great deal of research is to be hoped for, perhaps EU funded.

Sokli-Kovdor: the $20bn missed opportunity for cross-border business?

There is something of a risk in writing about Sokli development in 2009 because it is a work in progress. Its history to date gives little reason for cheer and underlines important, but, to my mind, quite unfamiliar facts about Finland and her relationship with her great Russian neighbour:

- Finland was amongst the last of countries to recover from Cold War thinking
- Previously, knowledge of 'Russia' was restricted to a very few people
- Business knowledge of the Russian language is (still) extremely limited
- In the 1990s, Finns were pre-occupied with Nokia, the EU and the Euro

All these factors meant that no natural business connections were developed by Finland with Russia. However, if you look at the history of Finland, you will see that the nation developed herself, her language and culture, during Tsarist Russian rule, and not under the Swedish control of previous centuries. Oddly enough, it was largely Swedish speaking Finns who developed *suomalaisuus*, Finnishness, with the first poetry and prose (*Kalevala*, a folklore collection, or the *Seven Brothers*) not to mention music, coming from people who did not speak Finnish as their mother tongue. It is notable that today, less than 1% of Finnish schoolchildren are even offered Russian as a foreign language.

Sokli phosphates

Sokli is the most extraordinary case. It is one of many plugs of mineralization from deep in the earth which came up in the Cretaceous,[30] punching through the ancient Precambrian shield of FennoScandia. They are called carbonatites because of the carbon dioxide which comes up with them from mantle rocks, but those of interest here are rich in phosphorus minerals. At Sokli, you could have spotted the deposit just by noticing the sudden change towards thick undercover of juniper which likes phosphate-rich soil.

My mining colleague, Steve Henley, and I posted internet sites to describe Sokli and the nearby Kovdor phenomenon. Each of these carbonatite plugs has different mineral characteristics. For example, that which makes up the mountain of Kirovsk by Apatity, is nearly pure mineral phosphate, with minor quantities of a zirconium mineral (blood of the lapps; eudialite) and nepheline, used as a source for aluminium, though it is rich in iron. The Kirovsk deposit produced up to 20m tonnes *per annum* of refined phosphate in Soviet days, when it was exported to Belarus, Ukraine and the South of Russia regardless of transport costs. Or out through Murmansk port in large dry bulk shipping, to compete with Moroccan phosphate (whose cadmium now pollutes EU fields).

Sokli lies just 50 kilometres from Kovdor over on the Finnish side. For reference, the tiny settlement of Tulppio is easily found on map searches. Crucially, too, it is right next to a National Park in an area of great beauty, making environmental considerations key. There is a minor road to it from the nearest large place, Salla. But there is no railway until 120 kilometres south, along the west-east trace of the old railway which Stalin terminated by moving the Finnish border westwards after the war. Allied in

[30] The Cretaceous period is usually noted for being the last portion of the 'Age of Dinosaurs'. The end of the Cretaceous brought the end of many previously successful and diverse groups of organisms, such as non-avian dinosaurs and ammonites. This laid open the stage for those groups which had previously taken secondary roles to come to the forefront. The Cretaceous was thus the time in which life as it now exists on Earth came together. See: <http://www.ucmp.berkeley.edu/mesozoic/cretaceous/cretaceous.html>.

Lapland with Germany, a railway had been built to attack the Oktober Railway and Murmansk, from the present border at Salla (to be exact, Kelloselkä) to the old limit of Finland at a village called Alakurtti.

During the 1980s, there were various plans to develop the phosphate mineral at Sokli by Kemira, the Finnish state fertilizer company. They foundered because reigning prices just were not good enough. The licence was neglected after quite substantial exploration expense, after it was transferred from one Finnish state company, Rautaruukki, to another. Then, Kemira spun out its fertilizer business as Kemira GrowHow. This was not a great stock market success, as fertilizers were not popular. Kemira GrowHow was not really big enough to develop the licence alone, and had its own successful pure phosphate mine further south in Finland at Siilinjärvi.

There had been ideas for a slurry pipeline from Sokli to the sea at Kemi, but it, like the idea of a whole new railway spur to reach Sokli from the south, came to nothing on grounds of cost. Besides, the track from Rovaniemi, capital of Lapland, to the border was in not too good condition and not electrified. The only industrial plant needing the railway was the Kemijärvi pulp mill, always at risk of closure since it was taken over by the Finnish state company Enso Gutzeit. It is now shut, although it was making a profit. The idea of relaying the old German railway line to Alakurtti was never popular and despite road improvements on both sides, a new customs post and frontier crossing point at Salla did not have much effect on traffic.

Two things have operated to change the situation. The local authorities, enraged by the constant failure to develop Sokli in any way, helped us to promote matters through two websites[31] which showed the link between Sokli and Kovdor. Secondly, Yara, Norsk Hydro's old fertilizer division, took over Kemira GrowHow, making a company big enough to handle Sokli development which was clearly in the one billion Euro category. At the same time, the price of mineral phosphate of this quality doubled and trebled, so that the old economics probably did not apply. Even iron ore became important, whilst specialized minerals, such as a world class deposit of niobium, rare earths and minor uranium and thorium, came into better focus. However, dealing with radioactive elements and separating specialized minerals from their ores was not a job for a fertilizer company, but was perfectly suited to skills developed just over the border at Kovdor in Russia.

Kovdor minerals

The mine in Kovdor (Russia) is, to put it mildly, more complex than that in Sokli. It includes commercial quantities of iron ores, whose separation from the phosphate is not so easy. It also has huge resources of lithium mica and mica crystals big enough to make window panes. Kovdor, as our Impivaara websites would show you, is a Russian town in the Murmansk Oblast, a *cul de sac* lying close to the Finnish border, but connected to the Oktober Railway by its own line.

---

[31] See: <www.sokli.info> and <www.kovdor.info>.

Infrastructure requirements

So we proposed a short transport link (mine railway, slurry pipeline or even a winter road) from Sokli (Finland) to Kovdor (Russia) as a first stage. Then, the products, mainly phosphate concentrate, could be exported via Murmansk, and possibly the southern Kola port of Kandalaskha (in Russia), which is not ice free, however. As a second stage, if the volumes of phosphate grew beyond 2m tonnes *per annum*, as they certainly could, it would make sense to bring them back round into Finland via a railway loop, Alakurtti/Salla.

Such an east-west link would transform the economy of the entire region. However, the appropriate environmental impact assessments and precautions should of course be taken if this link is developed. There are other major reasons for thinking about such a railway – developments in the Barents Sea, the enlargement and 'greening' of the aluminium smelters at Kandalaksha and even the idea of a new platinum metals refinery at Apatity. However, to make it happen would need one major project which proposes a railway track to be relaid for those missing 80 kilometres.

Technical problems

There is also a very peculiar technicality in the licensing arrangements in Finland. Between the 5 years exploration licence extension sought by Kemira GrowHow and a full 25 year mining licence, now lies an awkward limbo of two years in which they are right in the middle. They cannot apply for the full mining licence in 2009 without completing an environmental impact assessment (EIA) where the impact on the environment of a mine plan is made (without the benefit of its actual existence). Without knowing where the mineralization of niobium, radioative elements and rare earth elements actually is (all needing more exploration) one cannot just move phosphate and iron ore without environmental risk.

Not entirely surprisingly then, old Finnish state railway plans for a dream railway across virgin land north to Sokli from Kelloselkä have been dusted off and given prominence in the EIA. The Finnish state is willing to electrify and improve the existing railway all the way from Rovaniemi, and perhaps buy back the new *cul-de-sac* to Sokli on completion. That would put the cost of transport alone at something close to 1 billion Euro before a single bucket of ore had been dug at Sokli.

By contrast, our idea is to get material moving across to Kovdor on a trial basis as soon as possible, so that Sokli Ltd, a new company, would have some cash flow to finance its activities. Later, with government and other partners in mining, it could develop into something like the Antofagasta Railway in Chile – a penny railroad stock that found a huge copper mine on its land, and then had the railway to take it away.

From the point of view of environmental investment and socially responsible business activity, our plans for Sokli, via Kovdor and the Alakurtti-Salla railway, stand up to scrutiny better than an electrified red carpet to the Finnish-Russian border and a whole new *cul de sac* to the north. Besides, if the NATO lobby won, Russia would be looking at a new military transport threat parallel to the border where none is needed.

Ironically enough, putting back the old German railway, Alakurtti-Salla, would be a more friendly proposition, apart from its huge effect on the local economy on both sides of the border. Naturally, any investment, such as this one, in the minerals and mining sector would require expert scrutiny from an environmental impact perspective and need to incorporate stakeholder consultations.

Food for thought: Cross-border language teaching in remote holiday areas?

Now comes a special problem, touched on at the start of this chapter. Western (including Finnish) knowledge of Russia is often slight, even in Lapland where there is excellent university contact across borders.

To my mind, Finland, Russia and Norway should be making huge efforts to improve language knowledge. It is an illusion to think that English will cover business needs in Pomor. If you want to expand your business into Russia, or into any Slavonic land, you cannot do without this world language.

As a consumer of language teaching, I know that there comes a stage when you need one to one speaking practice; as well as encouragement to read novels for the first time. Why not do that using all those holiday cottages which dot northern Finland, within easy reach of everywhere, by sea, by land and by air (Oulu[32] is an important airport) out of season? Such a set up, web based, would have to be directed by skilled Russian language teachers and university collaboration in several countries, but the Pomor Region is the place to do it. You do not just need specialized and skilled language teachers; you also need a great deal of ordinary contact in which people trade languages with each other. I put forward this 'Little Idea' at an EU conference last April in the north of Finland at Saariselkä, close to the place where I first rode into Finland on my push bike from Finnmark

As secondary in importance as it may sound, Russian language training and knowledge is one additional way to encourage knowledge of these Northern regions and would no doubt promote investments.

Concluding remarks

As is clear from my own experiences in Murmansk Oblast and the Pomor Region, this part of Russia still seems relatively remote to outsiders, both in geographical and economic terms. However, successes are possible and investments are starting to flow, albeit slowly. The opportunities are vast, both in context of the mining and minerals sectors, and with respect to environmental investments. For example, a very important new EU grant application will focus on environmental and social problems which arise

---

[32] Oulu is a town in the province of Oulu and the region of Northern Ostrobothnia, in Finland. It is the largest and most important city in Northern Finland and the sixth largest city in the country.

on the closure of old mines or the opening of new ones. This follows an outstanding TACIS programme over the last 2 years which provided an umbrella for collaboration amongst Finnish, Russian and Swedish geologists to set up a database on the worldscale platinum metals which span 800 kilometres from the White Sea to the Baltic. One deposit in Finland holds one year of world production of these metals, but requires environmentally sensitive investment as well as US$1 billion. On the Russian side, in the same trends, there are similar huge opportunities which would change the economics of the Kola Peninsula and fund much needed clean-up work in places like Monchegorsk, where old technology has caused low life expectancies and laid waste the landscape.

Part III

Practical Trends and Synergies

B. CSR & philanthropy

Chapter 19
DEVELOPING CORPORATE SOCIAL RESPONSIBILITY IN RUSSIA: THE ROLE OF THE UNITED NATIONS DEVELOPMENT PROGRAMME[1]

Alexey E. Kostin[2]

From the beginning of the 21st century, corporate social responsibility ('CSR'), i.e., responsibility of corporations to the community and society on the whole, has been steadily developing in the Russian Federation. A strategic approach towards corporate social responsibility has spread among Russia's companies. This is a result of the globalization of business and an increased awareness amongst Russian companies of the various risks and opportunities they face. When developing their social strategies, many companies became increasingly aware of the needs of society. Russia, thanks to economic growth, has seen the active development of principles of CSR, social accountability, sustainable development, public-private partnerships ('PPP'), related to businesses' social obligations. At the very least, these principles have been applied in the practice of Russian companies; facilitated the development of the market of consultancy services on CSR and the sustainable development of business and society; and supported the foundation of voluntary initiatives and associations that contribute to the spread of CSR in Russia (e.g., by way of 'The Social Charter of Russian Business' of the Russian Union of Industrialists and Entrepreneurs ('RSPP')).

This chapter sets out to describe how CSR is progressing in Russia, investigates non-financial reporting and the role which the United Nations Development Programme ('UNDP') and other organisations have had in promoting CSR in Russia, in particular through the UN Global Compact Principles.

CSR has been developing in Russia in various forms and along various lines of corporate activities. Within the last 5 years the main trend has gradually revealed itself: CSR is being introduced as a form of sustainable corporate development that benefits companies' shareholders and employees, the communities they work in and society as a whole. This notion is extremely important since, at first, CSR in the Russian Federation began in the form of public relations ('PR') events relating to corporate charity.

---

[1] The analysis and policy recommendations in this chapter do not necessarily reflect the views of the UN system and the institutions by which the experts and consultants are employed.

[2] Executive Director, Non-Profit Partnership 'Corporate Social Responsibility – Russian Centre' and Consultant to UNDP Russia; Email: *alexey.kostin@mail.ru.*

W.Th. Douma and F.M. Mucklow, eds., Environmental Finance and Socially Responsible Business in Russia
© 2010, T·M·C·ASSER PRESS, The Hague, The Netherlands and the Authors

CSR and sustainable development

CSR is linked to the concept of sustainable development[3] – which was most famously defined as 'development that meets the needs of the present without compromising the ability of future generations to meet their own needs'. This definition, coined in 1987 by the World Commission on Environment and Development – the Brundtland Commission – still applies today and is used by politicians, experts, and entrepreneurs all over the world.

The term 'sustainable development' has become quite a staple in public and corporate operations in developed countries and in a number of developing countries. There are not merely national state-run programmes and strategies for sustainable development – but frontrunners among companies and corporations also have their own corporate plans for sustainable development. The social and environmental aspects of sustainable development establishes, a.o., a new progressive management philosophy whereby any managerial decision takes into account economic, environmental, and social effects and consequences. An important milestone in the development of the practice of sustainable corporate development was the establishment of the international standard of the Global Reporting Initiative ('GRI') based on the principle of a Triple Bottom Line: corporate economy, environmental performance, and social policy.

Russian business, CSR and sustainable development

Corporate sustainability is based not only on consistent and effective manufacturing operations that bring profits and ensure extended production. It also means that the economic, environmental, and social realms of corporate operation gradually come together in a joint self-organized system. In that sense, sustainable corporate development requires, at the very least, economic effectiveness, compatibility with the biosphere, and social justice combined with a decreased general anthropogenic impact on the biosphere. Sustainable development of a community (a country, a region, a town) is impossible without sustainable business development. And vice versa: sustainable business development is impossible without sustainable development of a community.

Corporate sustainability achieved by way of applying the concept of sustainable development and CSR produces at least the following outcomes for a company:

- Enhanced image and business reputation;
- Stronger team spirit;
- Increased transparency;
- Better attractiveness for investors;
- Social contribution to sustainable development of the country as a whole;
- Increased capitalization.

---

[3] See also chapter 3 (*Guseva*) with respect to the concept of sustainable development.

The business community – including the Russian business community – does not share a single interpretation of the concept of sustainable development and sustainable performance. This is not surprising considering the diversity of conditions in which countries and companies operate. However, the essence, or goal, of businesses engaged in operating sustainably, is always the same: to disrupt the proportional relationship between economic growth and negative impact on the environment (i.e., the process usually referred to as 'de-coupling'). Only those enterprises that effectively reduce negative environmental impact while increasing their production of goods and services, and that manage to do this year after year, are considered 'sustainable' and, therefore, environmentally and socially responsible. This is where there is a link to CSR. If a company relocates 'dirty' production outside of its own jurisdiction, this does not have a material impact on its corporate sustainability indicator. The requirements for its environmental performance and social development remain, albeit in modified form, due to locally applicable requirements (with respect to any internationally established subsidiaries).

Russian business and expert circles still vigorously debate the essence, scope, and goals of CSR. Most Russian business leaders, to judge from their non-financial reports, as well as specialists that adhere to front-running international approaches, tend to view CSR as incorporating two interconnected components: the realm of social and environmental responsibility required 'by law', and the realm of additional voluntary public responsibility which is 'beyond the requirements of the law'.

An analysis of the main definitions and systems of corporate social responsibility and management in Russia has led the author to formulate the following description of CSR which appears to be the most comprehensive and most correct: CSR is a system of consistent economic, environmental, and social measures taken by a company in constant cooperation with stakeholders and aiming to decrease non-financial risks, enhance corporate image and business reputation in the long term perspective, and to increase capitalization and competitive advantages to ensure profitability and sustainable development of an enterprise.

Thus, CSR means that business is conducted in such a manner that a company meets or exceeds the ethical, legal, commercial, and social expectations of all stakeholders (aside from its owners). CSR describes those corporate actions that target reinforcement and improvement of interactions between the company and society. Usually, those actions are related to the core operations of the company. CSR also means the anticipation of the problems the company might face in the immediate and more distant future, and preparation to solve or neutralize them. 'Social charity' or philanthropy is only one part of the social 'pillar' of CSR, which in Russia often has a pronounced image-enhancing nature. In summary, CSR covers the following interconnected areas:

- Quality, safety, and consumer availability of goods and services;
- Environmental policies, clean manufacturing, and sparing use of nature's resources;
- Personnel development;
- Workplace health and safety;
- Cooperation with local authorities, state structures, and public organizations to solve common social problems;

- Cooperation with suppliers and consumers of goods and services;
- Corporate development – restructuring and organizational changes;
- Corporate governance;
- PR-support of all of the above and other areas.

In the context of actual Russian business practice, CSR is demonstrated in Russia through growing transparency of Russian companies, establishment of new systems of internal and external social programmes, increased significance of publicizing corporate social and environmental responsibility, and publication of open non-financial (social, environmental, and sustainability) reports.

It would be too early to say that the majority of Russian companies engage in the above mentioned activities, but Russian business leaders are definitely engaged in CSR. The number of such business leaders, according to the RSPP Register[4] is low – as of November 2009 only 74 companies published 195 non-financial reports. 114 of those were social reports, 48 – sustainability reports, and 33 – environmental reports. Approximately one third of those reports used methods and indicators from the voluntary international standards, such as GRI and AA1000S. These social reports were published by: Norilsk Nickel; LUKOIL; RAO UES of Russia; Tatneft; Gazprom; Rosneft; Euras Holding; EuroKhim; and other companies. Russian Railways, RUSAL and some others have based their non-financial report on their practice of upholding the 10 Global Compact principles. All of the companies mentioned above include Russian business leaders in the oil, gas, and chemical industries; metallurgy; paper manufacturing; mining; food manufacturing; electric power generation; utilities.

For most of these Russian companies CSR is becoming a component of corporate governance rather than merely a part of public relations. This is what is new about CSR in Russia: companies are increasingly complying with international practice and with 'soft' international standards, specifically those proposed by GRI and AA1000 Stakeholder Engagement Standard.

Socially responsible business conduct which is reflected in a system of economic, environmental, and social performance indicators, and which is realized through a regular dialogue with the community, is becoming a part of strategic corporate governance. Therefore, decisions regarding CSR are made by all interconnected parties to corporate governance: company owners, boards of directors, and management. Some advanced companies establish special working groups and committees for and to address CSR and non-financial risks, include independent directors responsible for those matters in boards of directors and corporate committees, and introduce systems for non-financial risk management. Importantly, most of their operational and economic decisions are made with regard to their social and economic consequences for the companies and the regions where they operate. That helps turn CSR into a powerful tool for strategic development, enhancement of image and business reputations, increasing competitive advantage, and promoting gradual growth of corporate market capitalization. CSR is, in fact, a policy tool and a way in which to implement a sustainable

---

[4] See <http://rspp.ru/Default.aspx?CatalogId=2257>.

development strategy at the corporate level within a modern system of corporate governance.

Non-financial reporting in Russia

As of November 2009, standardized 'non-financial' reporting, which reflects the degree and content of a company's CSR and sustainable development activities, is applied by 5,709 major international companies, and many of them based in Russia. These companies have collectively published 23,408 reports,[5] evidencing that non-financial reporting is no longer a fashion, but rather a response to significant changes in the business environment: non-financial risks are becoming more material and potentially significant alongside traditional operational, technological, ecological and financial risks.

Russia is no exception in this respect. In fact, the non-financial risks in Russia are higher than in most countries due to poorer predictability of the major stakeholders: the state, shareholders, consumers, environmental organizations, civil society, and others. CSR and non-financial reporting help Russian business leaders effectively manage non-financial risks and demonstrate the growth of social and environmental responsibility to the Russian and foreign audience. That, in turn, has a direct impact on their corporate governance quality and ratings, which is important both in view of stock markets ratings and with respect to strategic investors.

CSR and attracting investments

CSR, as shown by frontline Russian corporate experience, is a policy tool to implement a strategy of sustainable corporate development. CSR and sustainable development are two sides of the same coin. It is hard to overestimate the impact of CSR on the attractiveness of a business to investors: every strategic or portfolio investor, when making a serious decision concerning acquisition of shares of a Russian company, evaluates the whole spectrum of risks. A company may be attractive from the viewpoint of its current profitability, and at the same time be unstable from an environmental and social perspective. Since the beginning of this century, CSR reports have been bridging that 'informational gap' for investors interested in whether a particular company pays serious attention to environmental and social aspects of its operations and, thereby, minimizes its risks of internal or external social conflicts and environmental sanctions. Thus, CSR 'management' through implementation of a corporate sustainable development strategy is, in fact, a system for managing and minimizing non-financial risks. Thus, all areas of CSR are essentially also coordinated through corporate PR departments with PR-support within and outside a company.

From the very beginning, corporate CSR operations in Russia considered 'beyond the requirements of the law' had a strictly voluntary nature. Major Russian corporations have come to realize that the current business environment requires social (and

---

[5] See <http://www.corporateregister.com/>, accessed in February 2009.

environmental) responsibility and, most importantly, that social responsibility 'pays off' as it mostly results in increased capitalization, sustainability, and attractiveness of a business. In addition, Russian businesses are realizing that PR events that are not supported by actions, facts, and figures, are insufficient for the contemporary business community. Thus, preparation of CSR Reports and Sustainability Reports has started to receive special attention in the Russian business community. Russian companies are beginning to learn best practices from each other, as well as through the vast international network of websites, seminars, and conferences.

This approach is highly reminiscent of management studies, consisting of two components: management theory and management practice (case studies). By sharing information about both positive and negative practices and by learning from other Russian or international experiences of applying CSR, Russian companies can avoid repeating mistakes made by their contemporaries. The communication network which has developed and continues to evolve in the area of CSR, enables Russian companies to get involved at a low cost.

Changing direction – CSR principles

Why do companies increasingly switch from the position of stating that 'we make the product for consumers and pay taxes, and that's the end of our responsibility' to the position of open and voluntary civic responsibility? Russian and international experience shows that there are two main reasons for this.

The first one is the objectively growing dependence of manufacturers on social conditions in regions of their operation. Regular research among international CSR leaders indicates that such companies voluntarily spend 0.5 to 1.5% of their annual turnover on CSR-related activities. In Russia, major companies spend even more – around 2% of their annual turnover – for those purposes, including social investments and corporate charity. This evidences that the Russian business community certainly does not underestimate the importance of social responsibility.

The second factor that motivates companies to invest into the 'non-manufacturing sphere' is the realization of the fact that a successful systemic introduction of CSR means an introduction of a system of corporate non-financial risk management which has management of stakeholder relations at its core. Those two main driving forces produce the consistent improvement of corporate governance systems, which has a positive effect on a company's image and business reputation, and ultimately on capitalization. That is, operationally, companies do incur additional non-manufacturing expenses, but ultimately they gain by way of increased sustainability and corporate value. In addition, capitalization normally grows much faster than CSR expenditures.

In the same manner, as any management concept is embodied in specific management tools, CSR is realized on the basis of the below principles that determine its structure, methods, and processes:

- *Systemic nature*, or integration, which ensures efficient realization of CSR's functions and trends in a corporate governance system;

- *Comprehensiveness* for the purpose of coordination of simultaneous actions in three areas of sustainable development: economy, environment, and social policy, along with ethical business conduct;
- *Social nature*, i.e., a thrust toward society as a whole and its most significant parts and groups;
- *Transparency* of all corporate operations in that area and of corporate non-financial reporting;
- *Accountability* to stakeholders and society as a whole, and compliance with obligations and standards;
- *Innovations* in CSR actions and technologies inside and outside the company;
- *Integration* with main manufacturing and sales operations of the company;
- *Engagement* of stakeholders in a regular open dialogue;
- *Materiality* of CSR actions and technologies that have a real and specific significance for stakeholders, local communities, and society as a whole;
- *Responsiveness* to stakeholders' evaluation, criticism, and expectations regarding all areas of social responsibility that are reflected, among other things, in transparent non-financial reporting on obligations according to and beyond the law.

It would be an overstatement to say that all these principles are fully implemented by Russian business leaders, but most of them are present in their mission and value statements, and in social reports. Russian business leaders are coming to realize that underestimating, or even ignoring, the principles of CSR leads to a distortion of a company's main requirements and goals, to a disruption of the Triple Bottom Line with respect to economic, environmental, and social activities, and to a lack of systematic accounting for opinions and requirements of main internal and external stakeholders.

Corporate governance and CSR

CSR principles apply to the whole corporate architecture. CSR becomes a multi-vector tool of influencing a company and, very importantly, the entire corporate governance system. When the significance of CSR is viewed in this manner, corporate governance of a modern socially responsible company is seen as a system and process of relationships, and an aggregate of principles, rules, and procedures for interactions between owners (shareholders) of a company, its board of directors, management and other stakeholders inside and outside the company. Over the last 10-15 years, corporate governance, worldwide and in Russia which has decisively led to transforming the management of a company as an 'open system' which is under permanent pressure from the outer business community.

Due to the high practical significance of this concept, best practices in corporate governance in Russia apply the whole hierarchy of CSR tools consisting of the following set of interrelated elements:

- Mission and values; Code of Corporate Ethics;

- Corporate strategy, including sustainability goals (economic, environmental, and social);
- Concept for managing non-financial risks and stakeholder engagement;
- CSR and sustainability activities (projects and programmes);
- Social (sustainability) reporting;
- Communications: stakeholder dialogues, PR, internet and intranet, and partnership between sectors.

In leading Russian companies, where CSR is a part of corporate governance, responsibilities and functions of parties to corporate governance pertaining to integration of the CSR principles and technologies are normally distributed as follows:

Owners (shareholders) make key decisions concerning the development of a socially responsible business taking into account all of its pluses and minuses (mostly the cost). When necessary, they participate in related PR events supporting and enhancing the socially responsible image of their company.

The *Board of Directors* adopts a corporate strategy including goals, priorities, indicators, and major CSR activities. The Board of Directors oversees CSR-related work of the management and approves the format and standard used for social (sustainability) reporting. Usually, large companies establish a CSR (or Sustainability) Committee consisting of the Board of Directors and chaired by one of the directors. The Chair of the Board usually attends prestigious national and international CSR and sustainability events on a regular basis. Occasionally, some Boards of Directors invite to work with the independent directors for CSR and sustainable development, often representing civil society and NGOs.

The *Management* (often together with the Governing Board) of a company implements CSR strategies and tactics; ensures the achievement and realization of social, environmental, and ethical goals and CSR (sustainability) indicators; introduces CSR technologies for non-financial risk management; designs and prepares social (sustainability) reports; and organizes related PR communications.

Corporate governance and non-financial reporting in Russia

The active development of non-financial reporting in Russia is displayed through the large variety of its forms. The type of non-financial reporting usually applied in Russia depends on the level of corporate governance of a particular company. A low level of corporate governance results in the use of the simplest forms of reporting, such reports are mostly created for the purposes of PR rather than for disclosing material information concerning CSR with the help of international standards – GRI and AA1000 – and corresponding performance indicators and stakeholder procedures. A high level of corporate governance is reflected in more advanced forms of non-financial reporting. Below is a list of principal types of non-financial reports which are produced in Russia.

Low level of corporate governance

- Social report – a booklet on corporate social and charitable programmes;
- Report on compliance with the Global Compact Principles (without reference to GRI G3);
- Environmental report (without reference to ISO 14001 and other environmental standards).

High level of corporate governance

- Social and environmental responsibility report (using GRI and AA1000);
- Corporate sustainability and social responsibility report (using GRI and AA1000);
- Sustainability report (using GRI and AA1000);
- Combined annual and non-financial report (using GRI and AA1000);
- Report on compliance with the UN Global Compact principles (using GRI G3);
- Brief sustainability report with specialized addenda on various activities for different stakeholders (using GRI, AA1000, and other standards).

International organizations and UN Global Compact

Various international organizations, including the UN, the Organization of Economic Co-Operation and Development, the European Union, and others, actively support the efforts of businesses to increase their social and environmental responsibility which ultimately result in their sustainable development and in social significance of their additional voluntary efforts and initiatives. There is simple logic at work here: global sustainable development is impossible without a sustainable development of businesses, and vice versa.

The UN Global Compact (the 'Global Compact' or 'UNGC') is an initiative that encourages companies to abide by the ten universal principles in the spheres of human rights, workers' rights, environmentalism and anti-corruption. The then UN Secretary-General, Kofi Annan, first advanced the idea of the Global Compact initiative at the World Economic Forum at Davos (Switzerland) on 31 January 1999. Officially, Global Compact was formed at the headquarters of the United Nations in New York in July 2000. Since then, Global Compact, which is now made up of thousands of participants from over 100 countries, has become the largest and most important corporate citizenship and sustainability initiative in the world. At the beginning of 2009, more than 6,000 companies from all over the world responded to the call for the development of a more stable and open global economy.[6] By joining the Global Compact initiative, companies declare their adherence to the conviction that a business practice, by aligning their operation and strategies with ten universally accepted principles which promote greater sustainability, justice, and social responsibility, help establish thriving and successful communities.

---

6 See <www.unglobalcompact.org>.

The Global Compact is a strictly voluntary initiative that pursues two main goals:

1. Integration of the ten principles into business practice on the international level (the 'Global Compact Principles');[7]
2. Promotion of actions in support of broader UN development goals, such as the Millennium Development Goals.

UNDP Russia and CSR

The UNDP Office in Russia plays a leading role in the UN efforts to support the promotion of the Global Compact Principles in Russia. At the end of 2001, the first presentation and discussion of the Global Compact premises was conducted. UNDP Russia ensured the support of the Russian business community and governmental structures, and also initiated public-private partnership in that area in Russia. The Global Compact initiative provoked significant interest among Russian businesses, as well as among civil society organizations and the mass media.

UNGC-Russia Network 2008

On 10 April 2008, in Moscow, more than 30 senior representatives of Russian companies met with UN Secretary-General Ban Ki Moon, and declared their intention of starting a Russian local network of the UNGC. '*I welcome today's formation of the Russian Local Network as a sign that the business community of Russia, one of the largest economies in the world, is ready to fully support the universal values of the United Nations*,' explained Ban Ki-Moon. There was an annual general meeting of participants of the UNGC – Russia network on 17 December 2008. The meeting ratified the 'Statutes of the UNGC Network in Russia' and 'Priority Aims and Objectives of the UNGC – Russia in 2009-11' and elected the Steering Committee of the UNGC – Russia for 2009. The Steering Committee of the UNGC – Russia elected the chairman of the UNGC – Russia network for 2009 and, in January 2009, signed memorandum about the mutual understanding between the UNGC –Russia network and the UNGC office in New York.

Reports on Social Investment in Russia

Earlier, in 2004, the UNDP Office in Russia signed a joint project with the Association of Managers with respect to 'Supporting Public-Private Partnership in Russia Based on the Global Compact Principles and on the Strategy for Corporate Social Responsibility'. The main goal of the project was to continue promoting specific standards of CSR among Russian companies. The project included expertise and information sharing, as well as promoting the understanding of the principle of CSR by Russian private

[7] See Annex IV (*UN Global Compact Principles*).

companies, state structures, public organizations, and NGOs. The project also targeted enhancement of cooperation between businesses and the state in solving broader problems related to social responsibility, as well as more active involvement of civil society. Well-known specialized non-profit organizations, such as the RSPP, the Agency for Social Information, the 'CSR-Russian Centre', and others joined the project.

In the same year (2004), a pioneer research paper entitled the 'Report on Social Investments'[8] was published in connection with the UNDP Russia project to launch a 'Social Investment Index' ('the Index') in collaboration with the Association of Managers. This paper was a milestone in the promotion of the concept of CSR and the Global Compact in Russia. The Index is based on the analysis of primary data from 1,000 Russian companies representing different sectors. The research demonstrated that Russian companies equal leading international corporations in terms of their social expenditures, and many of them even exceeded the international level. The costs of social programmes were evaluated using qualitative and quantitative indicators, including data on the system of social reporting, financial situation of companies, and their share in the Russian economy. Investments into personnel development, safety, manufacturing, environmental and charitable programmes were taken into account.

This first nation-wide 'Report on Social Investments' in Russia included data on annual turnover allocated by different companies for social programmes (by weight, area, and sector), a forecast for increasing or decreasing corporate social expenditures, and best practices in social programmes run by Russian companies. This empirical study was the first of its kind with respect to Russian businesses, and it played a key role in provoking a public discussion concerning CSR and participation in the Global Compact.

The UNDP provided financial support with respect to developing these reports on the issues of corporate social responsibility and social investment. The 2008 'Report on the Social Investments in Russia' was a continuation of the 2004 study of social investments in Russian business, and the report, 'From Russia with Love: the National Contribution to the Global CSR Context', compared Russian CSR practice with that of multinational companies. These reports systemized the existing experiences and innovative practices of the responsible conduct of business, developed unified approaches to understanding CSR and social investments, analyzed the dynamics of the changes in the quantitative and qualitative indices of the social investments, and they also developed a number of recommendations regarding the improvement of CSR principles in the following years.

International conferences and CSR in Russia

International conferences for the promotion of the Global Compact principles were very significant for the dissemination of CSR practices in Russia.

In 2005, the Eurasian Conference on the Global Compact, where the first open dialogue on Russian participation in the initiative took place, was held in Saint Peters-

[8] See <http://www.undp.ru/index.phtml?iso=RU&lid=1&cmd=publications1&id=40>.

burg (Russia). This conference was attended by representatives of the Global Compact Office in New York and members of the Russian business and non-profit communities. All participants discussed CSR issues and reflected the growing interest in the topic of CSR in Russia.

In 2006, UNDP Russia together with the RSPP and the UN Global Compact Office conducted an international conference on 'Social Investments in the Economic Development of Russian Regions', in Rostov-on-Don (Russia). The purpose of the conference was to develop inter-regional and international cooperation, provide an opportunity for experience sharing, and present examples of initiatives concerning responsible investment. The conference was attended by representatives of major ministries and governmental agencies of the Russian Federation, heads of large Russian and international companies, as well as representatives of regional businesses, civil society, international organizations, and agencies within the UN system. Participants adopted a resolution that outlined actions required by Russian and international partners to promote corporate responsibility, support social and economic development of the Russian regions, and create a Russian network of the Global Compact. An Agreement on Cooperation was signed between Vnesheconombank and UNDP during the conference. The agreement dealt with joint projects to promote sustainable economic and social development of Russian regions.

Principles for responsible investing

Other UN organizations also participate in promoting components of the CSR concept in Russia. In May 2006, the UN approved, at its supreme executive level, the Principles for Responsible Investing ('PRI') – an initiative proposed by a large group of banks, as well as insurance and investment funds and companies – the UNEP Finance Initiative.[9] The PRI are, in effect, a form of introducing CSR and sustainable development practices into the finance sector. The PRI are grouped into six areas of responsibility in institutional investment and include a commitment to follow progressive approaches in environmental, social, and corporate governance. Before that notable event, UNEP FI conducted its first presentation in Moscow, but it received little public response.

As of April 2008, the group of international banks and investment companies that have joined UNEP FI and adhere to the PRI included around 170 participants among largest finance organizations of the world with assets totalling over USD 9 trillion. Russian banks and investment companies do not participate in UNEP FI yet, even though seven Russian finance institutions have already published 19 social reports. Obviously, there are vast possibilities for joint initiatives by UNEP and the Russian finance sector, but that would require a more energetic promotion of the Finance Initiative in Russia.

---

9 See <http://www.unepfi.org>.

Global Compact Leaders Summit

The Global Compact Leaders Summit on 5-6 July 2007, in Geneva[10] was a notable international event to advance the initiative. The Declaration adopted at this Summit stated: 'It is unprecedented in history to have the objectives of the international community and the global business community so aligned. Common goals, such as building sustainable markets, combating corruption, safeguarding human rights and protecting the environment, are resulting in new levels of partnership and openness among business, civil society, labour, governments, the United Nations, and other stakeholders.' The Declaration determined the directions for action and the roles of the business community and governments, as well as of companies and organizations participating in the Global Compact.

Of significance for Russia were also the results of the Summit discussion of the role of governments of UN member countries in supporting responsible corporate citizenship (and CSR) domestically. Formally speaking, CSR in Russia is only a direction for development declared by the President, a slogan, while there is still no concept for its development that would be formulated and approved by the federal government. The role of executive state authorities in that process has yet to be determined officially, even although authorities at the regional and local levels are actively involved in it. A round table of governments of the countries that participated in the Summit recommended the following directions for the promotion of responsible corporate citizenship by national governments:

1. Creating an enabling environment;
2. Raising awareness;
3. Promotion;
4. Tools development;
5. Funding.

Impact of financial crisis on CSR

There is no particular difference between anti-crisis rationalisation and routine rationalisation, which has been largely neglected by big Russian businesses as they, until recently, enjoyed high profits and cheap foreign loans. The financial crisis in Russia, similar to that in other nations, means economic and cost reduction: personnel optimisation, promoting energy efficiency, cutting salaries and bonuses (especially for top management), non-production cost reduction, and so forth.

As for CSR, the situation is more complicated. On the one hand, CSR is a type of 'litmus test' which shows to what extent social responsibility has been 'integrated' into Russian companies and whether the modern ethical principles have, in fact, transformed the corporate governance principles and business practices in Russia. On the other

[10] See <http://www.globalcompactsummit.org/>.

hand, CSR is a tool with which one can see the effects of the financial crisis on Russian companies and local communities.

Heated discussion of this issue in expert and business communities at the beginning of 2009 clearly indicated that Russian CSR and sustainability leaders undoubtedly realise that any sharp departure from the social and environmental responsibility principles and programmes will inevitably undermine their business reputation both in Russia and abroad – and will grossly aggravate their relations with most stakeholders (from local and regional governmental authorities to investors and consumers). At the same time, they have recognised the need for anti-crisis resource saving and cost-cutting measures which objectively call for CSR 'reformatting', mainly relating to personnel (both employed and those who have been made redundant).

In these times of financial crisis, many Russian companies plan to wind down numerous social and charity programmes, save for the most sensitive ones that cannot be terminated that easily: orphanage and specialized hospital support, additional payments to their own retired employees, etc.

CSR must be fully 'reoriented' in order to focus on cooperation and support programmes and projects aimed at local communities in the Russian regions. Rise in unemployment, expected high inflation and general impoverishment of the population in 2009-2010 may substantially affect public safety and the social climate in the regions. Therefore, starting 2009, Russian companies are likely to wind down most of their social programmes of public and charity nature (experts estimate the annual volume of corporate philanthropy efforts in Russia at 1.5-2.0 billion US Dollars) and redirect these funds to the programmes of unemployed adaptation and retraining, support for small business and the poor, etc. For Russian companies, this is not just an adjustment of their charity efforts, but a conscious attempt to ensure the necessary level of local public safety for their own businesses purposes.

Concluding remarks

The Russian economy has, of course, its own unique features, both structural and institutional. But then all the countries with vigorous CSR, corporate citizenship, and sustainability policies vastly differ from each other, which does not prevent them from conducting active and consistent state policies based on 'encouragement rather than on punishment'. It would obviously be senseless and counterproductive to exercise state control over voluntary CSR work of businesses, which is shown by the entire international experience. The author believes that further promotion of the Global Compact, and structuring and introducing the concept of CSR and corporate citizenship based on new organizational and economic forms and methods in the Russian Federation, require the following priorities:

- to facilitate and promote public and social responsibility in the business community, through cooperation with Russian businesses;

- to encourage the formulation of national priorities and regulation CSR and non-financial reporting at the federal level (to a federal minister) without establishing any new administrative forms;
- to conduct annual forums on social and environmental responsibility of businesses, how businesses can cooperate with authorities, and on participation of Russian businesses and NGOs in the Global Compact;
- to design and distribute regular public communications (declarations) on the development of the system of corporate social and environmental responsibility, and on participation in the Global Compact;
- encouraging the leading Russian business associations (RSPP, Business Russia, and OPORA) and sector associations of manufacturers to develop recommendations on voluntary introduction on international sustainability and environment governance and reporting standards, and to participate in the Global Compact;
- to conduct specialized forums on sustainable development of the regions;
- to promote involvement of Russian businesses and non-profit organizations in the Global Compact;
- to develop and implement public-private partnerships in the social, environmental, and cultural sectors, co-funded by business and authorities;
- involving representatives of federal, regional and municipal authorities in business events, including so-called dialogues on corporate social responsibility, in order to develop specific recommendations for businesses on how to develop social programmes;
- conducting seminars and specific events to develop inter-sector social partnerships between the authorities, businesses, and civil society;
- to assist the authorities in improving the procurement system of social goods and services.

The year 2009 could become a turning point for UNDP with respect to the promotion of the Global Compact Principles in Russia and for active involvement of Russian companies in using those principles in their business practice.

UNDP Russia and the UN Global Compact Russia Network launched a multi-year project for 'Engaging Russian business in GC-driven sustainable development' aimed at the joint development and the implementation of sustainable socio-economic development projects by UNDP and Russian and foreign multinational companies, who are either currently participants, or anticipate to be participants, of the UNGC – Russia. The core priority of this project is the development and implementation of projects that aim to minimize the economic, social and environmental fallouts of the financial and economic crisis for the company-participants of the UNGC – Russia in the territories where they operate.

The main purpose of the UN Global Compact Russia Network is to ensure progress of the companies that have signed the Global Compact in implementing the initiative's ten principles in human rights, labour, environment, and anti-corruption, as well as to create opportunities for involving all interested parties and new members from Russia in that process and in collective actions. At the time of writing, 35 Russian companies and organizations in the Russian Global Compact Network are able to follow the example of many other countries in consistent integration of CSR and the Global Compact principles in their companies and organizations with the help and coordination from UNDP, and to actively recruit new members through mutual contacts.

The development of CSR practices and the promotion of the Global Compact Principles in Russia coincide with, and in some cases they have already resulted in, the creation of synergies: Russian companies are coming to realize that engaging in responsible business practices is not a fashion, but rather an objective necessity which benefits businesses, the Russian state, as well as the Russian and the international community.

Chapter 20
RUSSIA: CORPORATE SOCIAL RESPONSIBILITY IN A POST-SOCIALIST STATE

Sergey Dayman[1]

Business practices and even business ethics are increasingly standardised throughout the world. There are of course known differences between 'Western' and 'Eastern', 'developed' and 'developing' country practices, but it is expected that general business operations and conduct are globally similar. Social responsibility of businesses is an area which, regardless of the growing standardisation of approaches, will remain ever dependent on the local, regional, and national settings. In this respect, transitional post-socialist industrialised countries, such as Russia, have their peculiarities that affect the way in which social responsibility of businesses is perceived and put into practice.

Social and economic background

The Russian Federation is a successor of the Union of the Soviet Socialist Republics (or 'USSR'). The USSR, now almost forgotten, was one of the world's political poles at the times. Its differences from the post-war Europe and, in particular, the USA were manifold – for example, although socialism during the times of the USSR was flawed in the area of human rights, strong achievements were made in the areas of education, health and medicine, as well as the support of the elderly. Another feature of the USSR, which is important in context of corporate social responsibility ('CSR') in Russia, was its planned economy structure and significant size. The results of these were, amongst others, the construction of huge industrial plants for which entire cities were built. These cities were integral parts of the plants, embedded within the communal and social infrastructure of the USSR.

When the Soviet Union collapsed, various Russian business people acquired the major part of the existing industries. A good many enterprises, however, remained state-owned or inherited the Soviet administration style and method of doing business. In the next few years following the collapse of the Soviet Union, many changes took place and virtually all such changes had negative consequences in numerous social spheres. Enterprises struggled to find a strong foothold in their respective markets – or (nearly) went bankrupt as there was no demand for many of the products produced in

[1] Senior consultant of the Sustainability Assurance and Advisory Services group in Ernst & Young (CIS) branch in Moscow; Email: *Sergey.Dayman@ru.ey.com*.

W.Th. Douma and F.M. Mucklow, eds., Environmental Finance and Socially Responsible Business in Russia
© 2010, T·M·C·Asser press, The Hague, The Netherlands and the Authors

the USSR. Managers tried to improve efficiency and reduce costs. Social infrastructure, including water supply and wastewater treatment, heat supply, residential building maintenance, solid waste management, medical services and so on, quite reasonably became the focus point of such efforts – and also an inevitable victim of these. In most cases, the facilities were passed into the ownership of the respective municipal governments. However, the shift of management to the municipalities did not always impact positively on the administration of the social infrastructure. This was so due to the municipal government's budgetary limitations, in many cases, the lack of motivation and management experience, insufficient control over funds spending and transparency, and many other hampering elements commonly found in public administrations of various sizes around the globe. As a result, in many regions, towns and settlements, all types of infrastructure – from cultural (libraries, cinema, theatres, youth groups etc.) to transportation and communication – began to collapse.

First steps towards CSR

Increasingly, Russian businesses have come to recognise that their support of society is a vital ingredient to their commercial survival and development, and by so doing, they contribute to Russia's development as a whole. This recognition commenced at the turn of the century, when Russian companies, including large and small entities, started to adopt CSR practices, although in most cases unknowingly (and sometimes even unwillingly). Businesses were not necessarily eager to assume the role of local government by investing in infrastructures and communities, but often they have been left without any other option but to do so. In order to maintain and develop their businesses, companies have had to build (or assist in building) public roads to access their sites and nearby settlements, to arrange natural gas supply to their business sites and, thus, also providing it to the homes of the local population, to improve supplies to the local stores, and so on. Along with providing jobs, and so increasing the income of the surrounding population, companies have trained employees and developed their skills and social positions. Larger companies have contributed significantly to regional and national universities, high and technical schools, providing support for the establishment of new courses and even for entire faculties. Furthermore, sponsorship has become a part of every major company's budget – not only to increase their visibility and improve their public image, but also to respond to the most urgent, or most appealing, community requests.

For example, a medium-sized Russian company invested into a plant which was practically abandoned for three years. Inhabitants of the relevant village were generally surviving on forest products, their gardens, unqualified labour, and theft; spending their minimal income on alcohol. Not only did this company provide new jobs and income to the local community, it also improved goods supply to the village shop and introduced free alcoholism treatment to everyone willing to undergo such treatment. The response to the proposal was nearly 100%. A few years following the recommencement of business, the plant has become very competitive, attracting large corporate clients. This example of engaging in socially responsible business, however, was done without actual knowledge by the directors or the company owners of the very concept of CSR. Furthermore, an audit conducted by one of the company's clients, which cared for its CSR image revealed child labour requirement violations – young people were working at the factory in order to increase their family's earnings. Following the audit such practices were, of course, stopped.

With respect to the environmental and economic components of CSR, the evolution towards adopting environmentally responsible practices has not happened quite as dramatically. Nevertheless, the transition towards adopting environmental elements of CSR has been challenging for all players. In the wake of the emerging free market economy, most businesses in Russia tried to avoid paying taxes, often had no way to escape bribery, and kept investments to a minimum, trying to avoid risks and obtain maximum returns as quickly as possible. Attempts to save costs were made at the expense of health and the environment, including by way of switching off existing ventilation equipment to save on electricity, setting up facilities without obtaining environmental licenses and so on. After the first decade of such chaotic development in the 1990s, the Russian government tried to establish effective enforcement in all fields, combat corruption and increase tax collection, with the instruments it considered necessary. After a number of years, the effect of such changes have become visible: The introduction of environmental or tax inspection has made it difficult for businesses to avoid major law requirements, and Russian legislation has, in general, been further developed. In addition, businesses themselves have learned their lessons. Those businessmen who took a long-term view, sooner or later recognised that creating a reliable future for their businesses requires not only good marketing and financial management, but – of course – process management, material and energy efficiency improvements, and a safe working environment. In this respect, first steps have been made regarding investments in environmental and safety measures – and not only technical, but also managerial, measures, for leaders – by way of adopting ISO 14001- and OHSAS 18001-based management systems.

Governmental promotion of CSR

The above would have been an adequate description of CSR in Russia until the autumn of 2003 when the then President Putin declared that Russian businesses should be 'socially responsible'. His speech was not very specific as to what was meant by 'socially responsible', but it prompted a swift reaction from large Russian business. The President's speech was made in context of discussing the case of YUKOS, whereupon YUKOS immediately responded by publishing a 'Social Report' that discussed its social importance on the country level – as a taxpayer, employer, sponsor and community developer in many regions.

Social reports

In order to respond to the governmental push for Russian businesses to be socially responsible, Russian businesses drew on international experience. British American Tobacco Russia, which by that time had already published its first social report and was preparing its second social report, provided Russian businesses with an interesting example. Additional hints were found in the glossy reports of major international companies; ready instruments such as Global Reporting Initiative ('GRI')[2] guidelines and the AccountAbility AA1000 assurance standard[3] were uncovered and further publicised by consultants. These standards and practices have now become hugely popular in Russia – at least in speeches and publications.

Furthermore, during this peak of interest in the topic, conferences and large-scale presentations discussing various aspects of social or sustainability reporting, audit, and CSR in general, were held in Moscow every month if not more often. Companies promised that they would soon publish their own social reports. First reports which were released in 2004 were not very comprehensive, but bore more resemblance to PR booklets or a section of an annual report. A large, if not major, part of these reports covered sponsorship and philanthropy activities, and only a small amount of information was provided on actual economic, social and environmental policies, actions and impact(s) of the companies in society.

Large companies in Russia have only just started to prepare social reports and mostly outsource their preparation to external organisations. The following wave of social reports displayed more detailed reporting, but again reflected the fact that the management structure underlying CSR declarations is lacking in most aspects, except perhaps with respect to human resources. In general, CSR is mainly viewed by Russian top managers as a tool to managing non-financial risks and improve capitalisation through better public image.

In this respect, the following example is indicative of the current situation in Russia: A large industrial corporation which has significant environmental impacts prepared its first 'Social report' in 2005. The report included reference to the company's environmental, health and safety policies, and contained a few paragraphs on a cleaner production programme, but gave no account of the company's environmental issues or performance. Yet, in practice, the company made significant efforts to improve its environmental performance and gain a more environmentally friendly image. This contradictory situation was due to the company wanting to prepare a nice report, but, at the same time, it was not ready to discuss any of its environmental problems with the wider public. However, after the publication of their first report and attempt to approach the stakeholders, the company gained confidence. Its CSR reports, which are published on an annual basis since then, discuss company issues and performance with increasing detail.

---

[2] At the time, Guidelines version 2002, see <http://www.globalreporting.org>.
[3] See <http://www.accountability21.org>.

Concluding remarks

The question begs as to whether the CSR practices of Russian businesses are now the same as they exist in Europe, North America or Japan? The answer is: yes and no. The management and, more importantly, the directors and shareholders of Russian companies have yet to understand (as is the case in other regions of the world), firstly, the concept of sustainability and to distinguish it from the 'sustainability of the company' (what many Russian companies understand 'sustainability' is about), and secondly, their role in the development of Russia and the community and dependence of it, not just with respect to the tax budget and its own risk management. There is still a long way to go before Russian companies will automatically incorporate the concept of 'sustainability' into their regular business processes.

However, if one cuts out the fashionable words from the reports of most international companies, or from the speeches of public persons and, instead, one looks at their actions, one arrives at the same point: CSR is, worldwide, not yet a generally accepted practice. For all but a few leaders it represents, although a necessary activity, only a side activity. Russia is no different. This means that, effectively, Russia, too, needs to develop its own 'sustainability leaders' that are brave enough to take on this role and to direct and show the way for those who might follow towards universal implementation of CSR.

Chapter 21
SOCIALLY RESPONSIBLE BUSINESS: HELPING CHILDREN IN RUSSIA

Kit Lawry[1] and Joanna Rogers[2]

Corporate Social Responsibility ('CSR') has been a buzzword on the Moscow corporate scene for the past five years but interpretations of the concept differ enormously. One company may consider it fulfils all social responsibilities by paying its staff regularly and paying taxes; for another company, this would be merely a starting point and the company will look to play an active role in supporting the work of one or more charities over a period of years.

The Russian Business Social Charter, adopted by the Russian Union of Industrialists and Entrepreneurs in November 2004, helped clarify the concept for Russian companies.[3] A major report by the Russian Managers Association in 2004 recommended that '*Corporations should strive to establish partnerships with ... civil society organisations in order to develop ... systematic approaches to corporate responsibility.*'[4] However, businesses with a well-defined social policy and established accounting, managerial and information procedures for social investments are still rare in Russia. A random approach to social investment predominates: non-systematic philanthropy based on the Chief Executive's or Head of Human Resources' personal charity interests or the off-chance of a local charity managing to put their case to a senior manager.

How do children's support organisations cooperate with businesses?

Non-governmental children's support organisations and charities are getting better at making their case to corporates and raising funds for their work. In Moscow, a handful of small non-governmental organisations ('NGOs') raise funds successfully from mainly Western corporates, for programmes that are mainly specific service delivery projects (e.g., children with Downs syndrome, or disabled children in orphanages) based in Moscow. They are often supported by members of the international community in Moscow who provide access to high-level corporate individuals. Their fund-raising efforts focus on grants from larger companies, institutional grants from international

[1] At the time of writing, Moscow Coordinator, EveryChild Russia; Email: *kit.lawry@gmail.com*.

[2] Country Director, EveryChild Russia; Email: *jo_rogers@mail.ru*.

[3] M. Polyakova, *Russia Profile*, October 2005.

[4] S. Litovchenko, ed., *Report on Social Investments in Russia 2004* for the Russian Managers Association.

W.Th. Douma and F.M. Mucklow, eds., Environmental Finance and Socially Responsible Business in Russia
© 2010, T·M·C·Asser press, The Hague, The Netherlands and the Authors

donors, collecting donations from staff, gifts in kind and fund-raising events (e.g., charity dinners, auctions). Sometimes, staff from donor companies are personally involved in the fundraising activity (e.g., sponsored skiing or bike riding) but most interaction between donor organisations and children's support organisations is limited to the handing over of cash donations or gifts in kind. Long-term relationships with beneficiaries, or any kind of education programmes run by charities for their donors, are rare.

Children's organisations in other parts of Russia have a much harder time gaining access to potential corporate donors, particularly the big corporate entities based in Moscow. They tend to develop relationships with local Russian businesses which usually favour *ad hoc* donations of gifts in kind (food products, clothing, etc.). It is rare to find instances of systematic long-term strategic support from a business to a children's charity or organisation.

The Charities Aid Foundation ('CAF') Russia positions itself as a general charity broker helping corporates direct their funds into social investment, as well as supporting wealthy private individuals to establish trust funds. There are some local 'umbrella' children's charities that work mainly on bringing together local organisations to advocate with the government on a range of issues pertaining to children rather than linking into the business community.

EveryChild, a British based international charity, is the only major international children's charity currently based and working in Russia.

How do businesses typically work with charities?

When companies start thinking about investing in the community, they often – following the random approach mentioned above – gravitate towards helping local orphanages. There is usually a local orphanage in need of help, conditions can be easily improved by a lick of paint or some new equipment, or the children's lives brightened by taking them on an outing. This short-term approach to social investment offers quick, easily perceived benefits for both the orphanage and the company; it does not cost too much and is relatively simple to organise. The problem is that nothing really changes: children continue living in orphanages, the paint gradually chips and cracks over time, the new equipment may be sold off or broken, the children's memories of a good day out will fade and the bleakness of institutional life is still their only reality. Some businesses help locally based children's charities in their work with children in the community, but these relationships tend to be formed on an *ad hoc* basis and are also characterised by a predominance of gifts in kind.

Brief historical background to child institutionalisation in Russia

Institutional childcare in Russia has a long and unhappy (for the children) history. Catherine the Great's government established two large foundling homes in Moscow

(1764) and St Petersburg (1770) with schools attached 'to inculcate enlightened morality, the work ethic, civic mindedness and respect for constitutional authority, and thus to create from unwanted children the educated, urban estate that Russia then lacked'.[5] However, conditions in the homes were very bad and proportionally very few children survived into adulthood. In the twentieth century, the Bolsheviks approved of children's homes, rather than families, as places where the new 'Soviet Man' could be reared by professionals, but conditions were still dreadful. The Soviet family managed to survive as an institution despite harsh criticisms but doubts about the family's adequacy to bring up children remained deep in people's consciousness. There was a belief that 'the professional' in the institution knew best – particularly common for children with disabilities or from socially at risk families.

Similarly, and because Soviet society was by definition perfect, the perception of poverty as personal failure lingers on today. The Soviet mindset could not accept that it was appropriate to work with problem families to solve their problems: far better to take children away from such inadequate families. Consequently, Russia in the early nineties had no support systems (e.g., social workers) to deal with the huge social problems that overwhelmed so many people when the economy collapsed.

In some areas, matters have not improved much since the early nineties. EveryChild's report 'Family Matters' of November 2005[6] researched extensively conditions in children's residential institutions in Russia and the former Soviet Union. Findings included:

- *The poor physical state of buildings*. Many institutions have serious structural problems. Equipment is also in a poor state of repair: plumbing is bad and washing and toilet facilities are very substandard. Heating is often poor to non-existent.
- *A lack of financial resources*. Food supplies are inadequate, cooking methods and conditions are poor, and clothing supplied for the children is substandard.
- *A lack of individual attention.* It is hardly surprising that children do not receive the love and attention they need to thrive when 30 or more children are frequently kept in one room with one or two poorly-trained carers.Childcare policies are also frequently out of date.
- *Frequent abuse*. Abuse in residential institutions appears to be common across many different cultures and settings.[7] Research has identified staff, relatives and minors as perpetrators. Severe physical and verbal bullying and humiliation by both staff and other children is also common. In Russia, Human Rights Watch reported ten

[5] D.R. Ransel, *Mothers of Misery: Child abandonment in Russia* (Princeton, Princeton University Press 1988) p. 31.

[6] R. Carter for EveryChild: *Family Matters:A study of institutional childcare in Central and Eastern Europe and the former Soviet Union.*

[7] For a very useful review of abuse of children in institutions, see C. Barter, *Abuse of Children in Residential Care* (London, National Society for the Prevention of Cruelty to Children 2003) <www.nspcc.org.uk/inform/Info_Briefing/AbuseofChildreninResidentialCare.pdf>; for a shorter review, see A. Kendrick *Abuse of Children in Residential and Foster Care: a brief review* SIRCC research papers (Glasgow, Scottish Institute for Residential child Care 2003) <www.sircc.strath.ac.uk/research/kendrick.html>.

years ago on the systematic and casual abuse of the children in their care by staff in the institutions (see box below).

Abuse in Russian institutions

- slapping or striking children
- shoving their heads in the toilet
- squeezing a hand in a vice
- squeezing testicles while interrogating them
- stripping their clothes off in front of peers
- locking them in a freezing, unheated room for days
- engaging them in sexual relations
- sending them to a psychiatric institute to punish them for misdeeds such as attempting to run away

  HRW also reported that older or stronger orphans, goaded by the adult staff, had maliciously abused younger or weaker ones by such measures as:
- beating them on the neck, forehead and cheeks
- throwing them out the window in a wooden chest
- wiring a metal bed to electricity and shocking a child forced to lie on it
- forcing a child to beg or steal for them

Human Rights Watch (1998)

- *No right of contact*. In EveryChild's experience there are no regulatory requirements to sustain contact between child and parent. Staff frequently impose stringent conditions on parents and Child Protection Departments when a child is admitted.

EveryChild's work

EveryChild advocates family-based care for disadvantaged children, in foster families if not with their birth family, as a far preferable alternative to institutional orphanages. Children in Russia can be disadvantaged or marginalised from mainstream society for a wide range of reasons: loss of parents, child protection reasons, disability either mental or physical. There is a problem of over-diagnosis of disability, where children are diagnosed 'disabled' for minor problems such as a cleft palate or as in the example in the box below.

'Natalia, who has a congenital deformity of the fingers of her right hand, was sent to a new special kindergarten for disabled children, and thus was unable to enter regular school when she was seven. Her parents are divorced and her mother does not have the time or energy to fight the authorities for a new diagnosis. However she is worried about Natalia's future; being labeled as disabled and not having access to a decent education reduces her future opportunities.'

Psychologist, Moscow, Russia (Sammon 2001[8])

[8] E. Sammon, *Children and disability in the context of family breakdown in Central and South Eastern Europe and the Former Soviet Union: Defying Prejudice, Advancing Equality – 2* (London, EveryChild 2001).

Once these children are labelled and admitted into the 'special, disabled' system, the lack of stimulation soon delays their development so that they effectively become disabled.

Research has shown conclusively how damaging institutional life is for children emotionally, educationally, socially, even physically. Younger children, aged from 0 to 3 years, in baby homes are especially traumatised. A far preferable alternative – and far cheaper for local authorities – is family-based care, either within a child's birth family or, where that is impossible for the best interests of the child, within a local foster family.

EveryChild works at both government and grassroots levels. With government departments, we promote and prepare for family-based childcare: recruiting and training child and family social workers and potential foster parents. The social workers, we train then work at the grassroots level, helping and advising families in crisis on issues like debt, unemployment or homelessness. The earlier a family with difficulties can get support and advice, the more likely they are to overcome their problems so their children will not be removed and placed in institutions.

All our projects aim to achieve sustainability. Any new services we set up eventually become fully run and financed by local structures. We model new ways of working and then encourage them to be replicated widely by regional and, in the longer term, national, authorities. Thus our projects eventually impact on a large number of children, not just the few thousand who are direct project beneficiaries.

EveryChild has been working in Russia since the early 1990s, so has a long-term, well-developed commitment to disadvantaged children and their families and a wide experience of their problems and possible, well-tried, solutions. To highlight these problems and possible solutions, EveryChild made a film in 2005, in collaboration with the Early Intervention Institute of St Petersburg and HealthProm (UK). This is currently being shown to audiences of government policy makers and childcare staff across Russia. EveryChild also facilitates a network of organisations working to prevent children entering institutional care and promote family-based care as an alternative. The network, which is for government and non-governmental organisations, exists to share experience and spread best practice, so each region does not have to 'reinvent the wheel' in order to reduce the number of children living in institutions.

A family in crisis

Tamara, a young mother, was living at home with her mother when her son Sasha was born. Tamara had to work day and night to provide her son with food and clothing; she didn't get any support and so struggled to care for him properly. Sasha's grandmother quite often stayed at home to look after Sasha while Tamara was at work, but she had a problem with alcohol and didn't always give him the care and attention he needed.

When Sasha was four, there was a disturbance at the house involving Sasha's grandmother. The police were called and they decided to take him to an orphanage. Tamara tried to find out where her son had been placed, but the local officials wouldn't tell her. They had decided that Tamara was a bad mother and that it was in Sasha's best interests to remain in state care. Because of the lack of national standards in Russia, Tamara did not know what was happening to her own son – it was a traumatic time for her, and left her feeling isolated and hopeless.

Tamara was then summoned to court several times, but there were no social services available nearby to advise her that she needed to attend. She didn't have any knowledge about the law and didn't understand what would happen if she didn't go. And so the court hearing took place without her and it was decided that her parental rights should be removed. She would be allowed to see Sasha only with the permission of the Fostering and Guardianship Department. But now that she was seen as a bad mother, nobody wanted to take responsibility for giving her this permission and so she sadly lost all contact with her son.

Five years passed. Having learnt about the new EveryChild social services, Tamara came to us for help. Our social worker provided her with the support and guidance she needed, and together they planned how they could bring Sasha back home. With the help of our social worker, Tamara managed to convince the director of Sasha's orphanage of her good intentions and he finally gave her the chance to meet with Sasha, who had been waiting for her for all these years. And much to Sasha's delight, she is now able to visit him twice a week and the director has really started to support her and the renewed relationship she has with her son. With EveryChild's help, the local officials are also considering restoring her parental rights, and things are beginning to look much better all round.

There is plenty of work ahead, but we firmly believe that when the time is right, Sasha will return home to live with his mother and be part of a loving, secure family once again.

EveryChild, Russia (2006)

How does EveryChild promote socially responsible business in Russia?

EveryChild encourages long-term, strategic investments. This reflects our long term goals of changed attitudes and actions on the part of social service policy makers and providers in Russian government, away from institution-based childcare towards family-based and socially-integrated care. We seek to build long term relationships with individuals and corporate investors in Russia, both Russian and international, through various kinds of interactions as well as raising funds for our projects.

Companies can be involved with EveryChild in many different ways, some of which do not involve the exchange of money at all. For example:

- Offering access to lists of clients / customers / members whom we can contact for individual sponsorship
- Offering access to staff lists for individual sponsorship purposes
- Giving gifts in kind (intangibles, e.g., staff time and expertise in a particular area, printing facilities, meeting rooms for activities, as well as tangible gifts, e.g., equipment)
- Organising events (e.g., bazaars, sports events, musical events) in our favour
- Paying for consultancy services (see below)
- Giving grants or donations for projects or specific items
- Giving general donations (unrestricted funding)
- Doing cause related marketing with us (see below).

EveryChild has pioneered the practice of offering consultancy services for businesses in Russia. With offices and projects in Ekaterinburg and St Petersburg and our long experience of networking with contacts in children's organisations all over Russia, EveryChild is well placed to help companies develop effective social investment strategies, whether at planning, implementing or evaluating stages. Companies who wish to invest socially, but do not have in-house in-depth knowledge of the possibilities on the current Russian children's charity scene engage us to advise, run competitive tenders and draw up short lists of possible projects for investment. EveryChild can also monitor projects and report back regularly on progress. This way, businesses know their money is being well-spent for the long-term benefit of Russia's disadvantaged children.

Cause related marketing ('CRM') is when a particular item is sold with a message on the label about a charitable organisation, project or cause which receives a small amount from each item sold. This enhances sales of the item, raises awareness and raises funds for the cause.

Barriers, historical and contemporary

Because social deprivation and economic failure are still frequently perceived as failure on the part of the individual rather than as weaknesses of the system or society, 'charity' has quite a negative connotation in Russia. Successful Russian businesses can be reluctant to link themselves with the underdogs of society. They prefer to back potential future winners with scholarships for further training or to engage in prestigious infrastructure projects, like building hospitals and orphanages, rather than supporting the complex task of changing systems so that fewer people will need such support in future.

Very wealthy individuals have established foundations or trust funds over the past ten years to support their own charitable interests. However, public perception of a business person who has made so much money that they can afford to spend it on social projects can also be negative: where did they get that money from in the first place? Many of Russia's national assets are perceived as having been 'stolen' in shady deals by a few now very wealthy people during the early nineties.

So both business and the general public in Russia have concerns about transparency which can result in ambivalent attitudes about links between the corporate and charitable worlds. 'Foundations set up on the personal initiative of businessmen are perceived as being driven more by self-interest, and …it can even be harmful for a businessman to get involved with charity', the vice president of RSPP is quoted as saying.[9]

Some local authorities have exploited the possibilities of business funding for areas of their own social services. Instances of 'charity racketeering' have occurred in regions where local authorities impose additional 'taxes' on companies 'to develop the social sector', force companies to contribute to funds they have set up or even 'dream

[9] Interview with *Money and Charity* magazine, quoted in *Russia Profile*, October 2005.

of having businesses finance the entire social system', Olga Alexeyeva, former director of CAF Russia reported.[10]

None of these historical facts foster trust, goodwill, or even understanding between the public and private sectors.

Paths forward

As Russia and Russians engage more with worlds outside their own borders, both commercial and non-commercial, broader perceptions of what is good for people and good for business are percolating through historical filters and barriers.

One method of avoiding some of the pitfalls outlined above is to involve partners in a process of dialogue – business, government and community groups talking and listening to each other – to develop programmes which meet the needs of all sides. Tenders can be held for corporate grants, aimed at supporting local groups to resolve local social issues. LUKoil's charity fund director, Igor Beketov, commented: 'This is an effective procedure; people don't just receive the money but put forward specific projects and develop financing plans for them.'[11] This is one of the models EveryChild has pioneered in our consultancy services to corporates. The corporate benefit, of course, is that eventually more affluent consumers will be predisposed by their experience of the corporate's involvement in their community towards buying that business' products.

Observers of the Russian business scene are cautiously optimistic about the future of socially responsible business in Russia. As the market develops and society matures, they say, local communities, consumers and other stakeholders will become involved and bring the necessary civil society pressure to bear on moving business forward from its present declared interest in investing in the community to actually implementing its declarations. The Russian Managers Association expects Russian business social investments to increase by 10% and concludes there is 'likely to [be] a long term trend of social investment development in Russia'.[12]

In the end 'at some point', says Elena Topoleva of the Agency for Social Information, 'your social responsibility becomes an economic asset, and if you don't measure up, you risk losing out'.[13]

---

[10] At a conference on corporate social responsibility in March 2004, reported in *Russia Profile*, October 2005.

[11] Quoted in *Russia Profile*, October 2005.

[12] *Report on Social Investments in Russia 2004*, Russian Managers Association.

[13] Quoted in *Russia Profile*, October 2005.

LIST OF CONTRIBUTORS

Trevor Barton, Senior Vice President and Managing Director, Europe, is responsible for the company's Europe-wide operation. A lawyer by profession, he previously spent 14 years with the international law firm Clyde & Co, where he was a partner and managed the firm's St Petersburg office for 6 years. Trevor is also Chairman of the CIS Advisory Group of International Financial Services London and has been a Non-executive Director of the Russo-British Chamber of Commerce ('RBCC') for 7 years.

The PBN Company is a consultancy specialising in advising Western investors into Russia and the former Soviet Union on market entry and expansion strategies, public affairs, government relations, communications and crisis management. The company has offices in Moscow, Kyiv, Almaty, London and Washington DC.

Contact details:
Email: *trevor.barton@pbnco.com*
Website: *www.pbnco.com*

Dr. Michael Begak is the Key Expert on the Harmonisation of Environmental Standards (HES) II project in Moscow, Russia. He is also based at the St.Petersburg Scientific Research Institute for Ecological Safety RAS (since 1995), working as the Head of Department. From 1990 to 1993, Dr. Begak was the Deputy Head of the Standing Commission for Municipal Economy and Environment of St.Petersburg City Soviet (St.Petersburg Parliament) and from 1969 to 1990, he worked in the Krylov Shipbuilding Research Institute (Full State Councilor of St. Petersburg (III rank)). His professional interests nowadays include the legal regulation of environmental safety issues.

Education: Leningrad Shipbuilding Institute (now – State Marine Technical University of St. Petersburg) from 1963 to 1969. M.Sc. in Hydrodynamics Diploma. Doctor of Science (Engineering, first degree) in 1979.

Publications: More then 70 scientific publications in the fields of fluid dynamics and the environment. Latest publications include:

- The prospects of harmonisation (together with Dietrich Hahn) – Business environmental magazine N4 (28), 2008
- The factor of availability (together with Tatiana Guseva) – Business environmental magazine N1 (29), 2009, pp. 46-48
- Charges for environmental pollution and environmental taxes in Russia and CIS member-states (together with Alexandra Manvelova) – Nature management economy, VINITI, Review information N3, 2009, pp. 90-100
- Management systems providing for energy efficiency improvement: major principles of development and implementation (together with T. Guseva and S. Mironova) – Management in Russia and abroad N3, 2009
- Harmonisation of environmental standards (together with Tatiana Guseva) – chapter in the book "Russia in the Surrounding World" – 2009. Sustainable development: ecology, policy, economy: Analytical series /Ed. N.N. Marfenin; ed. board N.N. Marfenin, S.A. Stepanov. – Moscow: IIUEPS Press, 2009, ISBN 978-5-7383-0303-6

Contact details:
St.Petersburg Scientific Research Centre for Ecological Safety RAS
Email: *mbegak@gmail.com*

Website: *www.ippc-russia.org*
Tel.: +7 (812) 230-7894; +7 (812)373-9003
Fax: +7 (812) 235-4361

Godfrey Cromwell has spent 20 years working in and with Russia and other countries in the CIS region. This has included running businesses as well as NGOs working on environmental and social reform issues. He was Executive Director of the Russo-British Chamber of Commerce ('RBCC') for three years (2004-2007) and retains involvements across the CIS.
He now works at Barclays Wealth as a private banker for individuals, businesses and charities.
Contact details:
Email: *godfrey.cromwell@barclayswealth.com*

Dr. Sergey Dayman is a Senior Consultant in Ernst & Young Sustainability Assurance and Advisory Services based in Moscow. Ernst & Young is one of the four worldwide leading auditing and advisory companies, with very large practice in CIS (14 offices and 3,000 employees), serving clients in attested services, auditing, corporate finance, tax and business advisory areas. In the CIS, Ernst & Young is consistently ranked among top 2 companies in terms of professional fees, growth and personnel. Core fields of Sustainability Assurance and Advisory are sustainability reporting assurance and advisory, sustainability strategic advisory, environmental, health and safety risks and liabilities assessment, climate change assurance and advisory.

Dr. Dayman has over 10 years of consulting experience in sustainability strategy, sustainability reporting, climate change, energy efficiency, environment, health and safety, and quality management, environmental and EHS due diligence, knowledge management.

Sergey worked with such industry sectors as automotive, consumer products, diversified industrial products, governmental, public sector & non-for-profit organizations, mining, oil and gas, and utilities.

Dr. Dayman has received Ph.D. in environmental protection and M.Sc. in environmental sciences. He also received trainings on business risk, environmental management, climate change and energy efficiency.

Dr. Dayman has published a number of books and various articles in business and environmental magazines and journals covering different topics of sustainability.
Contact details:
Email: *Sergey.Dayman@ru.ey.com*
Tel: +7 495 755 9700
Fax:+7 495 755 9701

Dr. Wybe Th. Douma is Senior Researcher European Law and International Trade Law at the Asser Institute in The Hague, The Netherlands. He worked on numerous research and applied research projects in the Netherlands, Asia, South America and Central and Eastern Europe over the last 15 years, notably as long term senior legal expert in the TACIS project Harmonisation of Environmental Standards, Russia in the period 2002-2004. Dr. Douma also provided advice to the Russian-European Centre for Economic Policy ('RECEP') and lectured at Moscow State Institute for International Relations ('MGIMO'), the Institute for State and Law (Moscow) and the Universities of Irkutsk and Kazan (Russia).
Education: Dr. Wybe Th. Douma studied law at the University of Groningen (The Netherlands) and the Eberhard Karls University, Tübingen (Germany) and wrote his Doctorate thesis on "The Precautionary principle. Its application in International, European and Dutch Law". He also passed his propaedeutic year in Russian language studies at the University of Groningen.

Publications: Dr. Wybe Th. Douma is editor-in-chief of the European Environmental Law website (http://www.eel.nl) and editor for several journals on environmental law. He has published extensively on EU law, the EU-Russia relationship, environmental law, climate change issues and WTO law. Some of his publications related to the Russian Federation are:

- Russia and the present and future international climate change regime (Wybe Th. Douma, Michael Kozeltsev and Julia Dobrolyubova), in: Sebastian Oberthür and Marc Pallemaerts (eds.), *The Climate Policies of the European Union: Internal and External Dimensions*, Brussels, 2010
- *The Kyoto Protocol and Beyond: Legal and Policy Challenges of Climate Change*, The Hague, 2007 (co-editor)
- The European Union, Russia and the Kyoto Protocol, in: M. Peeters and K. Deketelaere, *EU Climate Change Policy*, Cheltenham, 2006, pp. 57-66
- Russia and the enlarged European Union. Mending fences whilst shaping a Wider Europe´s environment, in: J. de Zwaan (ed.), *The European Union, a Gradual Process of Integration*, The Hague, 2004, pp. 319-336
- *Economic aspects of environmental policy in Russia* (ed.), Moscow, January 2004

Contact details:
Email: *W.T.Douma@asser.nl*
Websites: *www.eel.nl* (European Environmental Law website including a page on Russia) and *www.asser.nl*

Dr. Ivetta Gerasimchuk, is a Scholar-in-Residence/Lecturer in Environmental Economics and Management at the Cyprus International Institute of Management ('CIIM') and Senior Advisor, Trade and Investment Programme, WWF Russia. Dr. Gerasimchuk has over 10 years professional experience on issues related to international development, economic policy and the environment. Prior to joining WWF-Russia, she has done research and provided advice for the Institute for African Studies of the Russian Academy of Sciences, Moscow State Institute of International Relations ('MGIMO'), Centre for Current Politics of Russia, Australian Embassy in Moscow and leading business media (Kommersant Daily, Oil & Gas Eurasia, Russia Investment Review, etc).
Education: Ivetta Gerasimchuk has a PhD degree in international economics from the Moscow State Institute of International Relations ('MGIMO').
Publications: She is author of over 30 academic publications on the role of environmental factors in the world economy, including three monographs: 'International Environmental Cooperation in Southern Africa' (2003), 'Environmental Practice of Transnational Corporations' (2007), 'Re-Think Russian Investment in Southern Africa' (2009). She is also contributor and editor of the WWF report 'Pure Profit for Russia: Benefits of Sustainable Finance' (2009). Dr. Gerasimchuk is the first prize winner of International Essay Prize Competition (Weimar, 1999), Hugo Grotius Award in International Law (Moscow, 2000), and ONS International Media Scholarship (Stavanger, 2004).

Contact details:
Email: *igerasimchuk@wwf.ru* or *ivetta@ciim.ac.cy*
Tel: + 7 (495) 727 0939
Fax: + 7 (495) 727 0938
Address:
WWF Russia,
Nikoloyamskaya St., 19, bld. 3
Moscow, 109240
Russia

Irina D. Gorkina is a former Deputy Director General of the Executive Directorate of NPAF at the National Pollution Abatement Facility, Russia ('NPAF') where she undertook project preparation management, environmental assessment, control on project implementation, etc.

She is now based at Basic Element Company as an environmental expert. Since 1995, she has been based at the Center for Preparation and Implementation of International Projects on Technical Assistance/National Pollution Abatement Facility (CPPI/NPAF). At CPPI/NPAF she is the Expert on Environmental Assessment and Technology and she conducted various environmental assessments and evaluated 23 investment projects in the pulp-and-paper sector, paint-and-varnish sector; oil and gas sector; as well as of the oil refining and chemicals sectors.

From 1990-1995, Irina was based at the State Committee of the Russian Federation on Environmental Protection ('SCEP') where she Chief Expert of the Development of Mining and Metallurgy Enterprises Review Department and of the Environmental Disaster Zones Review Department. She also worked as the Head of Environmental Legislation Application Section at SCEP. Prior to then, she was at the Moscow State University named after M.V. Lomonosov from 1978-1990.

Publications: about 40 publications – articles and monographs – in particular:

- *Comments to the Law on Subsoil* (Item 23), 2001.
- *The Russian Mining Legislation. Documents and Comments* (with co-authors) 1996
- "*Environmental Impact Assessment. The Guide for Practitioners*" (with co-authors), 1996
- *Calculation of MAE volume and environmental impact assessment* (with co-authors), 2000
- *On the role of Environmental Impact Assessment at preparation of investment projects*, 2000
- *Technique for the development of prospecting forecasts of geological environment changes* (with co-authors), 1988
- *EIA as a mechanism for sustainable social-economic development*, 1999
- *Development of SEA methodology in Russia*, 1998
- *On the ways of formation of guidance/regulatory base of realization of environmental impact assessment at designing. //Environmental Impact Assessment.* Practical experience (1992)

Education: Moscow State University named after M.V. Lomonosov, PhD of the Geographical Sciences; and Postgraduate study at the Moscow State University (master's thesis "The impact of mining industry on the geological environment (Moscow Coal-basin as an example)"). She has undergone training at UNEP/UNITAR/UNCHS (Habitat) TRAINING PROGRAMME "In environmental Law and Policy, held at the United Nations environment programme (International Environmental Convention, National Environmental Legislation and Institutions, Legal Aspects of Strategies for Integrating Environment and Development, and New Developmental in Environmental Law and Institutions)" in Nairobi, Kenya. As well as training at USAID/AED/ELI training course "Developing and Implementing Effective Environmental Laws", Washington, D.C., San Francisco, USA, as well as at the Sir Willion Halcrow Partners Ltd, training program "Environmental Audit" (UK), Azov, Russia, 1996.

Contact details:
Email: *gorkina@npaf.ru*
Address:
National Pollution Abatement Facility
Block 19, Leninskiy Avenue,
Moscow, Russian Federation
Tel./fax: +7-495-955-31-14, 730-40-97
Email: *npaf@npaf.ru*

Michele Hege Grønbech is the Renewable Energy Adviser in the Russian Department of the Bellona Foundation in Oslo, Norway. Previously, she worked in the field of Arctic environmen-

tal protection at the Norwegian Ministry of Environment and the World Wide Fund for Nature ('WWF') in Oslo (Norway). Mrs. Grønbech has a Masters Degree in Political Science from the Ohio State University in Columbus, Ohio. Mrs. Grønbech has worked at the Bellona Foundation since January 2007.
Contact details:
Email: *michele@bellona.no*
Websites: www.bellona.ru / www.bellona.org

Tatiana V. Guseva is a professor working at the Department for Management and Marketing of D. Mendeleyev University of Chemical Technology of Russia (in Moscow). Since the early 90s she has been involved in a number of national and international projects aimed at the development and practical application of instruments of environmental regulation, such as environmental impact assessment, environmental monitoring and environmental management systems. Currently, she is involved on three projects, as follows: "Climate Change Strategies for Russian egopolises" (funded by SPF (UK)) in Russia, as co-ordinator; the "Harmonisation of Environmental Standards II, Russia" (funded by the EU), as an expert; and a project on "Resource and Energy Efficiency", a training and research programme funded by BP plc, in Russia, as co-ordinator.
Publications: She has published over 200 articles, monographies, handbooks and text-books, used both by practitioners and students. These include a textbook on *Environmental Management Systems* (2008, in Russian), a handbook on *Hydrochemistry* (*Hydrochemical Parameters of the State of Environment*) (2007, 5th edn., in Russian), *Best Available Techniques for Large Combustion Plants* (2006, in Russian, adopted from EU BREF and edited), and *A Practical Guide to the Implementation of Environmental Management Systems* (2005, in Russian).
Tatiana V. Guseva has extensive experience in Russia and the NIS countries, including Armenia, Azerbaijan, Belarus, Kazakhstan, Tajikistan, Turkmenistan, Ukraine and Uzbekistan.
Contact details:
Email: *tguseva@muctr.edu.ru*

Dr. Dietrich Hahn is currently Team Leader on the Harmonisation of Environmental Standards ('HES') II project in Moscow, Russia (*www.ippc-russia.org*). For the past 15 years, he has been active as a team leader and project leader in EU supported accession and harmonisation driven projects. He sees himself as an environmental generalist, who has over 20 years of professional experience in working with environmental development issues on the national and local level, in the administrative, legal and technical context. He has advised various levels of the public sector, as well as private enterprises and their associations. Dietrich possesses a German university degree in Environmental and Spatial Planning. His work focuses on environmental law and regulations, norms and standards – which enables him to work in the full spectrum, from technical, normative to legal environmental matters.

Dietrich Hahn has over 12 years of experience in government and public sector consulting, 10 of which he spent on EU projects in transition countries where he has been working as an expert in or for authorities in Bulgaria, Estonia, Germany incl. a transition period in East Germany, Latvia, Lithuania, Montenegro, Romania, Russia, Serbia, and Thailand.

He has thorough experience working with topics of the EU Environmental Acquis and the underlying regulations, directives and norms. He succeeded with his team recently to develop for the Commonwealth of the Independent States ('CIS') the IPPC model-law and prepared for Russia concrete proposals to implement IPPC elements and related institutional and administrative concepts including steps towards the establishment of Russian BREFs.

Dietrich has a strong involvement in accession and approximation related projects in which he has been responsible for a wide range of issues from the development of EU compliant legislation, related institution and capacity building to the implementation of related sub-projects. He has experienced the harmonisation process from two sides, the implementation and the assessment, including the development of adequate advice for improved compliance. Due to his experience in consulting private enterprises in permitting procedures and to reach technical environmental compliance, he is aware of the potential cost and investment implications of normative settings, as well as of the necessary administrative structures and efforts for their enforcement.

Contact details:
Email: *D.Hahn@ippc-russia.org* or *Dietrich.Hahn@web.de*
Tel.: +49 (0) 170 48 77 018

Dr. Philip Hanson is currently an Associate Fellow at the Royal Institute of International Affairs, London. He is also an Emeritus Professor of the Political Economy of Russia and Eastern Europe, and honorary senior research fellow at the Centre for Russian and East European Studies ('CREES'), The University of Birmingham, UK. He is currently researching fiscal federalism in Russian conditions; business-state relations in contemporary Russia; Russian energy policies; and the effects of EU accession on ex-communist Central European economies.

Past positions held by Dr. Hanson are Economic Assistant, HM Treasury, London; lecturer in economics, University of Exeter, UK; First Secretary, British Embassy in Moscow, and senior research officer, Foreign & Commonwealth Office (1971-2); visiting professor of economics, University of Michigan (1977); Mellon Senior Fellow at Harvard Russian Center (1986-7); senior economic affairs officer, UN Economic Commission for Europe, Geneva (1991-2); Distinguished Visiting Fellow, Radio Free Europe/Radio Liberty Research Institute (1992); Visiting Professor, Kyoto Institute of Economic Research (2000); Director, CREES, 2001-02; Guest Professor, Baltic and East European Graduate School, Södertörns högskola, Stockholm (2004); Addy Speaker, Rollins College, Florida, February 2005.

He has undertaken consultancy work for Radio Liberty, Oxford Analytica, the Economist Intelligence Unit, AIG, and other companies. He has provided technical assistance work for UK Know How Fund and Tacis (EU). Dr. Hanson has made 30+ visits to Russia over the years and was expelled from the USSR in 1971.

Publications:

- *Trade and Technology in Soviet-Western Relations* (Columbia U.P., 1981);
- (with Keith Pavitt), *The Comparative Economics of Research, Development and Innovation in East and West: a Survey* (Harwood Academic Publishers, 1987);
- (co-ed with Michael Kirkwood), *Alexander Zinoviev as Writer and Thinker* (Macmillan, 1988);
- *From Stagnation to Catastroika. Commentaries on the Soviet Economy, 1983-1991* (Praeger/CSIS, 1992);
- (co-ed with Michael Bradshaw), *Regional Economic Change in Russia* (Elgar, 2000); *The Rise and Fall of the Soviet Economy* (Longman, 2003).
- Articles: *Soviet Studies*; *Europe-Asia Studies*; *Post-communist Economies*; *Journal of Comparative Economics*; *European Economic Review*; *Economy and Society*, *International Affairs* and elsewhere. Occasional op-ed pieces in *The Times* and *Financial Times* (mostly in the 1980s)

Education: trained as military interpreter in Russian in the British Army (1955-7); BA in Economics, Cambridge, England (1960); PhD in Economics, University of Birmingham (1971).

Contact details:
Email: *HansonPhl@aol.com*

Dr. Ekaterina N. Khmeleva is the Environmental Law Adviser at WWF-Russia. Since 1992 she has been working in environmental NGOs, including 12 years of professional experience in environmental law and litigation practice protecting environmental citizens rights. Another sphere of interest – ingidenous peoples rights for traditional way of life and matural resources.

Dr. Khmeleva is one of the founders of public interest law firm Legal Center Rodnik (1999).
Education: Ekaterina Khmeleva has a PhD degree in environmental law (1997) from the Institute of State and Law of Russian Academy of Sciences.
Publications: She is an author over 20 academic and educational publications on public interest environmental law issues. She is co-author of manuals for citizens and NGOs on court defence of environmental rights, on access to information, on protection of indigenous peoples rights (2003-2004). Ekaterina is also the compiler and editor of *Library of Environmental Rights Protection Cases* (2006).

Contact details:
Email: *ekhmeleva@wwf.ru*
Tel: + 7 (495) 727 0939
Fax: + 7 (495) 727 0938
Address:
WWF Russia,
Nikoloyamskaya St., 19, bld. 3
Moscow, 109240
Russia

Dr. Alexey E. Kostin, graduated the Economics Department of the Moscow State University in 1976. In the 1980s, he worked as a deputy dean of the Department. He has a doctorate degree in public and business administration and is a leading Russian expert in CSR and corporate sustainable development. During 1990s and 2000s, he has worked on many international projects for Russia, including the EU-TACIS project on environmental monitoring for the post-Soviet countries at the European Environment Agency in Copenhagen in 2002-2003. Dr. Kostin works as an expert on CSR and non-financial reporting for UNDP and UNECE. He is the Executive Director of the “Corporate Social Responsibility – Russian Centre”. This Center promotes CSR and sustainable development principles, the training of Russian and foreign businesses in Russia on how to apply CSR principles in corporate governance and to introduce new social and economic technologies, including non-financial reporting and modern forms of intersectoral cooperation between the state, business and civil society.

Contact details:
Email: *alexey.kostin@mail.ru*
Mobile: +7 916 675 5241
UNDP Russia contact:
Mr. Evgeny Levkin
Sustainable Livelihoods and Economic Development
Russian Federation, 125003, Moscow, UNDP
9, Leontyevsky Lane, Building 1
Tel: +7 495 787 2176
Fax: +7 495 787 2101
Website: *www.undp.ru*

James Lanman contributed to the article while an intern at The PBN Company (*www.pbnco.com*). The PBN Company is a consultancy specialising in advising Western investors into Russia and the former Soviet Union on market entry and expansion strategies, public affairs, government relations, communications and crisis management. The company has offices in Moscow, Kyiv, Almaty, London and Washington DC.

Kit Lawry has extensive international work experience in Finland, Tanzania, Kuwait, Malaysia, Swaziland, Malawi, Kazakhstan and Russia (Moscow). She has taught English, trained English teachers, wrote, edited and broadcast schools' English textbooks, radio and video programmes, facilitated organisation development for DFID projects in Malawi, lectured, edited and broadcast in Kazakhstan. Mrs. Lawry also worked, on a voluntary basis, with disadvantaged children in Africa and Kazakhstan.

In Moscow, Kit was Moscow Coordinator for EveryChild Russia, establishing EveryChild's profile, fundraising programme and office in Moscow (*www.everychild.ru*).

Since November 2008, Kit has been working as a Fundraiser for CORD (*www.cord.org.uk*), an international relief and development organisation, working with people affected by violent conflict in Africa and Asia.

Education: Durham University honours degree in English Literature and Language (1976); Teacher training qualification in Teaching English as a Foreign Language at London University Institute of Education (1978-79); Masters degree in Management Learning at Lancaster University (2000-2002), dissertation in cross-cultural training in international NGOs.

Contact details:
Email: *kit.lawry@gmail.com*

Nina Lesikhina is the Coordinator of Energy Projects of the Bellona Foundation in Murmansk, Russia. Miss Lesikhina is educated at the Murmansk State Pedagogical University, and was active in the nongovernmental environmental organisation Murmansk Nature and Youth for many years. She has been working in Bellona Murmansk since 2005.

Contact details:
Email: *nina@bellona.ru*
Websites: *www.bellona.ru* / *www.bellona.org*

Dr. Yurij L. Maksimenko currently holds the position of Principal (Health, Social, Environmental, policy advice, methodology, data base, reporting, etc.) at Basic Element Company.

Between 1999-2008, Dr. Maksimenko was Executive/General Director at the Russian Federation National Pollution Abatement Facility ('NPAF') (advising on NPAF policy, creation of NPAF project-investment cycle, project preparation, methodology of environmental impact assessment, technological and environmental project assessments). Dr. Maksimenko also was the Deputy Executive Director at the NPAF and Manager on Technology and Environmental Issues (advising on NPAF policy, creation of NPAF project-investment cycle, project preparation and environmental evaluation, methodology of environmental impact assessment, technological and environmental project assessments, 1995-1999).

Prior to this, from 1993-1995, Dr. Maksimenko held the position as Director of the Department at the Russian Federation Ministry on Environmental Protection & Natural Resources in Department of Environmental Legislation & Regulations (drafting of environmental laws & regulations; connections both with government and public; environmental enforcement; systematization of different kinds of legislation; implementation of environmental priorities in all kinds of legislation; international cooperation in this area).

Dr. Maksimenko has further significant experience in environmental protection issues, including due to positions held at:

- Moscow City State Committee on Environmental Protection, State Environmental Review Department (Director of the Department, 1992-1993);

- the RSFSR State Committee on Ecology & Integrated Planning of Natural Resources, State Environmental Review Department (Deputy Director of Department, 1990-1992);
- USSR State Committee on Protection of Environment, State Environmental Review Department (Deputy-Director of the Department, 1988-1990);
- Russian Federation Scientific-Research Institute on Water Supply & Wastewater Treatment (Deputy-Director of the Institute, 1987-1988);
- USSR State Committee on Civil Engineering, Construction Expertise Department (Main Environmental Specialist, 1981-1987);
- USSR State Committee on Science and Technology, Division of Environmental Protection & Integrated Planning of natural Resources (Senior Scientific Employee 1974-1981);
- Scientific wastewater treatment laboratory under the Moscow Civil Engineering Institute Engineer (Senior Engineer, 1972-1974).

Education: 1966-1971: Moscow Civil Engineering Institute Faculty of "Water supply and Sewerage" (Diploma of Civil Engineer); 1971-1972: Profound English language course under the Moscow Civil Engineering Institute; 1976-1978: Post-graduate studies under the Moscow Civil Engineering Institute. Master of Science (Ph.D) (1979); 1999: Doctor Degree, Ph.D (economics), the Moscow State University names after M.V.Lomonosov; Since 2005: Professor of the Moscow State University on Civil Engineering.

Contact details:
HSE Principal
Basic Element Company
30, Rochdelskaya Street, Moscow, Russia 123022
Email: *YuriyLM@basel.ru*
Website: *www.basel.ru*
Tel.: +7 985 226 09 61

National Pollution Abatement Facility
Block 19, Leninskiy Avenue,
Moscow, Russian Federation
Email: *npaf@npaf.ru*
Tel./fax: +7-495-955-31-14, 730-40-97

Fiona Mucklow is a research associate at the Asser Institute (since 2003) and was until recently practicing as a solicitor at the international law firm Chadbourne & Parke LLP (previously in Moscow and then in London). She has extensive experience of Public International Law and International Environmental Law. Her experience includes consultancies at the Lauterpacht Research Centre for International Law, University of Cambridge (1998-1999); the Royal Botanic Gardens, Kew (UK) (1998); the Legal Division of the World Trade Organisation ('WTO') Secretariat (Geneva, Switzerland) (1998/1999); the Foundation for International Environmental Law and Development ('FIELD') (London, UK) (1997; 1999); the Global Environment Facility ('GEF') Secretariat in Washington DC (USA) (1999); Milieu Ltd (Brussels, Belgium) (2004); and at United Nations Development Programme ('UNDP') Russia in Moscow (2006).

Ms. Mucklow practiced as a solicitor in finance and banking law as part of the Russia/CIS Group of Chadbourne & Parke (in Moscow and London). She advised borrowers and lenders with respect to their investments and projects in Russia, Ukraine and Kazakhstan. She previously practiced at Clyde & Co (London, UK) (1999 to 2003) and at Herbert Smith LLP (Moscow, Russia) (2003-2004).

She is pursuing a part-time PhD at Queen Mary, University of London (under the guidance of Professor Malgosia Fitzmaurice) and is focusing her thesis on International Environmental Law in Russia.

Education: Bristol University (Geography degree, BSc (Soc Sci)) from 1991-1994; University of Westminster (Diploma in Law, LL Dipl) from 1994-1995; College of Law (Legal Practice Course) from 1995-1996; London School of Economics and Political Sciences (Master in Laws, LLM, Public International Law) from 1996-1997.
Publications:
- *Joint Implementation in Russia – an update* (Chadbourne & Parke, CIS Newswire) [12/2008] with R. Mikhailov
- *Biofuels in Russia and Ukraine – Legal pitfalls* (Chadbourne & Parke, CIS Newswire) [12/2008] with R. Mikhailov and V. Dovhan
- *Biofuels in Russia and Ukraine – An Introduction* (Chadbourne & Parke, CIS Newswire) [10/2008] with J. Kogan and A. Bloomfield
- *Joint Implementation in Russia* (Chadbourne & Parke, CIS Newswire) [07/2008] with R. Mikhailov and V. Andrianov
- Emissions trading in the UK – a case study, in *American Bar Association International Environmental Law Committee Newsletter, Vol. 7(1)* [December 2004]
- "Mechanisms for Regulating Environmental Barriers to Trade within the WTO" with Ruttley, P., Weisberger, M., in *Handbook of World Trade (2001)*
- "*Integration of environmental principles into the World Bank*", in *RECIEL* Vol 9:3 [2000]
- *The Commercial Use of Biodiversity* (1999) by K. ten Kate and S. Laird, research on regulation of crop protection and seeds, GMOs sector and biosafety

Contact details:
Email: *f.mucklow@asser.nl*

Soili Nystén-Haarala has been Professor of Civil Law in the Faculty of Law, Economics and Business Administration at the University of Eastern Finland since 2004 (tenured as of 2006). She completed her doctorate in the Faculty of Law of the University of Lapland in 1998. She has held a number of different positions at the University of Vaasa and University of Lapland. She has also worked as a visiting fellow in the Forestry Project at IIASA (International Institute of Applied Systems Analysis) in Laxenburg, Austria and as a visiting professor at Umeå University, Sweden. Ms Nystén-Haarala has led several research projects since 2003. Her main field of interest is contract law, international trade law and Russian law. In contract law she focuses on 'contracting', the use of contracts in business practice. She has publications on Russian law and transition as well as on contracting and proactive law.

Contact details:
Email: *soili.nysten-haarala@uef.fi*

Daria N. Ratsiborinskaya is a staff lecturer at the Erasmus University of Rotterdam (section of European law). She is also an associate researcher at the Asser Institute and responsible for the National Page of the Russian Federation of EEL (*www.eel.nl*).

Daria has broad international work experience, including a stage at DG Environment of the European Commission (2001), DG Environment, legal expertise for several European Union-funded projects of technical assistance in the field of approximation of environmental legislation, an UNEP project on sustainable transport and an environmental NGO in the Netherlands (Stichting "Natuur en Milieu"). From 2003-2006, Daria was a research associate at T.M.C. Asser Institute, Editor of the EEL.nl, and co-editor of the European Environmental Law newsletter. Between 2003-2004, Daria was working in Moscow on the TACIS project "Harmonization of environmental standards, Russia" as a legal expert. From 2002 to 2004 she was a guest lecturer at MGIMO-University, European Law department (Lecturing on European environmental law, national environmental law, international trade law (environmental aspects), energy and environment regulation in Russian Federation).

Daria read law at MGIMO-University of Moscow and was awarded an LL.M. in International and European Environmental law from University of Amsterdam.

Currently, Daria lives with her husband and two children in the Netherlands and is active, *a.o.*, as invited lecturer at MGIMO-University and Leuven Catholic University.

Some publications:

- "The Russian Federation and the Kyoto Protocol", contribution to *The Kyoto Protocol and Beyond, Legal and Policy Challenges of Climate Change*, W.Th. Douma *et al.*, eds., T.M.C. Asser Press 2007 (in English)
- *Contemporary waste regulation in Russia*, US Bureau of National Affairs' Russia Environmental Outlook 2005, January 2005 (in English)
- *European Environmental law*, chapter in textbook on European law, ed. L.Entin, MGIMO, Moscow, September 2004 (in Russian); revised for 2nd edition April 2009
- *Let Polluter Pay! Environmental liability regime in EU*, EVROPA magazine, Moscow, February 2004 (in Russian)
- *TACIS in Russia*: *approximation of Russian and European environmental legislation*, Ekooko [ecological magazine], St.Petersburg, January 2004 (in Russian)
- *Energy and problems of environmental protection in Russian Federation and Foreign States*, methodological materials for lecture course, MGIMO, Moscow, September 2003 (in Russian)
- *Environmental Rights Protection in EU Legislation*, Ekologicheskoe pravo [Environmental law journal], Moscow, April 2003 (in Russian)
- *Northern Dimension*: *Briefing for Mrs. Wallström, Commissioner for Environment*, together with P. Schellekens, DG Environment, Brussels, January 2002 (in English)

Presentations:

- "*Russian Environmental law*", a lecture course for master students, Leuven Catholic University, Belgium, 1-2 March 2008, 30-31 January 2009
- "*The EAP Task Force Toolkit for Environmental Inspectorates in Eastern Europe, Caucasus and Central Asia on behalf of the EAP Task Force Secretariat*", Russian-Dutch Conference on Compliance With Environmental Regulations, St. Petersburg, 29 September - 1 October 2004
- "*Perspectives for harmonization of EU-Russia environmental legislation*", Conference on Environmental protection in aluminum industry in Russia, Moscow, May 27, 2003
- "*Issues of Environmental liability*", Conference "Day of Environmental law in Moscow State Institute of International Relations", Moscow, May 16, 2003
- "Proposed changes of Russian legislation within the frames of harmonization process with EU Directive on Integrated Pollution Prevention and Control", Joint Conference (EU, OECD, State Duma of Russian Federation) on Perfection of Environmental permitting system in Russia and other countries of Eastern Europe, Moscow, April 10-11, 2003

Contact details:

E-mail: *ratsiborinskaya@frg.eur.nl*
Tel.: +31 10 40 82 94

Olga A. Razbash is an environmental lawyer and is the Environmental Law and Public Awareness Expert at the Russian Regional Environmental Centre ('RREC'). She is co-founder and Vice-President of an NGO "Ecojuris-WLED" (Women-Lawyers for Environment and Development), the first Russian public interest environmental law organization.

She provides legal consulting assistance to citizens, local municipal councils, NGOs; business on EIA and public participation aspects of projects; assisted in compiling the first compendium of Russian legislation; participated in public hearings on the environmental disputes and in hearings regarding the Moscow landfills reconstruction project; and participated in various environmental legal researches. She has also been a member of NGO Coalition in 10 negotiation sessions the drafting working group on UNECE Convention on Access to Information, Public Participation in Decision-making and Access to Justice in Environmental Matters.

She is a founding member and Chair of the Regional Public Center "For Human Rights and Environmental Defense". Olga has more than 10 years of experience protecting Russia's environment. Olga served as a legal expert at the State Duma Environment Protection Committee and participated in drafting laws and environmental policies.
Publications: "EIA and governmental ecological expertiza"; "Manual for citizens on legal issues of covering health damages caused by environmental violations.
Contact details:
Russian Regional Environmental Centre
1 Volkonsky per., 13, building 2
127473 Moscow, Russia
Email: *razbash@rusrec.ru*
Website: *www.rusrec.ru*
Tel./fax: + 7 (495) 737 64 48 and + 7 (495) 737 64 49

Dr. Jeffrey Roberts is a Director of Impavaara Securities Ltd and Pomor Petroleum Ltd. He did his Ph.D. in geology in Finnmark, northern Norway, then worked briefly as a wellsite geologist in the North Sea before joining Wood Gundy's international oil team in 1973 to cover the North Sea and Middle East. He soon had two careers competing with each other in the 1970s and 1980s – one in the City as an analyst in oil & gas; the other as a local councillor in the London Borough of Hackney. In 1987, Dr. Roberts went to work in Finland as Director of Research for the then leading Finnish stockbroker. That led to an interest in Estonia and nearby Russia, including Murmansk and Apatity (*www.kovdor.info*).

Impivaara Securities (regulated by Financial Services Authority), was formed in 1991 to offer investment advice on Finland and 'Hansa' stock markets which would be influenced by Russia. That advice is increasingly directed to specific client demand against soft dollar arrangements as large institutions try to avoid overdependence on banks. Impivaara Securities organized the first conference on Russian Frontier Natural Resources in London in 2005 (details *at www.impivaara.com*).

Pomor Petroleum Ltd (registered in England, 1991) is an English company equally owned by Murmansk directors, to aid Russian regional development and integration with international markets: (*www.pomor.com*). Pomor has provided significant advice to Russian concerns, and forms a natural partner to Impivaara Securities in organizing opportunities in Russia. Originally for oil, the emphasis has now shifted to industrial minerals, through their partner O.A.O. Pana in Apatity. Pana is the operator on major platinum metals licences for Barrick Gold, BHPBilliton and others (*www.panapgm.com*).
Contact details:
Email: *jrobertsa@cix.co.uk*

Joanna Rogers is the Country Director for EveryChild Russia (full time since 2007) (part-time 2002-2006). Other experience she has had is as a Team Leader for an EU project on the Support to Child Welfare Reform in Georgia (2006-2007). She has also undertaken various consultancy assignments (since 2002), including with Birks Sinclair, DFID SPRILO programme (2005-

2007) and HealthProm (2005). Jo has further extensive experience in the field of child care and protection, and was Head of Programmes with EveryChild from 2001-2002, Regional Director FSU for the European Children's Trust (1997-2001), Programme Manager for the Christian Children's Fund of Great Britain (1995-1997) and the Country Director for the Russia Programme at Christian Children's Fund of Great Britain (1993-1995).
Selected recent publications/conference papers:
- '*Gatekeeping*' – '*A Family for Every Child International Conference*' Moscow, 2-4 December 2008 – UNICEF, EveryChild, MoES, MoLHSD
- '*Community Based Family Support Services*' – EU Food Security Programme Regional Conference, Tbilisi, March 2007
- 'Decision-making in the sphere of Child Protection in St Petersburg' – Rogers, Smyikalo, 2007
- 'Participation of parents of children at risk in Social Policy Making' – Rogers, Sundh, 2006

Education: Master of Social Work – Ersta Skondal University College, Stockholm (2006); BA English/Russian at University of London (1993).
Contact details:
Email: *jo_rogers@mail.ru*
Website: *www.everychild.ru* or *www.everychild.org.uk*

Dr. Evgeny A. Shvarts, is the Director of Conservation Policy of WWF Russia, and holds a Doctor of Science (Geography, 2003). He holds the title and honorary mark of "Distinguished (Honorary) Public Figure of Nature Conservation", granted by the Ministry of Nature Resources (2006).

From 1990 until 1998, Evgeny Shvarts was a Senior Research Fellow, Department of Biogeography, and Member of Scientific Council of Institute of Geography at the Russian Academy of Sciences.

He is founder and Chairman of the Board of Biodiversity Conservation Center (BCC, 1992-1998). He was an international researcher of the Swedish University of Agricultural Sciences (1993).

In 1996-1998, he was the Manager of the Protected Areas Component of "Russian Federation Biodiversity Conservation Project" of the Global Environment Facility/World Bank (GEF/WB) project. In 1987-1988, 1997-1999 he was Co-chairman of the International Socio-Ecological Union. He has been a member of the World Commission on Protected Areas of IUCN (since 1996) and a Fellow of the Program for "Leadership for Environment and Development" (1998). In 2008, he was awarded an Erasmus Mundus fellowship by the European Commission with the Masters of Environmental Sciences, Policy and Management (MESPOM) Consortium (Lund University – Central European University – University of the Aegean – University of Manchester).

He is co-founder and an editorial board member of the "Environment planning and management" (Russian Academy of Sciences), "Sustainable forestry", "Wild nature conservation", and "State Reserves and National Parks" journals.
Education: Evgeny graduated from the Moscow State University, Biological faculty (1982). In 1987, he defended a PhD thesis in geography.
Publications: Evgeny Shvarts is an author of three books and more than 120 articles, mainly published in leading scientific journals in Russia (e.g., "Doklady of the Academy of Sciences", "Proceedings of the Academy of Sciences", "Advances of modern biology", "Zoological journal", "Polis" ("Political studies"), "Social sciences today") and abroad (e.g., "Journal of Biogeography", "Ecoscience", "Biodiversity and Conservation", "Journal of Zoology", "Journal of the Royal Society of New Zealand").

Contact details:
Email: *EShvarts@wwf.ru*
Tel.: + 7 (495) 727 0939
Fax: + 7 (495) 727 0938
Address:
WWF Russia,
Nikoloyamskaya St., 19, bld. 3
Moscow, 109240
Russia

Jonti Small, Senior Account Manager at The PBN Company (London), provides financial communications, media relations and corporate communications support to PBN's Western and CIS clients. Prior to joining PBN, Jonti was Managing Editor of Russian Investment Review – the leading regular English-language publication focused on Russia's economic and political development.

The PBN Company is a consultancy specialising in advising Western investors into Russia and the former Soviet Union on market entry and expansion strategies, public affairs, government relations, communications and crisis management. The company has offices in Moscow, Kyiv, Almaty, London and Washington DC.
Contact details:
Email: *jonti.small@pbnco.com*
Website: *www.pbnco.com*

Michael Yulkin is the Director General of Climate Change Global Services (CCGS). Previously, Michael was director of the Environmental Investment Center, a Russia based autonomous non-commercial organization. Michael Yulkin is an internationally recognized climate change expert with a vast experience in GHG emissions accounting, management and improvement as well as in JI projects preparation and implementation. He has participated in the GHG inventory at a number of Russian industrial enterprises, in development of Russian National Strategy Study on climate change, and also in development of climate change strategies at both regional and corporate levels in Russia. Michael Yulkin has more than 35 publications on climate change and Kyoto Protocol related issues in Russian and international media. Michael speaks fluent English. Having graduated from Moscow State University, faculty of economics, Michael also studied project analysis at the World Bank Institute.
Contact details:
Email: *profi@aha.ru or yulkin.ma@gmail.com*

Sir Andrew Wood GCMG was a member of the British diplomatic service from 1961 to 2000. He was posted to three foreign capitals: Moscow 1964-6, 1979-82 and 1995-2000 (the last as Ambassador); Washington DC, 1967-70 and 1989-92 (as Deputy Head of Mission); and Belgrade, 1975-9 and 1985-89 (as Ambassador). His various London postings concluded as Deputy Under Secretary in charge of the finance and administration of the Diplomatic Service.

Since leaving Moscow in early 2000 Sir Andrew has returned regularly to Russia, as well as to Ukraine, Uzbekistan, Kazakhstan and Azerbaijan. He has acted at various times on behalf of a number of Western companies, including Ernst &Young, Unilever, Diageo, and GlaxoSmith Kline. More recently he has assisted PriceWaterhousCoopers, Toyota Motor Europe, Citibank and Montrose Associates. He served on the Boards of the Foreign and Colonial Investment Trust (2000-2007), the PBN Company (now as Chairman), and (2000-2008) the Russo-British Chamber of Commerce (including as Chairman). He was on the Board of OAO Mechel (a

Russian coal and steel company listed on the NYSE) between 2004 and 2006. He was a member of the Advisory Council of Renaissance Capital from 2003-2007 and is a member of the Advisory Council of Air Products Inc. He is a Special Representative for BP, an Adviser to ITE Group plc, a Trustee for the Victor Zorza Hospice Trust, and the President of the King's College (Cambridge) Association. He has been an Associate Fellow of Chatham House since 2008.

Andrew Wood was born in Gibraltar on 2 January 1940, and has an MA in history from King's College, Cambridge. He is married to Stephanie (née Masset). They have three children, and live in London and Jersey (Channel Islands).

BIBLIOGRAPHY

2003 Environmental Policy Review. Achieving the Bank's environmental mandate through direct investments, European Bank for Reconstruction and Development, January 2008

Agenda 21. The United Nations Programme of Action from UN Conference on Environment and Development. Rio de Janeiro. 1992. Chap. 30.1, 30.2 and 30.3. See also <http://www.un.org/esa/sustdev/documents/agenda21/index.htm>

The Arctic Nuclear Challenge, Bellona Report, Vol. 3-2001, pp. 40-43

Baker & McKenzie, '*New Code of Administrative Violations of the Russian Federation: Increased Responsibility of Commercial Organizations and Their Managers*', 2 July 2002: <http://www.bisnis.doc.gov/bisnis/bisdoc/020710AdmCode.htm>

Bambulyak, A. and Frantzen, B., *Oil Transport from the Russian Part of the Barents Region* (Status 2005), Svanhord Environmental Centre: <http://www.wwf.ru/data/publ/oil_report_english_1.pdf >

Bank of Finland, *BOFIT Weekly*, 22 February 2008

Barter, C., *Abuse of Children in Residential Care* (London, National Society for the Prevention of Cruelty to Children 2003), <www.nspcc.org.uk/inform/Info_Briefing/AbuseofChildreninResidentialCare.pdf>

Bobylev, S.N., Sidorenko, V.N., Safonov, Y.V., Avaliani, S.L., Strukova, E.B. and Golub, A.A., *Macroeconomic Assessment of Environmental Pollution Health Effect in Russia* (Moscow, World Bank Institute, Nature Protection Fund 2002)

Brinchuk, M.M., *Ekologicheskoe pravo*, Uchevnick (Moskva, Vyshee obrasovanie 2005)

Внешний долг Российской Федерации в 2008 году по срокам погашения и финансовым инструментам, Central Bank of Russia, 6 January 2009, <http://www.cbr.ru/statistics/print.aspx?file=credit_statistics/debt_maturity_08.htm&pid=svs&sid=vdRF_fisp>

Bruntland, G., (ed.), *Our Common Future, World Commission on Environment and Development* (Oxford, Oxford University Press 1987) see <http://www.un-documents.net/wced-ocf.htm>

Carter, R., for EveryChild: *Family Matters: A study of institutional childcare in Central and Eastern Europe and the former Soviet Union*

Chowdhury, A. and Verbina, I., 'Reforming Russia's Education System', in *Transition, The Newsletter About Reforming Economies*, The World Bank, January/February/March 2003, <http://www.cdi.org/russia/johnson/7171-15.cfm>

Comments on the EBRD Environmental Policy Discussion Paper, Bank Information Center, 29 June 2007

Comments on the EBRD Environmental Policy Discussion Paper, CEE Bankwatch Network, June 2007

Commerce and Industry Chamber of the Russian Federation / ТПП РФ, 28 February 2006. *Russia and Africa in the Short- and Mid-term. / Россия и Африка в кратко- и среднесрочной перспективе*, Moscow

* It is noted that the websites and URLs mentioned herein (and in the chapters) have been accessed in November 2009. It is possible that access to these websites and URLs may not be guaranteed at a later point in time.

Commission of European Communities Country Strategy Paper 2002-2006, National Indicative Programme 2002-2003, Russian Federation (27 December 2001)

Communication from the Commission on EU-Russia environmental cooperation of 17 December 2001 [COM (2001)772 final – Not published in the *Official Journal*], p. 4.

Compliance review report relating to the Vlorë thermal power generation project, Document of the European Bank for Reconstruction and Development, 17 April 2008

Country Profile of Russian Federation to the UN Commission on Sustainable Development, 1997

Cuamba, B., *Sustainable Energy Pathways for the SADC Region.* Presentation at the DFID-DSA policy forum *International Development in the Face of Climate Change: Beyond Mainstreaming?*, Greenwich, 2 June 2008

DeBardeleben, J., and Heuckroth, K., 'Public Attitudes and Ecological Modernization in Russia', in Massa, I. and Tynkkynen, V., (eds.), *The Struggle for Russian Environmental Policy* (Helsinki, Kikimora Publications) Series B 17.2001, pp. 49-76

Declaration by the Ministers of Environment of the region of the United Nations Economic Commission for Europe (UNECE), 25 October 1995, <http://www.unece.org/env/wgso/pre-kiev.declar/Sofia.E.pdf>

Doing Business, Russian Federation, 21 March 2008, <www.doingbusiness.org/Explore Economies/?economyid=159>

Doklad ob Ocobennoctyax Klimata ha Territoriy Rossiskoy Federatsiy za 2006 god

Du Plooy, P., *Re-Think Investment in (South) Africa* (Johannesburg, WWF South Africa 2008) pp. 36, 47-48

EBRD no longer considers current financing package for Sakhalin II, 11 January 2007, <http: //www.ebrd.com/new/pressrel/2007/070111.htm>

Eklof, B., Holmes, L.E. and Kaplan, V., (eds.), *Educational Reform in Post-Soviet Russia: Legacies and Prospects* (The Cummings Center Series, 2005)

Environmental and Social Policy, European Bank for Reconstruction and Development, 2008, <http://www.ebrd.com/about/policies/enviro/policy/2008policy.pdf>

Environmental and Social Policy Report on stakeholders consultation 2006-2008, Document of the European Bank for Reconstruction and Development, <http://www.ebrd.com/about/policies/enviro/policy/review/consult.pdf>

Environmental Component of the Samara Regional Investment Initiative, Final Report, USA, Chemonics International Inc. (December 1999)

Environmental Performance Reviews: Russian Federation, Organization for Economic Cooperation and Development (Paris, OECD 1999) pp. 3, 6

Environmental Policy, European Bank for Reconstruction and Development, 2003, <http:// www.ebrd.com/about/policies/enviro/policy/policy.pdf>

Environmental Policy Discussion Paper, Document of the European Bank for Reconstruction and Development, <http://www.ebrd.com/about/policies/enviro/policy/review/issues.pdf>

The Environmental Outlook in Russia, National Intelligence Council, January 1999

Esty, D. and Porter, M., *Ranking National Environmental Regulation and Performance: A Leading Indicator of Future Competitiveness?* The Global Competitiveness Report 2001-2002, (New York 2001) p. 78

EU Ambassador says Russia at Risk, Moscow Times, 27 July 2009

Executive summary of the Phase 2 environmental and social impact assessment process, Sakhalin Energy Investment Company, November 2006

Feshbach, M. and Friendly Jr., A., *Ecocide in the USSR: Health and Nature under Siege* (New York, Basic Books 1992)

Gerasimchuk, I., *World Economic Forum: Green Russia*, Vedomosti, 27 January 2009, <http://www.vedomosti.ru/newspaper/article.shtml?2009/01/27/178689>

Gerasimchuk, I., *Environmental Practice of Transnational Corporations* (Moscow, World Wildlife Fund (WWF) 2007) <http://wwf.ru/resources/publ/book/240/>

Hansen E., and Tønnessen, A., *Environment and Living Conditions on the Kola Peninsula*, Fafo-Report 260, FAFO Institute for Applied Social Science, Section 2.2 *Overall Population Development*, p. 28

Hanson, P., *The Rise and Fall of the Soviet Economy* (London, Longman 2003) chap. 8

Holliday Jr., C.O., Schmidheiny, S. and Watts, P., *Walking the Talk: The Business Case for Sustainable Development* (Sheffield, Greenleaf 2002)

Holme R., and Watts P., *Making Good Business Sense*, World Business Council for Sustainable Development (2003)

Ikonickaya, I.A., *Zemel'noe pravo Rossiiskoi Federacii* (Moscow, Iurist 2002)

Independent Recourse Mechanism, Eligibility Assessment Report, Complaint Sakhalin 2 Phase 2 Calypso. European Bank for Reconstruction and Development, <http://www.ebrd.com/about/integrity/irm/0501ear.pdf>

The ISO Surveys of ISO 9000 and ISO 14000 Certificates for 1998-2004

The ISO Survey of ISO 9001:2000 and ISO 14001 Certificates 2004 / International Standardization Organization. Geneva, 2007, <http://www.iso.org/iso/iso_catalogue/management_standards/certification/the_iso_survey.htm>

Kendrick, A., *Abuse of Children in Residential and Foster Care: a brief review*, SIRCC research papers (Glasgow, Scottish Institute for Residential child Care 2003) <www.sircc.strath.ac.uk/research/kendrick.html>

Kits, A., Novitskov, S. and Stroganov, A.G., *Paradoksy rossiskogo biznesa. Ili snova ob investitsijah* (Moskva, Kompania ASTSh 2001)

Knizhnikov, A., Pusenkova, N. and Solntseva, E., *Social and Environmental Aspects of Russian Oil Refining* (Moscow, World Wildlife Fund (WWF) 2008) <http://wwf.ru/resources/publ/book/302/>

Kolbasov, O. and Krasnova, I., *Russian Federation*, in *International Encyclopaedia of Laws* (London, Kluwer Law International 2003) p. 42

Kommersant Daily, 10 February 2004, p. 20

Kornegay, F., 'Africa's Strategic Diplomatic Engagement with China', in Edinger, H., Herman, H. and Jansson J., (eds.), *New Impulses from the South. China's Engagement of Africa* (Centre for Chinese Studies, University of Stellenbosch, South Africa 2008)

Kramer, A.E., 'In Russia, Pollution Is Good for Business', New York Times, 28 December 2005

Kriukov, V., Seliverstov, V. and Tokarev, A., 'Federalism and Regional Policy in Russia: Problems of Socio-Economic Development of Resource Territories and Subsoil Use', in Solomon Jr., P.H., (ed.), *The Dynamics of 'Real Federalism'. Law, Economic Development, and Indigenous Communities in Russia and Canada* (Center for Russian and East European Studies, University of Toronto 2004)

Labatt, S. and White, R.R., *Environmental Finance: A Guide to Risk Assessment and Financial Products* (Hoboken NJ, John Wiley 2002)

Labatt, S. and White, R.R., *Carbon Finance: The Financial Implications of Climate Change* (Hoboken NJ, John Wiley 2007)

Larin, V., Mnatsakanyan, R., Chestin, I. and Schwartz, E., *Environment Protection in Russia: from Gorbachev to Putin* (Moscow, KMK 2003) <http://wwf.ru/data/publ/okhrana_ot_i_do.pdf>

Leksin, V., 'The New Russian Federalism', in Solomon Jr, P.H., (ed.), *The Dynamics of 'Real Federalism'. Law, Economic Development, and Indigenous Communities in Russia and Canada* (Center for Russian and East European Studies, University of Toronto 2004) p. 56

Litovchenko, S., (ed.), *Report on Social Investments in Russia 2004, The Role of Business in Social Development* (Moscow, UNDP/Russian Managers Association 2004)

Living Planet Report 2008. WWF, Gland, 2008, <http://assets.panda.org/downloads/living_planet_report_2008.pdf>

The Little Green Data Book 2008 (World Development Indicators 2007) World Bank, Washington DC, 2008, see <http://siteresources.worldbank.org/ENVIRONMENT/Resources/LGDB2008.pdf>

Long-Term State Programme of the Research of Subsoils and Reproduction of the Mineral and Raw Materials Base (2005 – 2010 and through 2020) of the Russian Federation / Долгосрочная государственная программа изучения недр и воспроизводства минерально-сырьевой базы России на основе баланса потребления и воспроизводства минерального сырья (2005-2010 гг и до 2020 г.) Российской Федерации

Lugar, G., *European Bank for Reconstruction and Development (EBRD)*, London; A report to Members of the Committee on Foreign Relations, United States Senate, 109th Congress, Second Session, 16 June 2006

Ministry of Natural Resources, Государственный доклад 'О Состоянии и Охране Окружающей Среды Российской Федерации в 2003 году' [State Report 'On Condition and Protection of Environment of Russian Federation in 2003'] (2005) p. 83

Mitrofanov, F., *The mineral resources of the Kola*, presented at the conference 'Natural resources on the Russian frontier', 2005, London, <www.impivaara.com>

Moran, T., Graham, E. and Blomstrom, M., (eds.), *Does Foreign Direct Investment Promote Development?* (Washingon DC, Peterson Institute for International Economics 2005)

Murray, F., *Ecological Disaster: Cleaning up the Hidden Legacy of the Soviet Regime* (Twentieth Century Fund Inc. 1995)

Nikolai, F., et al., *White Sea: Its Marine System and Ecoenvironment Dynamics Influenced by Global Change* (Chichester, Springer 2005) pp. 250, 252

Northwest Russia: Facts and Events, The Stockholm Region, Newsletter by the Stockholm Region Office in St. Petersburg, Issue 91, 2008, week 6

Nystén-Haarala, S., 'Contract Law and Everyday Contracting', in Wahlgren, P., (ed.) *Scandinavian Studies in Law the Proactive Approach*, Vol. 49 (Stockholm 2006)

Nystén-Haarala, S., (ed.), *The Changing Governance of Renewable Natural Resources* (London, Ashgate 2009)

Nystén-Haarala, S., *Russian Law in Transition. Law and Institutional Change* (Helsinki, Kikimora Publications 2001)

OECD, *Mobilising Financial Resources for the Environment in Russia* (2008) see <http://www.oecd.org/dataoecd/23/5/41445931.pdf>

OECD, *Policies for a better environment. Progress in Eastern Europe, Caucasus and Central Asia* (2007)

OECD, *Environmental Policy and Regulation in Russia. The Implementation Challenge* (2006)

OECD, *Economic Survey of the Russian Federation* (2006)

Open Register of EMS Certification in Russia, <http://www.14000.ru/register/>

PDG developed information portal of North-West Russia region, PDG, <http://www.pdg.ru/news/2006/11/03/portalallnwrussia/en/>

Peterson, D.J., *Environmental Activism in the New Russia: Prospects for the Coming Decade, Report to the National Council for Eurasian and East European Research* (Washington DC, National Council for Eurasian and East European Research 1997)

Polyakova, M., *Russia Profile*, October 2005

Porter, M., and van der Linde, C., 'Green and Competitive: Ending the Stalemate', 73 *Harvard Business Review* (September-October 1995) pp. 120-155

Poussenkova, N. and Solntseva, E., *Russian Companies in the 21st Century – Towards Competitive Corporate Citizenship* (Moscow, WWF-Russia 2007)

Priess, H.-J., (ed.), *Getting the Deal Through: Public Procurement, An Overview of Regulation in 43 Jurisdictions Worldwide* (2009), see chapter on Russia by Falk Tischendorf, Kamil Karibov and Ekaterina Soboleva of Beiten Burkhardt, pp. 199-207, see also <http://www.bblaw.com/uploads/media/PP2009_Russia_chapter.pdf>

'*Prospects for Development of Non-conventional and Renewable Sources of Energy on the Kola Peninsula*', <http://www.bellona.no/rapporter/Clean_energy_Kola>; <http://www.bellona.ru/reports/Energy_Kola_Peninsula/Kola_Peninsula_report in Russian>

Ransel, D.R., *Mothers of Misery: Child abandonment in Russia* (Princeton, Princeton University Press 1988)

Report of the Western Gray Whale Advisory Panel at its fifth meeting, Western Gray Whale Advisory Panel 5th Meeting, 3-6 December 2008, <http://cmsdata.iucn.org/downloads/wgwap_5_report_final_040209.pdf>

Report '*On the state of industrial safety of hazardous enterprises, rational use of mineral resources and protection of entrails in the Russian Federation in 2003*' Federal Mining and Technological Surveillance Service (2004)

Revision of EBRD's 2003 Environmental Policy: Revised and Updated Stakeholder Consultation and Disclosure Plan (SCDP), <http://www.ebrd.com/about/policies/enviro/policy/review/scdp2.pdf>

Roberts, J., 'Late Precambrian Dolomites, Vendian Glaciation and Synchroneity of Vendian Glaciations', 48 *Journal of Geology* (1976) pp. 47-63

Roberts, J., 'Late Precambrian Glaciation: an Anti-Greenhouse Effect', 234 *Nature* (1971) pp. 216-217

Rose R., '*Getting things done in an antimodern society: social capital networks in Russia*', in Dasgupta, S., (ed.), *Social Capital- A Multifaceted Perspective* (Washington DC, World Bank 2000) pp. 147-171, <http://www.exclusion.net/images/pdf/778_fimod_rose.pdf>

Rosstat, 15 August 2008. On Foreign Investment in Russia in the First Half of 2008 / Об иностранных инвестициях в I полугодии 2008 г. Russian Federal Service of State Statistics, Moscow

Russian Ministry of Natural Resources, Государственный доклад 'О Состоянии и Охране Окружающей Среды Российской Федерации в 2003 году' [State Report 'On Condition and Protection of Environment of Russian Federation in 2003'] (2005) p. 83

Rutland, P., *Business – State Relations in Russia*, draft paper developed at Center for Security Studies, Zurich: Swiss Federal Institute of Technology, (19 March 2005)

SADC, 1997. SADC Mining Sector Five Year Strategy 1997-2001. Southern African Development Community, Gaborone

Sakhalin pubic consultation in London transcript, European Bank for Reconstruction and Development, 28 February 2006

Sakhalin pubic consultation in Moscow transcript, European Bank for Reconstruction and Development, 14 March 2006

Sammon, E., *Children and disability in the context of family breakdown in Central and South Eastern Europe and the Former Soviet Union: Defying Prejudice, Advancing Equality – 2* (London, 2001) EveryChild

Schmidheiny, S., *Changing Course: A Global Perspective on Development and the Environment* (London, MIT Press 1992)

Schrijver, N.J., *The Evolution of Sustainable Development in International Law: Inception, Meaning and Status* (Leiden, Martinus Nijhoff Publishers 2008)

Schwartz, E.A., 'Awaiting Changes. Environmental Responsibility of Finance Organizations', 159 *Banks and Business World* (2008) pp. 88-89

Schwartz, E.A., 'Environment and Transnational Corporations: from Myths to Pragmatic Approach', 78 *Ecology and Life* (2008) pp. 48-53

Sokolow, R.H., Andrews, C., Berkhout, F. and Thomas, V., (eds.), *Industrial ecology and global change* (Cambridge, Cambridge University Press 1994)

Sustainability Reporting Guidelines, Global Reporting Initiative 2002, <http://www.globalreporting.org/guidelines/2002/gri_2002_guidelines.pdf>

Tangen K., Korppoo A., et al., '*A Russian Green Investment Scheme – Securing Environmental Benefits from International Emissions Trading*', Climate Strategies: <http://www.climate-strategies.org/gisfinalreport.pdf>

Transcript of public consultation meeting in Korsakov on the Sakhalin II Phase 2 project, European Bank for Reconstruction and Development, <http://www.ebrd.com/country/sector/natural/projects/sakhalin/consult/korsakov.pdf>

Transcript of public consultation meeting in London on the environmental and social issues associated with the Sakhalin II Phase 2 project, European Bank for Reconstruction and Development, <http://www.ebrd.com/country/sector/natural/projects/sakhalin/consult/london.pdf>

Trutnev, Y.P., RF Minister of Natural Resources, '*Concerning the Enhancement of Efficiency of Hydrocarbon Resources Development on the Continental Shelf of the Russian Federation*', <http://www.geoinform.ru/?an=trutnev_en>

Tsygankov, A., and Yanchik L., *On the development of the Cleaner Production Program in Russia* (2004) p. 4

Tynkkynen, V.-M. and Massa, I., 'The Russian Environment – a Source of Risk or the Lungs of the Planet?', in Massa, I. and Tynkkynen, V., (eds.), *The Struggle for Russian Environmental Policy* (Helsinki, Aleksanten Institute, Kikimora Publications 2001) Series B 17

US Geological Survey, 2007. Mineral Commodity Summaries 2007. US Geological Survey, Washington DC

Weiner, D. R., *A Little Corner of Freedom: Russian Nature Protection from Stalin to Gorbachev*, University of California Press (London, 1999)

Willard, B., *The Sustainability Advantage: Seven Business Case Benefits of a Triple Bottom Line* (British Colombia, Canada, New Society Publishers 2002)

World Commission on Environment and Development: *Our Common Future* (1987)

World Development Indicators Database 2007, World Bank

WWF Russia, 'Basic Principles for Environmentally and Socially Responsible Policy of Mining Companies', at <www.wwf.ru/data/pub/energy/mining_eng.pdf>, September 2009

WWF Russia, 'Promoting Responsible Finance in Russia', <www.wwf.ru/data/pub/business/sustainable_financing_eng.pdf>, May 2009

The World's Worst Polluted Places – the Top Ten, Blacksmith Institute (October 2006) Appendix 1, pp 33-34, <http://www.blacksmithinstitute.org/top10/10worst2.pdf>

Yasin, E., *Rossiiskaya ekonomika* (Moscow, Higher School of Economics 2002) chap. 16.

List of WWW sources

http//:www.forest.ru/eng/sustainable_forestry/
http://chelyabinsk.rfn.ru/rnews.html?id=77335
http://document.kremlin.ru/doc.asp?ID=046255
http://ec.europa.eu/environment/emas/index_en.htm
http://en.baikalforum.ru/
http://en.rian.ru/russia/20080221/99794450.html
http://eng.genproc.gov.ru/
http://eng.www.kadastr.ru/field_of_activity/lands_statisics/northwest/
http://eng.www.rosoez.ru/
http://europa.eu/publicprocurement/index_en.htm
http://factiva.com
http://handlingsplan.nrpa.no/English/englishfrontpage.htm .
http://ji.unfccc.int/JI_Projects/DeterAndVerif/Verification/PDD/index.html
http://minenergo.com/
http://news.bbc.co.uk/2/hi/business/7358542.stm
http://nw.siora.ru/english/index.php
http://oboc.narod.ru/ovos1.htm
http://president.kremlin.ru/text/appears/2008/01/158674.shtml
http://procurement.e-management.ru/news-state47.htm
http://rspp.ru/Default.aspx?CatalogId=2257
http://wood.ru
http://www.accountability21.org/
http://www.amr.ru/
http://www.arcticshelf.ru/index.php?option=com_content&task=view&id=52&Itemid=61&lang=en
http://www.bellona.no/en/energy/39771.html
http://www.bellona.org
http://www.bellona.org/articles/articles_2008/murmansk_poll
http://www.bellona.ru/articles_ru/articles_2008/ebrd_annual_meeting_2008
http://www.businessday.co.za/articles/article.aspx?ID=BD4A579828
http://www.busrep.co.za/index.php?fSectionId=553&fArticleId=4370487
http://www.cms.int/documents/convtxt/cms_convtxt.htm
http://www.corp-gov.org/links/link.php3
http://www.corporateregister.com/
http://www.customs.ru
http://www.dfid.gov.uk/procurement/environmental-proc-policy.pdf
http://www.ebrd.com/country/sector/natural/projects/sakhalin/corres/060905.pdf
http://www.ebrd.com/new/pressrel/2005/175dec14.htm
http://www.eca-watch.org/problems/eu_russ/russia/sakhalin/WSJ_Sakhalin_11oct05.htm
http://www.economy.gov.ru
http://www.eib.org/infocentre/epe/index.htm
http://www.engineeringnews.co.za/article.php?a_id=114089
http://www.environmental-finance.com/onlinews/0306sak.html
http://www.eua.be/eua/en/policy_bologna.jspx
http://www.fas.gov.ru/english/
http://www.fas.gov.ru/english/legislation/21182.shtml
http://www.fcgsen.ru/

http://www.forest.ru/legislation/new
http://www.fpf.ru/eng/
http://www.fsc.ru
http://www.gefweb.org/interior.aspx?id=21354
http://www.geopolitics.ru/news/afr-004.htm
http://www.globalcompactsummit.org/
http://www.globalreporting.org/
http://www.gosnadzor.ru
http://www.government.ru/content/governmentactivity/rfgovernmentdecisions/archive/2008/11/17/2982752.htm
http://www.government.ru/content/governmentactivity/rfgovernmentdecisions/archive/2009/01/08/1528437.htm
http://www.halcrow.com/html/our_projects/projects/st_petersburg_barrier.htm
http://www.icett.or.jp
http://www.ippc-russia.org/content.html
http://www.iucn.org/wgwap
http://www.iucn.org/wgwap/news_archive/?3079/IUCN-welcomes-reprieve-for-whales
http://www.iucnredlist.org/details/8099/0
http://www.jbic.go.jp/en/about/press/2008/0616-01/
http://www.kommersant.com/p-10421/Sakhalin_Energy_Gazprom/
http://www.mcx.ru
http://www.mid.ru/ns-dmo.nsf/a1c87897b58a9d2743256a550029f995/432569f10031eb9343256c9a002f10ca?OpenDocument
http://www.minedu.fi/minedu/education/bolognaprocess.html
http://www.mineweb.net/mineweb/view/mineweb/en/page36?oid=37493&sn=Detail
http://www.minfin.ru
http://www.minprom.gov.ru
http://www.minprom.gov.ru
http://www.minzdravsoc.ru
http://www.mnr.gov.ru/part/?act=more&id=3093&pid=729
http://www.mnr.gov.ru/part/?pid=398
http://www.mosnews.com/world/2009/04/15/newngolegislation
http://www.niisp.ru/News/Events/art34
http://www.nord-stream.com/en/
http://www.nti.org/db/nisprofs/russia/forasst/intnatl/intnatl.htm#TACIS;
http://www.nti.org/db/nisprofs/russia/forasst/intnatl/intnatl.htm;
http://www.nti.org/db/nisprofs/russia/reactor/power/kola.htm;
http://www.nytimes.com/2005/08/21/magazine/21OIL.html?pagewanted=all
http://www.rbcdaily.ru/2007/03/19/tek/268514
http://www.rian.ru/economy/20090117/159469833.html
http://www.romir.ru/en/news/res_results/184.html
http://www.rosbalt.ru/2008/11/11/540541.html
http://www.rosleshoz.gov.ru
http://www.rsppr.ru/
http://www.ruscp.ru/
http://www.ruscp.ru/tsygeng.htm
http://www.sakhalinenergy.com
http://www.smoney.ru/article.shtml?2006/10/09/1476
http://www.socialinvest.org/areas/research/trends

http://www.sundaystandard.info/news/news_item.php?GroupID=1&NewsID=1499
http://www.transparency.org
http://www.ucmp.berkeley.edu/mesozoic/cretaceous/cretaceous.html
http://www.undp.ru/index.phtml?iso=RU&lid=1&cmd=publications1&id=40
http://www.unece.org/env/eia/welcome.html
http://www.v-c-s.org
http://www.vesti.ru/doc.html?id=222485
http://www.wood.ru
http://www.worldbank.org/transitionnewsletter
http://www.wwf.ru .
www.14000.ru
www.14001.ru
www.armscontrol.org
www.barentsoberver.com
www.ecoline.ru
www.ed.gov.ru
www.en.gtk.fi
www.forest.ru
www.gov.mnr.ru
www.impivaara.com
www.ineca.ru
www.kovdor.info
www.mnr.gov.ru
www.ndep.org/projects
www.panapgm.com
www.pomor.com
www.rusrec.ru
www.sokli.info
http://www.unece.org/env/eia/sea_protocol.htm
http://www.unece.org/env/eia/welcome.html
http://www.unece.org/env/pp/
www.unglobalcompact.org

ANNEX I: USEFUL WEBSITES

A

Agenda 21:
www.un.org/esa/sustdev/documents/agenda21/index.htm
APPM Archangelsk Pulp and Paper Mill:
http://www.appm.ru/ [in Russian]
http://en.appm.ru/ [in English]
Asser Institute:
www.asser.nl

B

Bellona Foundation:
www.bellona.org
Biosafety Protocol:
www.cbd.int/biosafety/
British Embassy, Moscow, Russia:
http://ukinrussia.fco.gov.uk/en/our-offices-in-russia/our-embassy-in-moscow/
Brundtland Report:
www.un-documents.net/wced-ocf.htm

C

CAF Charities Aid Foundation:
www.cafonline.org
Camco International:
www.camco-international.ru
CEB Council of Europe Development Bank:
www.coebank.org
Center for Business Ethics and Corporate Governance:
www.cfbe.ru
Center for Russian Environmental Policy:
www.ecopolicy.ru/eng/
CIEL Center for International Environmental Law:
www.ciel.org
CIS Commonwealth of Independent States:
www.cisstat.com
CITES Convention on International Trade in Endangered Species of Wild Fauna and Flora:
www.cites.org
Convention on Environmental Impact Assessment in a Transboundary Context:
www.unece.org/env/eia/
Convention on the Protection and Use of Transboundary Watercourses and International Lakes:
www.unece.org/env/water/text/text.htm or
www.unece.org/env/water/pdf/watercon.pdf
Convention on Wetlands of International Importance especially as Waterfowl Habitat
www.ramsar.org/

D

DanishCarbon.dk (Danish Energy Agency):
www.danishcarbon.dk
Dutch Embassy, Moscow, Russia:
http://russia.nlembassy.org/homepage

E

Eco-Accord – Center for Environment and Sustainable Development:
www.ecoaccord.org/english/index.htm
Ecoline:
www.eac-ecoline.ru/index-eng.html
Ernst & Young Russia:
www.ey.com/cis
Environmental Finance (magazine):
www.environmental-finance.com
Equator Principles:
www.equator-principles.com/documents/Equator_Principles.pdf
European Bank for Reconstruction and Development:
www.ebrd.com
European Investment Bank (EIB):
www.eib.org
EIB Sourcebook on EU Environmental Law:
www.eib.org/about/publications/sourcebook-on-eu-environmental-law.htm
European Environmental Law:
www.eel.nl
European Principles for the Environment (EPE):
http://www.eib.org/infocentre/epe/
EveryChild:
www.everychild.org.uk/

F

FAS – Russian Federal Antimonopoly Service (in Russian):
www.fas.gov.ru
FFMS – Russian Federal Financial Markets Service (in English):
www.fcsm.ru/eng
FSC Forest Stewardship Council:
www.fscus.org

G

Global Environment Facility (GEF):
www.gefweb.org
Global Reporting Initiative:
www.globalreporting.org
Greenpeace Russia:
www.greenpeace.org/russia/en/
GOST – Russian Federal Agency for Technology and Metrology:
www.gost.ru/wps/portal/pages.en.Main
GOST R ISO 14001-2007:
www.gost.ru/wps/portal/pages.Resource [Russian language only]

H

Harmonisation of Environmental Standards (HES II) project:
www.ippc-russia.org

I

IBRD – International Bank for Reconstruction and Development (IBRD or World Bank):
http://web.worldbank.org
Impivaara Securities Ltd:
www.impivaara.com or
www.pomor.com or
www.kovdor.com
Institute for Corporate Governance:
www.inkup.ru
International Business Leaders Forum:
www.iblf.ru or *www.iblf.org*
International Chamber of Commerce (ICC) Russia:
www.iccwbo.ru
International Finance Corporation:
www.ifc.org
International Labour Organization (ILO):
www.ilo.org
Investor Protection Agency:
www.corp-gov.ru
ISO 14001 and ISO Management Standards:
www.iso.org/iso/management_standards.htm

J

JBIC – Japan Bank for International Cooperation:
www.jibc.go.jp/en/
Johannesburg Plan of Implementation:
www.un.org/esa/sustdev/documents/WSSD_POI_PD/English/POIToc.htm

M

D. Mendeleyev University of Chemical Technology:
www.muctr.ru or
www.muctr.ru/en/
Millenium Development Goals:
www.un.org/millenniumgoals/bkgd.shtml
Ministry of Economic Development (MED), Russian Federation:
www.economy.gov.ru
Ministry of Natural Resources and Environment (MNR), Russian Federation:
www.mnr.gov.ru
Multilateral Carbon Credit Fund:
www.myebrd.org/country/sector/energyef/carbon/mccf/index.htm

N

National Pollution Abatement Facility (NPAF) (Russia):
www.npaf.ru
NDEP Northern Dimension Environmental Partnership:
www.ndep.org

NEFCO, Northern Nordic Environment Finance Corporation:
www.nefco.org
NEFCO Carbon Finance & Funds:
www.nefco.org/cff
NGO Non-Governmental Organizations search website:
www.ngo.org
Nord Stream project:
www.nord-stream.com/en/

O
OECD:
www.oecd.org
OECD Principles of Corporate Governance:
www.oecd.org/daf/corporate/principles
OPORA:
www.opora.ru/

P
Protocol on Water and Health:
www.unece.org/env/water/text/text_protocol.htm

R
Rio Declaration on Environment and Development
www.un.org/documents/ga/conf151/aconf15126-1annex1.htm
Royal Institute of International Affairs (Chatham House), Russia and Eurasia Programme:
www.chathamhouse.org.uk/research/russia_eurasia/
RSPP – Russian Union of Industrialists and Entrepreneurs:
www.rspp.ru
Russian Corporate Governance Code:
www.fcsm.ru/eng/catalog.asp?ob_no=9073
Russian Embassy, London, UK:
http://london.rusembassy.org/
Russian Managers Association:
www.amr.ru
Russian Regional Environment Center (RREC):
www.rusrec.ru or
www.rusrec.ru/homepage/index_en.htm
Russian SME Resource Center:
www.rcsme.ru/eng/
Russo-British Chamber of Commerce:
www.rbcc.com

S
SEIC – Sakhalin Energy Investment Company:
www.sakhalinenergy.com/en/
Social Charter of Russian Business:
http://rspp.ru/Default.aspx?CatalogId=2398

T

The PBN Company:
www.pbnco.com/

U

United Nations Development Programme (UNDP) Russia:
www.undp.ru
United Nations Environment Programme (UNEP):
www.unep.org
United Nations Environment Programme Finance Initiative (UNEP-FI):
www.unepfi.org
United Nations Economic Commission for Europe (UNECE):
www.unece.org/
United Nations Framework Convention on Climate Change (UNFCCC):
www.unfccc.int
University of Joensuu, Finland:
www.joensuu.fi/englishindex.html
UN Global Compact Principles: *www.unglobalcompact.org/AbouttheGC/TheTENPrinciples/index.html*
UN Principles of Responsible Investing:
www.unpri.org/

W

WGWAP – Western Gray Whale Advisory Panel:
www.iucn.org/wgwap
World Bank Group: See entry for 'IBRD'
World Bank Russia:
www.worldbank.org.ru
World Business Council for Sustainable Development:
www.wbcsd.org
WWF Russia:
www.wwf.ru - news and information about WWF-Russia's work.
www.panda.org - news and information about WWF's work around the world.

ANNEX II: THE EQUATOR PRINCIPLES*

The 'Equator Principles'

A financial industry benchmark for determining,assessing and managing social & environmental risk inproject financing

PREAMBLE

Project financing, a method of funding in which the lender looks primarily to the revenues generated by a single project both as the source of repayment and as security for the exposure, plays an important role in financing development throughout the world.[1] Project financiers may encounter social and environmental issues that are both complex and challenging, particularly with respect to projects in the emerging markets.

The Equator Principles Financial Institutions ('EPFIs') have consequently adopted these Principles in order to ensure that the projects we finance are developed in a manner that is socially responsible and reflect sound environmental management practices. By doing so, negative impacts on project-affected ecosystems and communities should be avoided where possible, and if these impacts are unavoidable, they should be reduced, mitigated and/or compensated for appropriately. We believe that adoption of and adherence to these Principles offers significant benefits to ourselves, our borrowers and local stakeholders through our borrowers' engagement with locally affected communities. We therefore recognise that our role as financiers affords us opportunities to promote responsible environmental stewardship and socially responsible development. As such, EPFIs will consider reviewing these Principles from time-to-time based on implementation experience, and in order to reflect ongoing learning and emerging good practice.

These Principles are intended to serve as a common baseline and framework for the implementation by each EPFI of its own internal social and environmental policies, procedures and standards related to its project financing activities. We will not provide loans to projects where the borrower will not or is unable to comply with our respective social and environmental policies and procedures that implement the Equator Principles.

---

* Source: <www.equator-principles.com/documents/Equator Principles.pdf>.

[1] **Project finance** is 'a method of funding in which the lender looks primarily to the revenues generated by a single project, both as the source of repayment and as security for the exposure. This type of financing is usually for large, complex and expensive installations that might include, for example, power plants, chemical processing plants, mines, transportation infrastructure, environment, and telecommunications infrastructure. Project finance may take the form of financing of the construction of a new capital installation, or refinancing of an existing installation, with or without improvements. In such transactions, the lender is usually paid solely or almost exclusively out of the money generated by the contracts for the facility's output, such as the electricity sold by a power plant. The borrower is usually an SPE (Special Purpose Entity) that is not permitted to perform any function other than developing, owning, and operating the installation. The consequence is that repayment depends primarily on the project's cash flow and on the collateral value of the project's assets.' Source: *Basel Committee on Banking Supervision*, International Convergence of Capital Measurement and Capital Standards ('Basel II'), November 2005, <http://www.bis.org/publ/bcbs118.pdf>.

SCOPE

The Principles apply to all new project financings globally with total project capital costs of US$10 million or more, and across all industry sectors. In addition, while the Principles are not intended to be applied retroactively, we will apply them to all project financings covering expansion or upgrade of an existing facility where changes in scale or scope may create significant environmental and/or social impacts, or significantly change the nature or degree of an existing impact.

The Principles also extend to project finance advisory activities. In these cases, EPFIs commit to make the client aware of the content, application and benefits of applying the Principles to the anticipated project, and request that the client communicate to the EPFI its intention to adhere to the requirements of the Principles when subsequently seeking financing.

STATEMENT OF PRINCIPLES

EPFIs will only provide loans to projects that conform to Principles 1-9 below:

Principle 1: Review and Categorisation

When a project is proposed for financing, the EPFI will, as part of its internal social and environmental review and due diligence, categorise such project based on the magnitude of its potential impacts and risks in accordance with the environmental and social screening criteria of the International Finance Corporation ('IFC') (Exhibit I).

Principle 2: Social and Environmental Assessment

For each project assessed as being either Category A or Category B, the borrower has conducted a Social and Environmental Assessment ('Assessment') process[2] to address, as appropriate and to the EPFI's satisfaction, the relevant social and environmental impacts and risks of the proposed project (which may include, if relevant, the illustrative list of issues as found in Exhibit II). The Assessment should also propose mitigation and management measures relevant and appropriate to the nature and scale of the proposed project.

Principle 3: Applicable Social and Environmental Standards

For projects located in non-OECD countries, and those located in OECD countries not designated as High-Income, as defined by the World Bank Development Indicators Database, the Assessment will refer to the then applicable IFC Performance Standards (Exhibit III) and the then applicable Industry Specific EHS Guidelines ('EHS Guidelines') (Exhibit IV). The Assessment will establish to a participating EPFI's satisfaction the project's overall compliance with, or justified deviation from, the respective Performance Standards and EHS Guidelines.

[2] Social and Environmental Assessment is a process that determines the social and environmental impacts and risks (including labour, health, and safety) of a proposed project in its area of influence. For the purposes of Equator Principles compliance, this will be an adequate, accurate and objective evaluation and presentation of the issues, whether prepared by the borrower, consultants or external experts. Depending on the nature and scale of the project, the assessment document may comprise a full-scale social and environmental impact assessment, a limited or focused environmental or social assessment (e.g., audit), or straightforward application of environmental siting, pollution standards, design criteria, or construction standards. One or more specialised studies may also need to be undertaken.

The regulatory, permitting and public comment process requirements in High-Income OECD Countries, as defined by the World Bank Development Indicators Database, generally meet or exceed the requirements of the IFC Performance Standards (Exhibit III) and EHS Guidelines (Exhibit IV). Consequently, to avoid duplication and streamline EPFI's review of these projects, successful completion of an Assessment (or its equivalent) process under and in compliance with local or national law in High-Income OECD Countries is considered to be an acceptable substitute for the IFC Performance Standards, EHS Guidelines and further requirements as detailed in Principles 4, 5 and 6 below. For these projects, however, the EPFI still categorises and reviews the project in accordance with Principles 1 and 2 above.

The Assessment process in both cases should address compliance with relevant host country laws, regulations and permits that pertain to social and environmental matters.

Principle 4: Action Plan and Management System

For all Category A and Category B projects located in non-OECD countries, and those located in OECD countries not designated as High-Income, as defined by the World Bank Development Indicators Database, the borrower has prepared an Action Plan ('AP')[3] which addresses the relevant findings, and draws on the conclusions of the Assessment. The AP will describe and prioritise the actions needed to implement mitigation measures, corrective actions and monitoring measures necessary to manage the impacts and risks identified in the Assessment. Borrowers will build on, maintain or establish a Social and Environmental Management System that addresses the management of these impacts, risks, and corrective actions required to comply with applicable host country social and environmental laws and regulations, and requirements of the applicable Performance Standards and EHS Guidelines, as defined in the AP.

For projects located in High-Income OECD countries, EPFIs may require development of an Action Plan based on relevant permitting and regulatory requirements, and as defined by host-country law.

Principle 5: Consultation and Disclosure

For all Category A and, as appropriate, Category B projects located in non-OECD countries, and those located in OECD countries not designated as High-Income, as defined by the World Bank Development Indicators Database, the government, borrower or third party expert has consulted with project affected communities in a structured and culturally appropriate manner.[4] For projects with significant adverse impacts on affected communities, the process will ensure their free, prior and informed consultation and facilitate their informed participation as a means to establish, to the satisfaction of the EPFI, whether a project has adequately incorporated affected communities' concerns.[5]

[3] The Action Plan may range from a brief description of routine mitigation measures to a series of documents (e.g., resettlement action plan, indigenous peoples plan, emergency preparedness and response plan, decommissioning plan, etc). The level of detail and complexity of the Action Plan and the priority of the identified measures and actions will be commensurate with the project's potential impacts and risks. Consistent with Performance Standard 1, the internal Social and Environmental Management System will incorporate the following elements: (i) Social and Environmental Assessment; (ii) management program; (iii) organisational capacity; (iv) training; (v) community engagement; (vi) monitoring; and (vii) reporting.

[4] Affected communities are communities of the local population within the project's area of influence who are likely to be adversely affected by the project. Where such consultation needs to be undertaken in a structured manner, EPFIs may require the preparation of a Public Consultation and Disclosure Plan (PCDP).

[5] Consultation should be "free" (free of external manipulation, interference or coercion, and intimidation), "prior" (timely disclosure of information) and "informed" (relevant, understandable and accessible

In order to accomplish this, the Assessment documentation and AP, or non-technical summaries thereof, will be made available to the public by the borrower for a reasonable minimum period in the relevant local language and in a culturally appropriate manner. The borrower will take account of and document the process and results of the consultation, including any actions agreed resulting from the consultation. For projects with adverse social or environmental impacts, disclosure should occur early in the Assessment process and in any event before the project construction commences, and on an ongoing basis.

Principle 6: Grievance Mechanism

For all Category A and, as appropriate, Category B projects located in non-OECD countries, and those located in OECD countries not designated as High-Income, as defined by the World Bank Development Indicators Database, to ensure that consultation, disclosure and community engagement continues throughout construction and operation of the project, the borrower will, scaled to the risks and adverse impacts of the project, establish a grievance mechanism as part of the management system. This will allow the borrower to receive and facilitate resolution of concerns and grievances about the project's social and environmental performance raised by individuals or groups from among project-affected communities. The borrower will inform the affected communities about the mechanism in the course of its community engagement process and ensure that the mechanism addresses concerns promptly and transparently, in a culturally appropriate manner, and is readily accessible to all segments of the affected communities.

Principle 7: Independent Review

For all Category A projects and, as appropriate, for Category B projects, an independent social or environmental expert not directly associated with the borrower will review the Assessment, AP and consultation process documentation in order to assist EPFI's due diligence, and assess Equator Principles compliance.

Principle 8: Covenants

An important strength of the Principles is the incorporation of covenants linked to compliance. For Category A and B projects, the borrower will covenant in financing documentation:

a) to comply with all relevant host country social and environmental laws, regulations and permits in all material respects;
b) to comply with the AP (where applicable) during the construction and operation of the project in all material respects;
c) to provide periodic reports in a format agreed with EPFIs (with the frequency of these reports proportionate to the severity of impacts, or as required by law, but not less than annually), prepared by in-house staff or third party experts, that i) document compliance with the AP (where applicable), and ii) provide representation of compliance with relevant local, state and host country social and environmental laws, regulations and permits; and
d) to decommission the facilities, where applicable and appropriate, in accordance with an agreed decommissioning plan.

---

information), and apply to the entire project process and not to the early stages of the project alone. The borrower will tailor its consultation process to the language preferences of the affected communities, their decision-making processes, and the needs of disadvantaged or vulnerable groups. Consultation with Indigenous Peoples must conform to specific and detailed requirements as found in Performance Standard 7. Furthermore, the special rights of Indigenous Peoples as recognised by host-country legislation will need to be addressed.

Where a borrower is not in compliance with its social and environmental covenants, EPFIs will work with the borrower to bring it back into compliance to the extent feasible, and if the borrower fails to re-establish compliance within an agreed grace period, EPFIs reserve the right to exercise remedies, as they consider appropriate.

Principle 9: Independent Monitoring and Reporting

To ensure ongoing monitoring and reporting over the life of the loan, EPFIs will, for all Category A projects, and as appropriate, for Category B projects, require appointment of an independent environmental and/or social expert, or require that the borrower retain qualified and experienced external experts to verify its monitoring information which would be shared with EPFIs.

Principle 10: EPFI Reporting

Each EPFI adopting the Equator Principles commits to report publicly at least annually about its Equator Principles implementation processes and experience, taking into account appropriate confidentiality considerations.[6]

DISCLAIMER

The adopting EPFIs view these Principles as a financial industry benchmark for developing individual, internal social and environmental policies, procedures and practices. As with all internal policies, these Principles do not create any rights in, or liability to, any person, public or private. Institutions are adopting and implementing these Principles voluntarily and independently, without reliance on or recourse to IFC or the World Bank.

Exhibit I: Categorisation of projects

As part of their review of a project's expected social and environmental impacts, EPFIs use a system of social and environmental categorisation, based on IFC's environmental and social screening criteria, to reflect the magnitude of impacts understood as a result of assessment. These categories are:

- **Category A** – Projects with potential significant adverse social or environmental impacts that are diverse, irreversible or unprecedented;
- **Category B** – Projects with potential limited adverse social or environmental impacts that are few in number, generally site-specific, largely reversible and readily addressed through mitigation measures; and
- **Category C** – Projects with minimal or no social or environmental impacts.

Exhibit II: Illustrative list of potential social and environmental issues to be addressed in the Social and Environmental Assessment documentation

In the context of the business of the project, the Assessment documentation will address, where applicable, the following issues:

[6] Such reporting should at a minimum include the number of transactions screened by each EPFI, including the categorisation accorded to transactions (and may include a breakdown by sector or region), and information regarding implementation.

a) assessment of the baseline social and environmental conditions
b) consideration of feasible environmentally and socially preferable alternatives
c) requirements under host country laws and regulations, applicable international treaties and agreements
d) protection of human rights and community health, safety and security (including risks, impacts and management of project's use of security personnel)
e) protection of cultural property and heritage
f) protection and conservation of biodiversity, including endangered species and sensitive ecosystems in modified, natural and critical habitats, and identification of legally protected areas
g) sustainable management and use of renewable natural resources (including sustainable resource management through appropriate independent certification systems)
h) use and management of dangerous substances
i) major hazards assessment and management
j) labour issues (including the four core labour standards), and occupational health and safety
k) fire prevention and life safety
l) socio-economic impacts
m) land acquisition and involuntary resettlement
n) impacts on affected communities, and disadvantaged or vulnerable groups
o) impacts on indigenous peoples, and their unique cultural systems and values
p) cumulative impacts of existing projects, the proposed project, and anticipated future projects
q) consultation and participation of affected parties in the design, review and implementation of the project
r) efficient production, delivery and use of energy
s) pollution prevention and waste minimisation, pollution controls (liquid effluents and air emissions) and solid and chemical waste management

Note: The above list is for illustrative purposes only. The Social and Environmental Assessment process of each project may or may not identify all issues noted above, or be relevant to every project.

Exhibit III: IFC Performance Standards on Social and Environmental Sustainability

As of April 30, 2006, the following list of IFC Performance Standards were applicable:

- Performance Standard 1: Social & Environmental Assessment & Management System
- Performance Standard 2: Labour and Working Conditions
- Performance Standard 3: Pollution Prevention and Abatement
- Performance Standard 4: Community Health, Safety and Security
- Performance Standard 5: Land Acquisition and Involuntary Resettlement
- Performance Standard 6: Biodiversity Conservation and
- Sustainable Natural Resource Management
- Performance Standard 7: Indigenous Peoples
- Performance Standard 8: Cultural Heritage

Note: The IFC has developed a set of Guidance Notes to accompany each Performance Standard. While not formally adopting the Guidance Notes, EPFIs or borrowers may use the Guidance Notes as useful points of reference when seeking further guidance on or interpretation of the Performance Standards. The IFC Performance Standards, Guidance Notes and Industry Sector EHS Guidelines can be found at <www.ifc.org/enviro>.

Exhibit IV: Industry-Specific Environmental, Health and Safety (EHS) Guidelines

EPFIs will utilise the appropriate environmental, health and safety ('EHS') guidelines used by IFC which are now in place, and as may be amended from time-to-time.

IFC is using two complementary sets of EHS Guidelines available at the IFC website (<www.ifc.org/enviro>). These sets consist of all the environmental guidelines contained in Part III of the World Bank's Pollution Prevention and Abatement Handbook ('PPAH') which went into official use on July 1, 1998 and a series of environmental, health and safety guidelines published on the IFC website between 1991 and 2003. Ultimately new guidelines, incorporating the concepts of cleaner production and environmental management systems, will be written to replace this series of industry sector, PPAH and IFC guidelines.

Where no sector specific guideline exists for a particular project then the PPAH's General Environmental Guidelines and the IFC Occupational Health and Safety Guidelines (2003) are applied, with modifications as necessary to suit the project.*

The table below lists both the World Bank Guidelines and the IFC Guidelines as of March 1, 2006.

Industry Specific EHS Guidelines:

World Bank Guidelines (PPAH)	IFC Guidelines
1. Aluminum Manufacturing	1. Airports
2. Base Metal and Iron Ore Mining	2. Ceramic Tile Manufacturing
3. Breweries	3. Construction Materials Plants
4. Cement Manufacturing	4. Electric Power Transmission and Distribution
5. Chlor-Alkali Plants	5. Fish Processing
6. Coal Mining and Production	6. Food and Beverage Processing
7. Coke Manufacturing	7. Forestry Operations: Logging
8. Copper Smelting	8. Gas Terminal Systems
9. Dairy Industry	9. Geothermal Projects
10. Dye Manufacturing	10. Hazardous Materials Management
11. Electronics Manufacturing	11. Health Care
12. Electroplating Industry	12. Life & Fire Safety
13. Foundries	13. Occupational Health and Safety
14. Fruit and Vegetable Processing	14. Office Buildings
15. General Environmental Guidelines	15. Offshore Oil & Gas
16. Glass Manufacturing	16. Polychlorinated Biphenyls (PCBs)
17. Industrial Estates	17. Pesticide Handling and Application
18. Iron and Steel Manufacturing	18. Plantations
19. Lead and Zinc Smelting	19. Port and Harbor Facilities
20. Meat Processing and Rendering	20. Rail Transit Systems
21. Mini Steel Mills	21. Roads and Highways
22. Mixed Fertilizer Plants	22. Telecommunications
23. Monitoring	23. Tourism and Hospitality Development
24. Nickel Smelting and Refining	24. Waste Management Facilities
25. Nitrogenous Fertilizer Plants	25. Wastewater Reuse
26. Oil and Gas Development (Onshore)	26. Wildland Management
27. Pesticides Formulation	27. Wind Energy Conversion Systems
28. Pesticides Manufacturing	28. Wood Products Industries
29. Petrochemicals Manufacturing	

* Exception (the following are World Bank Guidelines not contained in the PPAH and currently in use): Mining and Milling – Underground; Mining and Milling – Open Pit

Industry Specific EHS Guidelines (cont.)

World Bank Guidelines (PPAH)	IFC Guidelines
30. Petroleum Refining	
31. Pharmaceutical Manufacturing	
32. Phosphate Fertilizer Plants	
33. Printing Industry	
34. Pulp and Paper Mills	
35. Sugar Manufacturing	
36. Tanning and Leather Finishing	
37. Textiles Industry	
38. Thermal Power Guidelines for New Plants	
39. Thermal Power Rehabilitation of Existing Plants	
40. Vegetable Oil Processing	
41. Wood Preserving Industry	

ANNEX III: UN GLOBAL COMPACT'S PRINCIPLES*

THE TEN PRINCIPLES

The UN Global Compact's ten principles in the areas of human rights, labour, the environment and anti-corruption enjoy universal consensus and are derived from:

The Universal Declaration of Human Rights

The International Labour Organization's Declaration on Fundamental Principles and Rights at Work

The Rio Declaration on Environment and Development

The United Nations Convention Against Corruption

The Global Compact asks companies to embrace, support and enact, within their sphere of influence, a set of core values in the areas of human rights, labour standards, the environment, and anti-corruption:

Human Rights

Principle 1: Businesses should support and respect the protection of internationally proclaimed human rights; and

Principle 2: make sure that they are not complicit in human rights abuses.

Labour Standards

Principle 3: Businesses should uphold the freedom of association and the effective recognition of the right to collective bargaining;

Principle 4: the elimination of all forms of forced and compulsory labour;

Principle 5: the effective abolition of child labour; and

Principle 6: the elimination of discrimination in respect of employment and occupation.

Environment

Principle 7: Businesses should support a precautionary approach to environmental challenges;

Principle 8: undertake initiatives to promote greater environmental responsibility; and

Principle 9: encourage the development and diffusion of environmentally friendly technologies.

Anti-Corruption

Principle 10: Businesses should work against corruption in all its forms, including extortion and bribery.

---

* Source: <www.unglobalcompact.org/AbouttheGC/TheTENPrinciples/index.html>.

ANNEX IV: RUSSIAN CODE OF CORPORATE GOVERNANCE*

5 April 2002

The final version of the Russian Code of Corporate Conduct presented to the Russian and foreign business community, international organizations and federal authorities at The IV Coordination Council for Corporate Governance met on 4 April 2002.

Table of Contents

INTRODUCTION

"Corporate governance" is a term that encompasses a variety of activities connected with the management of companies. Corporate governance affects the performance of economic entities and their ability to attract the capital required for economic growth. Improvement of corporate governance in the Russian Federation is vital for increasing investment in all sectors of the Russian economy from both domestic sources and foreign investors. One means to foster such improvement is to introduce standards that are based on an analysis of best practices of corporate governance.

These standards of corporate governance apply to all economic entities, but are most important for joint stock companies. This is because it is in joint stock companies that the separation between ownership and management is the greatest, and thus conflicts related to corporate governance are most likely. Therefore, this Code has been developed primarily for joint stock companies seeking access to capital markets. This consideration, however, does not rule out the possibility of its use by other economic entities.

Standards of corporate governance should be applied to ensure adequate protection of the interests of all shareholders, regardless of the size of their holdings. The greater the level of shareholders' protection achieved, the more investment capital will be available to Russian joint stock companies (hereinafter referred to as "companies"), which will favorably influence the Russian economy as a whole.

* Source: <www.ecgi.org/codes/documents/final code english.pdf>.

Standards of corporate governance should be instrumental to the attainment of high ethical standards in relations between market participants.

The following sections describe the background of the development of the Corporate Governance Code (hereinafter referred to as the "Code"). It should be also taken into account that any company may develop its own code of corporate governance in accordance with the recommendations of this Code, or incorporate some of the Code's provisions into its internal regulations. Subject to a company's legal status, branch affiliation, capital structure, and other specific parameters, such a company may selectively use those recommendations of this Code suitable to it.

1. Russian law already incorporates the majority of the fundamental principles of corporate governance, however, in practice, their use, especially in court, and corporate governance traditions, are still at the formative stage

The existing Russian legislation governing the operation of economic entities is relatively new, however it already reflects most of the generally accepted corporate governance principles.

On the other hand, the main problems in the corporate governance area arise from the lack of durable corporate governance practices rather than the quality of legislation, which is why corporate governance traditions are currently in the formative stage.

2. High standards of corporate governance cannot be assured by legislative provisions alone

Legislation alone cannot be expected and is inherently unable, to regulate all issues related to the management of companies.

First, the law establishes and should establish only general mandatory rules. It cannot regulate, and should not have as its purpose to regulate in detail all matters of corporate operations. Abundance of detail in legal norms makes it difficult for companies to function since each company's business is unique, making it impossible adequately to reflect such uniqueness in the law. That is why the law often completely omits provisions regulating certain relationships (and the absence of such regulation is often not at all a legislative weakness), or establishes general rules leaving it to the parties involved in appropriate business relationships to choose a line of behavior.

Second, legislation is unable to react rapidly to changes in corporate governance practice, as amending laws is very time consuming.

3. Many issues associated with corporate governance lie outside the legal arena and have an ethical rather than legal nature

Many legislative regulations covering corporate governance are based on ethical norms. For example, civil law regulations, in particular, stipulate the possibility of applying requirements of good faith, prudence and equity in the absence of applicable legislation, as well as exercising civil rights in a reasonable and fair manner. Thus, moral and ethical standards of reasonableness, equity and good faith are part and parcel of the existing legislation.

At the same time, such legislative regulations are not always sufficient to ensure proper corporate governance.

Therefore, companies should act in accordance not only with statutory standards, but also with ethical standards which are often more demanding than the law's requirements.

Ethical standards present a set system of behavioral norms and customs of the trade traditionally applied by the business community, which are not based on the law, and which form positive expectations with respect to the anticipated behavior of participants in corporate relations.

Ethical standards of corporate governance form sustainable behavioral patterns common to all participants in corporate relations.

Compliance with these standards is not only a moral imperative; it also helps the company avoid risks, supports long-term economic growth and facilitates successful business activity.

Ethical standards and best practice, together with the law, form a company's policy of corporate governance based on respect for the interests of both the shareholders and management of the company and help to strengthen the company and increase its profit.

4. The Code sets forth recommendations with respect to the best practices of corporate governance, but is, however, not obligatory

The Code shall play a key role in the process of development and improvement of Russian corporate governance practices. It shall become an important educational tool that be extensively used to define Russian company governance standards and to promote the further development of the Russian stock market.

The Code has been developed in accordance with the current provisions of Russian legislation, with due respect to established Russian and foreign corporate governance practices, ethical standards, and the specific needs and business environment of Russian companies and Russian capital markets at the present stage of their development.

The provisions of the Code are based on the internationally recognized corporate governance principles developed by the Organization for Economic Cooperation and Development (OECD), in accordance with which a number of other countries have adopted their own corporate governance codes and similar documents.

The Code is a list of recommendations. The application of the Code should be voluntary for companies, and should be motivated by their desire to increase their attractiveness to present and potential investors.

The Code sets forth the underlying principles of the best corporate governance practices that may be used by Russian companies to build their own systems of corporate governance, and contains recommendations on the practical implementation of these principles and the related disclosure of information.

When shaping their own corporate governance policies, companies may themselves determine which rules and procedures recommended by the Code they should follow and/or whether they should develop new rules and procedures in accordance with the corporate governance principles set forth by the Code.

CHAPTER 1. PRINCIPLES OF CORPORATE GOVERNANCE

Corporate governance should be based on respect for the rights and lawful interests of all participants and improve the quality of a company's operations by means of, among other things, increasing the value of corporate assets, creating jobs and enhancing the financial stability and profitability of the company.

Trust between all those engaged in corporate governance is at the root of the effective operation of a company and the ability to attract investment. The Principles of Corporate Governance described in this chapter are aimed at the creation of trust in relations arising in connection with corporate management.

The principles of corporate conduct are fundamental guidelines underlying the formation, operation and enhancement of a company's system of corporate governance.

The principles of corporate conduct set forth in this chapter form the basis of the recommendations contained in the chapters of this Code that follow, and also serve as fundamental guidelines to be observed in the absence of specific recommendations. These principles have been drafted according to the Principles of Corporate Governance of the Organization for Economic Cooperation and Development (OECD), international corporate conduct practice, as well as experience accumulated in Russia since the enactment of the Federal Law 'On Joint Stock Companies'.

(…)

CHAPTER 3. BOARD OF DIRECTORS OF THE COMPANY

(…)

4.12. A company may wish to establish an ethics committee to ensure that the company complies with ethical standards and contributes to the creation of an atmosphere of trust within the company.

The ethics committee defines the ethical standards for the company's operations, taking into account its area of activity. It is recommended that the company develop internal regulations, which should be approved by the board of directors and contain a description of the ethical standards by which the company is guided in its operations (hereinafter referred to as "Ethical Standards").

These Ethical Standards should reflect the company's social responsibilities and, in particular, state its duty to maintain high standards of quality for its products, comply with environmental and safety regulations, as well as the company's awareness of the need to develop and implement new technologies and improve conditions of labor. In addition, the Ethical Standards should prevent the officers of the company from misusing their authority to the detriment of the company and third parties through, among other things, illegal use of confidential and insider information. The Ethical Standards should be designed to reflect fundamental social values, such as honesty, mutual respect, and equity, and business principles, such as profitability, customer satisfaction, quality of products, workplace health care, safety and efficiency of labor. It is recommended that the ethics committee should detect and prevent violations of the applicable legislation and ethical standards.

(…)

CHAPTER 7. DISCLOSURE OF INFORMATION ABOUT A COMPANY

(…)

3.3.5. It is recommended that the annual report should cover matters related to health care, vocational training, workplace safety and environmental protection.

ANNEX V: EUROPEAN PRINCIPLES FOR THE ENVIRONMENT*

Declaration on the European Principles for the Environment

May 2006

We, the Signatories of the Declaration, have a shared responsibility towards protecting but also improving the environment in the interest of sustainable development, which we believe can best be achieved working more closely together.

We hereby endorse and reinforce the European consensus on the values attached to the fundamental right for both present and future generations throughout the world to live in a healthy environment.

This Declaration, recognizing the comparable approach taken by the Signatories to environmental management and to the integration of environmental considerations in their respective operations and mandates, is based on the particular EU approach to the environment, which is as strong as any that exists.

The 'European Principles for the Environment' (EPE) consist of the guiding environmental principles enshrined in the EC Treaty and the project-specific practices and standards incorporated in EU secondary legislation on the environment. The principles include, in particular, the precautionary principle, the prevention principle, the principle that environmental damage should as a priority be rectified at source, and the polluter pays principle[1].

We underline the importance of the EPE, which promotes the EU approach to environmental sustainability, and we are committed, subject to our respective environmental policies, to applying EU principles, practices and standards to all projects financed by the Signatory institutions.

The geographical scope of the EPE covers, at least the respective regions of operations of each Signatory institution, or any other geographic area it deems appropriate, including the EU 25 and European Economic Area (EEA) countries, the EU Acceding, Candidate and potential Candidate Countries and the Countries that are covered in the 'European Neighborhood and Partnership Instrument', implemented according to the following modalities.

In the Member States of the EU, the EEA countries, the EU Acceding, Candidate and potential Candidate Countries, the Signatories hereby agree to provide financing to public or private sponsors of projects only where the projects comply with the above principles and the relevant secondary EU legislation[2]. Of the EU secondary legislation particular emphasis is given to:

The EU acquis related to environmental assessment;

The EU Directives related to industrial production, water and waste management, air and soil pollution, occupational health and safety, and the protection of nature, where these can be applied to specific projects.

---

* Source: <www.eib.europe.eu/infocentre/epe/declaration/index.htm>.

[1] Treaty Establishing the European Community 2002, Article174 (2).

[2] In some countries, a phased approach with certain requirements of the *Acquis Communautaire* may be adopted in accordance with negotiated accession agreements.

We also agree that projects in this region should comply with any obligations and standards enshrined in relevant Multilateral Environmental Agreements (MEAs), according to applicable EU law (e.g., biodiversity, climate change, the ozone layer, wetlands, persistent organic pollution, trans-boundary air pollution, endangered species and environmental information, and others that may be ratified from time to time).

In all other countries, projects financed by the Signatories are expected to comply with the appropriate EU environmental principles, practices and standards – and with regard to EU financing, due respect for the European Neighborhood Policy and the EU policy towards Russia –, subject to local conditions. In such financing, we will apply the EPE, with reference to such factors as the costs of application, the local conditions that prevail and the time frame for the phased application for implementing the EPE.

In the case of co-financed projects, we will work together to agree a common approach to the project, where possible, based on or consistent with the EPE and its methods of implementation as outlined above.

We also aim as part of the EPE to promote best EU practice in the fields of environmental management, transparency, public consultation and reporting[3] .

As we move forward with the EPE process, the Signatories of the Declaration will encourage other European-based institutions to share the common approach to environmental sustainability as well as work together on specific topics in the interests of greater coherence.

Signatories

- **The Council of Europe Development Bank,**
- **The European Bank for Reconstruction and Development,**
- **The European Investment Bank,**
- **The Nordic Environment Finance Corporation and**
- **The Nordic Investment Bank**

[3] The EU 'Environmental Management and Audit Scheme' (EMAS) is an example of such best practice.

DETAILED TABLE OF CONTENTS

Part III: Practical Trends and Synergies

A. Business, NGOs and environmental projects